# Build Math Basics

**The core math skills second and third graders need!**

- Number Sense
- Basic Addition and Subtraction Facts to 18
- Basic Multiplication and Division Facts
- Multiplying Multidigit Numbers by One-Digit Numbers
- Fractions
- Story Problems
- Geometry
- Measurement
- Data Analysis and Probability

And much, much more!


ISBN# 1-56234-628-8



Manufactured in the United States
10 9 8 7 6 5 4 3 2 1

**www.themailbox.com**

# Table of Contents

## Number and Operations



## Geometry

## Measurement



## Data Analysis and Probability



## Algebra





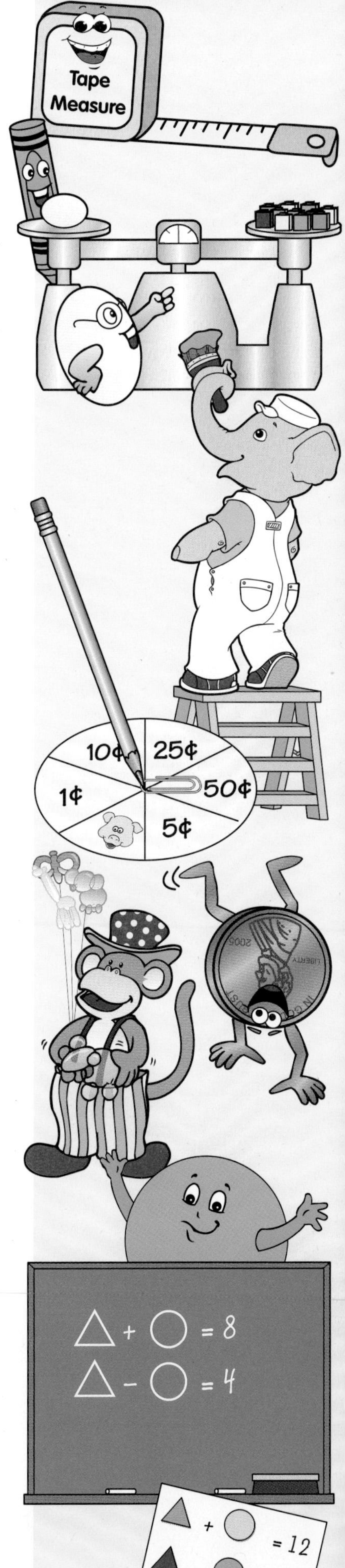

# Building Math Basics


**Managing Editor:** Hope Taylor Spencer

**Editorial Team:** Becky S. Andrews, Kimberley Bruck, Karen P. Shelton, Diane Badden, Thad H. McLaurin, Cindy K. Daoust, Leanne Stratton, Karen A. Brudnak, Sarah Hamblet, Hope Rodgers, Dorothy C. McKinney, Karen Anderson, Patricia Atkinson, Jenny Baker, Amy Barsanti, Bonnie Baumgras, Pat Biancardi, Jennifer Bohrer, Judith Brown, Peggy Morin Bruno, Lisa Buchholz, Connie Bugenhagen, Lynn Bump, Denine T. Carter, Mary Lou Corrigan, Lynda Currington, Stacie Davis, Billie Jo Deal, Sybil Derderian, Julie Douglas, Lori Emilson, Sherry Ezell, Betty Fast, Colleen Fitzgerald, Josephine Flammer, Kelli L. Gowdy, David Green, Kish Harris, Julie Hays, Ann Hefflin, Jeannie Hinyard, Cynthia Holcomb, Starin Lewis, Elizabeth H. Lindsay, Mary Manlove, Geoff Mihalenko, Laura Mihalenko, Kim Minafo, Jennifer M. Nelson, Kimberly Norman, Cheryl Radley, Amy Satkoski, Connie Semler, Shannon Singleton, Valerie Wood Smith, Eileen Tarconish, Pam Wheatley, Karen Widegren, Joyce Wilson, Eileen Wolejsza, Trisha Yates

**Production Team:** Lisa K. Pitts, Jennifer Tipton Cappoen (COVER ARTIST) Pam Crane, Rebecca Saunders, Jennifer Tipton Cappoen, Chris Curry, Sarah Foreman, Theresa Lewis Goode, Ivy L. Koonce, Clint Moore, Greg D. Rieves, Barry Slate, Donna K. Teal, Tazmen Carlisle, Amy Kirtley-Hill, Kristy Parton, Debbie Shoffner, Cathy Edwards Simrell, Lynette Dickerson, Mark Rainey, Clevell Harris, Karen Brewer Grossman

# Reading and Writing Numbers

## The Envelope, Please!

Put the spotlight on number-reading skills and listening skills with this simple activity. Begin by listing a different whole number for every three students in your class. Write a different form of each listed number (standard form, word form, and expanded form) on a separate slip of paper and place each slip in a separate envelope. Next, cut your original list into number strips and place them in a cup. To start, give one envelope to each child. Then pull a number from your cup and read it aloud. Invite each student to open his envelope to see whether he has the number that was called. Repeat the number as needed until all three of its forms have been identified. Have each student read his number aloud. Confirm each answer and then instruct students to switch envelopes. Return your number to the cup and repeat the activity. Now that's number practice that makes sense!

## Too High, Too Low

100 101 102 103 104 105 106 107 108 109
110 111 112 113 114 115 116 117 118 119
120 121 122 123 124 125 126 127 128 129
130 131 132 133 134 135 136 137 138 139
140 141 142 143 144 145 146 147 148 149
150 151 152 153 154 155 156 157 158 159
160 161 162 163 164 165 166 167 168 169
170 171 172 173 174 175 176 177 178 179
180 181 182 183 184 185 186 187 188 189
190 191 192 193 194 195 196 197 198 199
200

There's more than guesswork involved in this game—there's number-reading practice too! In advance, write the numerals from 100 to 200 (or another group of numbers that you wish to have students practice) on a large sheet of paper as shown. Laminate the page and then post it where it is easily visible. Secretly choose one of the numbers from the list and record it in a location where it will not be visible to students. Invite a student to guess the number. Tell the class whether the guess is too high or too low, and then use a dry-erase marker to cross out the eliminated numbers. Before the next guess, refer students to the chart and lead them in a discussion about which numbers are still available for choosing. Before long, the secret number will be revealed, and students' skills will be sharper!

## Getting Into the Act

Involving students in this class act really reinforces skills! To prepare, obtain a tennis ball and use a marker to draw a large comma on it. Also program ten 9" x 12" sheets of construction paper each with a different number from 0 to 9. Distribute four of the programmed cards each to a different child, and instruct the students to stand in front of the classroom and arrange themselves as desired to form a number. Next, tell students that the tennis ball will stand for the comma. Give another child the tennis ball and direct him to place himself correctly in the number. Guide seated students to read the number aloud. If desired, add extra skills to the lesson by posing questions such as "What is the value of Erika's number?" and "Which number is in the thousands place?" That's role-playing practice that helps students stay on the ball!

8 , 4 2 1

## House Hunting

Invite students to visit this nifty neighborhood and watch them feel more at home with number-reading skills. Write a four- to six-digit number on the board and ask students where to place the comma. Use brightly colored chalk to write the comma. Next, draw a simple house outline around each group of digits on either side of the comma. Tell students that the comma represents Thousands Street, the street that the first house of numbers lives on. Then guide students to read the whole number by reading each house number individually and the name of the comma in between. What a neighborly way to learn how to read numbers!

Find more student practice on pages 7–8.

Name____________________ Date____________

# Checking In

Read the number words.
Write the number on the line.
Color the key with the matching number.

1. seven hundred fifty-four ____________

2. eight hundred ninety-one ____________

3. one hundred thirty-two ____________

4. six hundred seventy-nine ____________

5. four hundred sixty-eight ____________

6. five hundred eighty-three ____________

7. three hundred thirty-five ____________

8. two hundred forty-six ____________

Name ______________________ Date ______________________

# King of the Number Jungle

Read each set of number words.
Write the matching numerals in the puzzle.

**Across**

3. four hundred three
5. six hundred twelve thousand, five hundred eighty
6. twenty-seven thousand, nine hundred forty-seven
8. five hundred forty-eight thousand, seven hundred ninety-two
10. one hundred sixty-eight

**Down**

1. five thousand, three hundred sixty-seven
2. sixty-one thousand, eight hundred thirty-seven
4. nine thousand, two hundred twenty-four
7. four hundred thirty-eight
9. two hundred thirty-six

# Counting to 1,000

## Lend a Hand!

Here's a math display that's sure to help students visualize 1,000. Give each child several sheets of construction paper. Direct her to trace each of her hands several times and cut out the resulting handprints. Stop when you have a total of 100 hands cut out. Then have each student tape her handprints along a wall to make one long row of prints. Students can count the paper fingers when counting by ones, fives, or tens. Very handy!

## Base Ten Buildings

Counting to 1,000 is fun, but building to 1,000 is even better! In advance, make a class supply of centimer graph paper in three colors and gather several sheets of 18" x 24" blue construction paper. Have students use the graph paper to create base ten sets of ones, tens, and hundreds (ones cut from the same color, tens from a different color, and hundreds from the third color). Challenge small groups to use the base ten sets to create a two-dimensional building that uses exactly 1,000 squares and includes ones, tens, and hundreds. Have teams glue their finished buildings onto the sheet of construction paper. Then encourage each team to add crayon details to the landscape. Display the buildings on a wall or bulletin board. In their free time, students can check out the buildings and verify that each has 1,000 squares. What an impressive skyline!

# Counting Pays!

Counting to 1,000 makes plenty of "cents" with this learning center! Gather a large supply of play nickels, dimes, quarters, half-dollars, and dollars at the center, along with five clean, empty containers with plastic lids. Label each container and its lid as shown. To make the center self-checking, label the bottom of the container with the appropriate number of coins or bills needed to make 1,000. (200 nickels, 100 dimes, 40 quarters, 20 half-dollars, or ten one-dollar bills). Display the cans and the money underneath a sign like the one shown. Challenge each student to fill each container with enough of the correct type of money to total $10. Students are sure to learn that there are many ways to count to 1,000!

$10.00 = 1,000 cents
How many ways can you count to 1,000?

## Have a Ball!

With this fast-paced game, counting to 1,000 is a ball! To make a die, wrap an empty square tissue box with white paper and label each side with a different one of the following directions: "count by 1s," "count by 2s," "count by 5s," "count by 10s," "count by 25s," "count by 50s." Have students sit in a circle. One student holds a ball and rolls the die. Depending on the roll of the die, announce a number to start counting from. The student who rolled the die states that number and the next number in the series and then passes the ball to the child on his left. That child states the next number and then passes the ball. Repeat until the group reaches 1,000. Then start the activity again. The numbers will just roll on!

Josh, please start the count at 680.

Find more student practice on pages 11–12.

Name ______________________________ Date ____________________

# Cleaning Crew

Write the missing numbers.

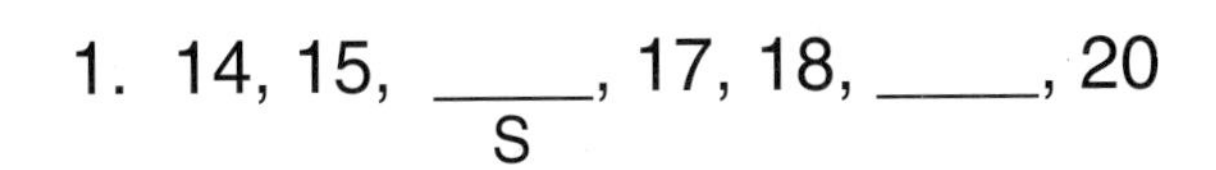

1. 14, 15, ____ (S), 17, 18, ____, 20
2. 136, 137, 138, ____, ____ (V), 141, 142
3. 261, 262, ____, 264, ____ (!), 266, 267
4. 398, ____, ____ (E), 401, 402, 403, 404
5. 524, 525, ____, 527, ____ (O), 529, 530
6. 675, 676, ____, 678, 679, ____ (T), 681
7. ____ (E), 740, 741, 742, ____, 744, 745
8. 852, 853, ____, 855, 856, 857, ____ (W)
9. 895, ____ (P), 897, 898, ____, 900, 901
10. 917, 918, 919, ____, ____ (R), 922, 923

**Why was the broom late?**

To solve the riddle, match the letters and punctuation above to the numbered lines below.

Because he ____ ____ ____ ____ ____ ____ ____ ____ ____ ____

| 528 | 140 | 739 | 921 | 16 | 858 | 400 | 896 | 680 | 265 |
|---|---|---|---|---|---|---|---|---|---|
| ____ | ____ | ____ | ____ | ____ | ____ | ____ | ____ | ____ | ____ |

Name________________________ Date______________

# Twists and Tangles

Read each rule.
Follow each ball of yarn.
Write the missing number in each box.

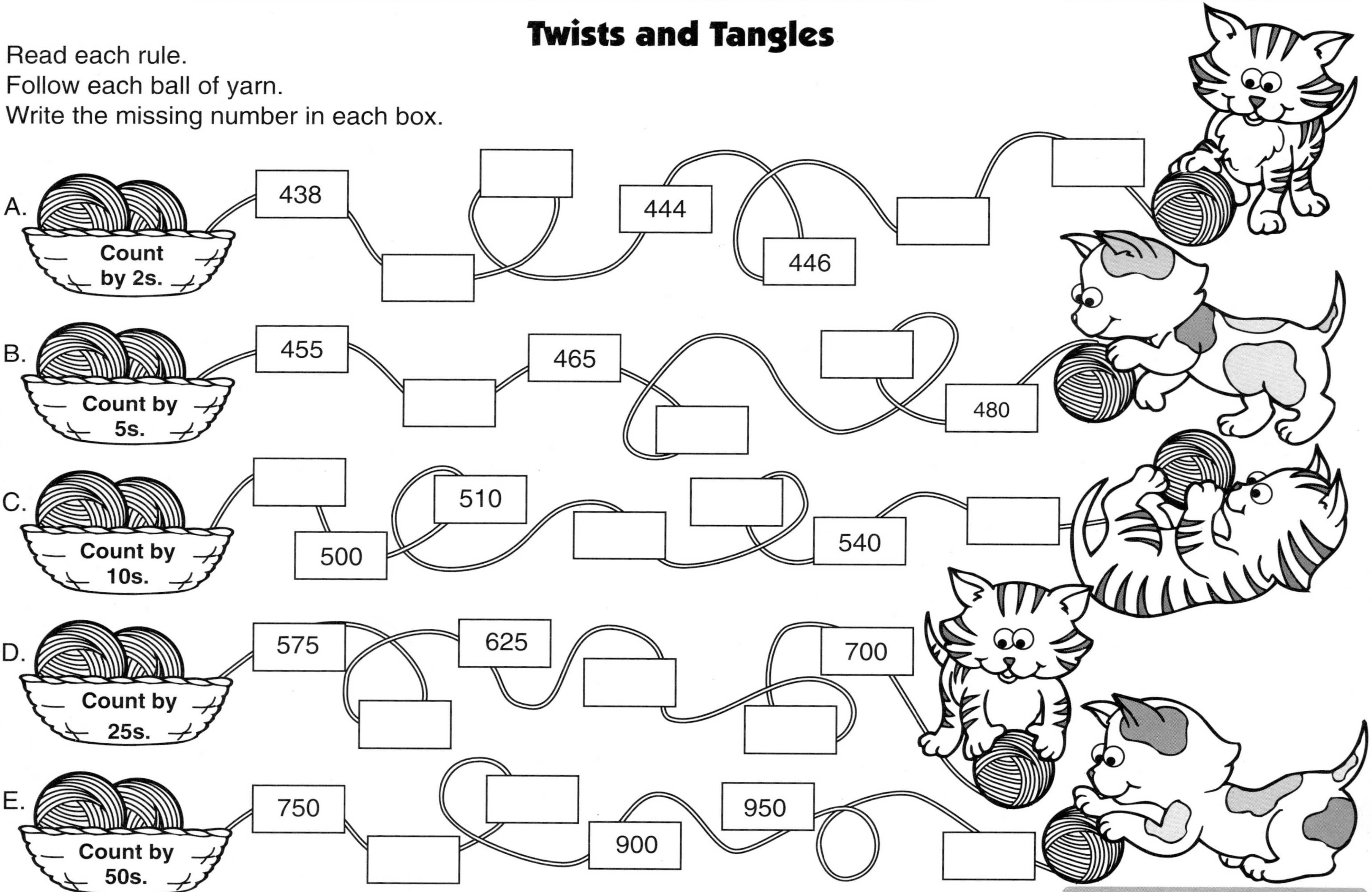

A. Count by 2s.
438 | ____ | ____ | 444 | 446 | ____ | ____

B. Count by 5s.
455 | ____ | 465 | ____ | ____ | 480

C. Count by 10s.
____ | 500 | 510 | ____ | ____ | 540 | ____

D. Count by 25s.
575 | ____ | 625 | ____ | ____ | 700

E. Count by 50s.
750 | ____ | ____ | 900 | 950 | ____

# Place Value

## Race to 100

Looking for a fun way to assess students' understanding of place value? Try this fast-paced partner game! To make a game mat, have each child fold a sheet of paper in thirds and program the resulting columns as shown. Give the child ten red Unifix cubes, ten yellow Unifix cubes, and one blue Unifix cube. Pair students and give each set of partners a die. Then guide each pair through the directions to play the game.

| hundreds (blue) | tens (red) | ones (yellow) |
|---|---|---|
| | | |

To play:

1. Each player places his game mat in front of him. He piles his Unifix cubes next to the mat.
2. Player 1 rolls the die. He counts out yellow Unifix cubes to match the number shown on the die and places them in the ones column on his game mat.
3. Player 2 takes a turn in the same manner.
4. Players continue taking turns. As each player gathers ten yellow cubes in the ones column, he trades them in for one red cube and places it in the tens column.
5. The first player to gather ten red cubes and trade them in for one blue cube is declared the winner.

## In the Cards!

Teaching place-value skills just got easier! In advance, remove the tens and face cards from a deck of playing cards. Divide the class into groups of six and provide each group with a laminated sheet of blank paper, a dry eraser marker and a paper towel. Give five players in each group a card. Direct the group to arrange its cards to make the largest possible number. Have the sixth person record the group's number on the paper. At your signal, instruct students to display the numbers. Award a point to each group that created the largest possible number with its cards. Next, direct each player with the laminated paper to pass it to someone else in her group. Collect the cards, shuffle, and redistribute them to play additional rounds.

Find more student practice on pages 14–15.

Name________________________ Date______________

# Bowling Along

Write each number from the bowling ball on the matching pin.

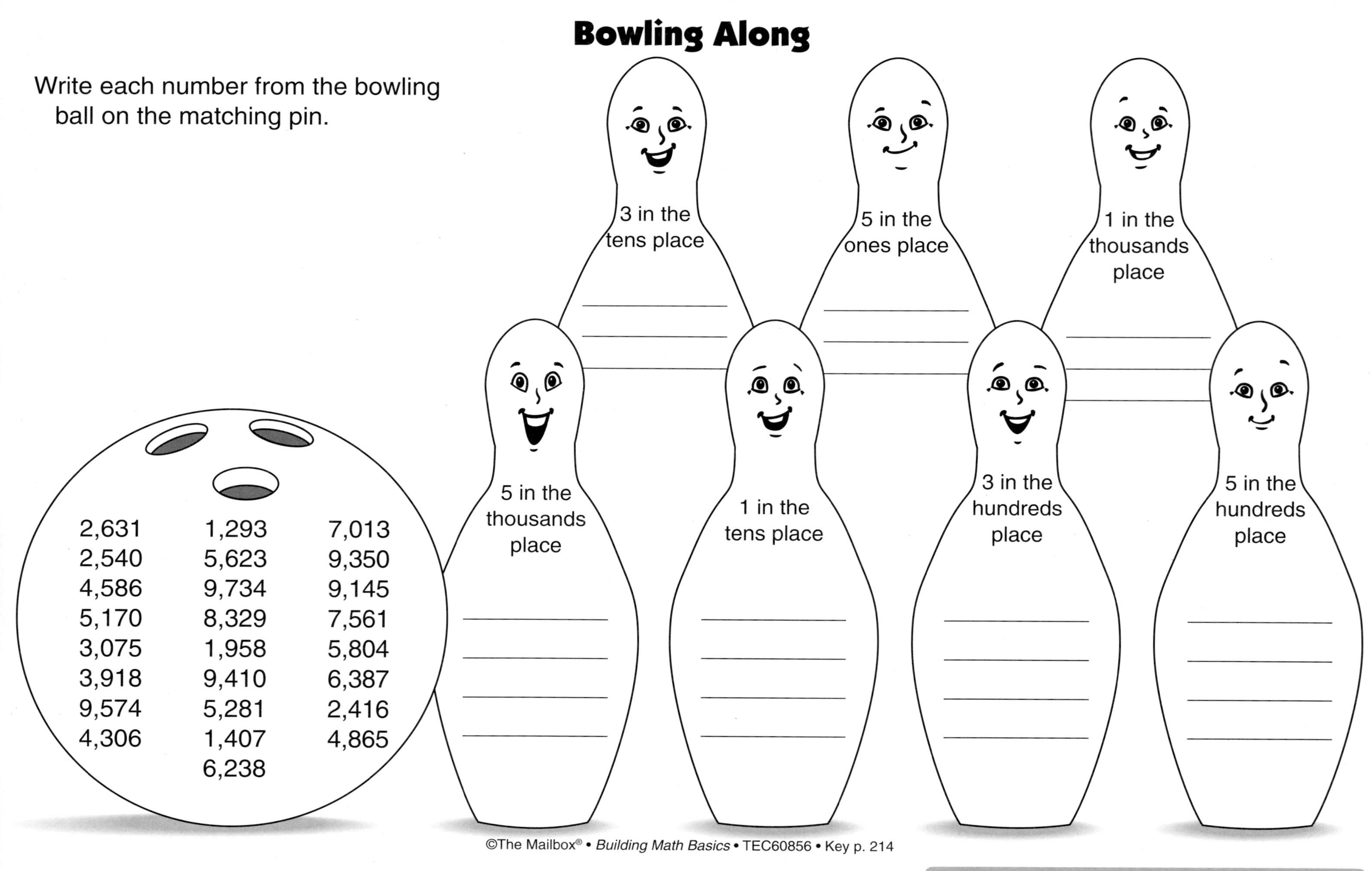

Name ______________________________ Date ____________________

# Happy "Buzz-day"!

Write a number for each clue.
Use the numerals on the matching candle in the number you write.

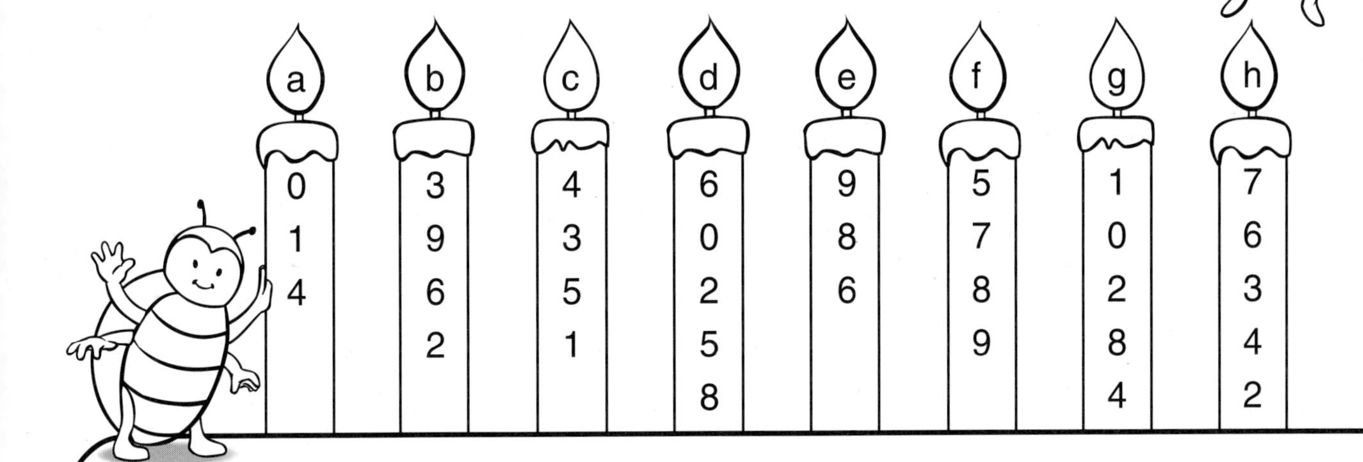


a. Form the greatest three-digit number with a 0 in the tens place. ___ ___ ___

b. Form the smallest four-digit number with a 6 in the thousands place. ___, ___ ___ ___

c. Form the smallest four-digit number with a 5 in the ones place and a 3 in the hundreds place. ___, ___ ___ ___

d. Form the greatest five-digit number with a 2 in the ten thousands place. ___ ___, ___ ___ ___

e. Form the smallest three-digit number with an 8 in the ones place. ___ ___ ___

f. Form the smallest four-digit number with a 9 in the hundreds place. ___, ___ ___ ___

g. Form the greatest five-digit number with a 2 in the thousands place and an 8 in the tens place. ___ ___, ___ ___ ___

h. Form the smallest five-digit number with a 7 in the ten thousands place. ___ ___, ___ ___ ___

# Rounding and Estimating

## Stepping Into Estimating

This quick tip helps students get their estimation skills in line. When students line up to walk to another new place in the building, have them estimate the number of steps it will take them to get there. Record their estimates on the board. As you lead students through the halls, appoint one student to be the official counter and count each step in a hushed voice. Once students return to the classroom, have them compare the official count to their estimates. After students have counted the steps to each place in the building, give them a new challenge by having them estimate and count the steps it takes to get from location to location on the playground. Ready? March!

## How Many Flakes?

When it's cold outside, this friendly, flaky pal helps students practice their estimation skills! Cut three large circles out of white paper and mount them on the wall to form a snowman. Display a package of small lace doilies and ask students to estimate how many doilies it will take to cover the snowman. Record their estimates on a sheet of paper posted near the snowman.

Next, begin taping the doilies on the top circle. Have a student volunteer make a tally mark on the board each time a doily is added. Once the circle is covered, encourage students to modify their estimations as you record their new ideas on the paper. Then have students continue to tape doilies and make tally marks. When the snowman is covered, direct students to count the marks and compare the results to their estimates. Brrr!

# Rounding the Mountain

Scaling this mountain gives students a visual reference for rounding numbers! On the board, draw the diagram shown below (omitting the numbers in red). Then demonstrate how to use the diagram. Write the number(s) you want to round, such as 73 and 78, under the mountain and the rounded down and rounded up numbers on either side (70, 80). Point to the number 73 and call students' attention to the digit in the ones column. Tell them to imagine that they are pushing a heavy ball up the mountain and must stop the ball at the given digit. For example, for 73, ask students to explain what will happen to the heavy ball when it stops on the 3. (It will roll back down the left side of the mountain.) Tell them that this means the number 73 should be rounded to 70. Next, point to the number 78. Explain that if the heavy ball is pushed over the top of the mountain and stopped on the 8, it will roll to the other side of the mountain. This means that 78 should be rounded to 80. Guide students through several more examples and then post the drawing in a clearly visible place. Encourage students to use it as an at-a-glance rounding reference!

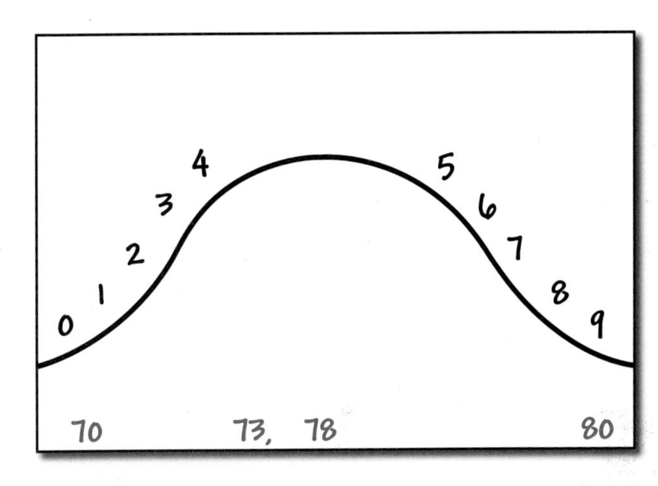


## Benchmark Bags

Closer estimates are in the bag with this center activity! Label ten resealable bags with the letters *A* through *J* and place between one and 60 beans in each bag. List the number of beans in each bag on an answer key. Also, pour a supply of beans into a bowl. Place these supplies along with three empty resealable bags and a class supply of blank paper at a center. Have each student estimate the number of beans in each labeled bag and record his estimates on a sheet of paper. Next, tell him to make a set of benchmark bags by placing ten beans in one empty bag, 25 beans in a second bag, and 50 beans in a third bag. Then direct the student to refer to the benchmark bags as he revises each estimate. Have him record his revised estimates on the paper. Direct him to check the answer key and record the actual number of beans in each labeled bag and tell him to circle the closer estimate for each bag. Finally, have the student empty the beans from his benchmark bags back into the bowl.

# Comparing and Ordering Numbers

## Puzzling for Points

Combine the fun of puzzles and the skill of ordering numbers with this self-checking partner game! To make the game, write a set of numbers on a sentence strip, leaving space between each. Then, to make puzzle pieces, cut the numbers apart so that they fit together in only one way. Place the puzzle pieces in a plastic bag. Repeat, cutting out the same number of pieces each time. Challenge partners to each choose a bag of numbers, set an egg timer, and race to see who puts his puzzle together more quickly. Once time is up, students count the number of pieces that they put in order. The student who ordered more puzzle pieces after five rounds of play wins!

## The Greatest Race

Keep young mathematicians on their toes with this quick game. Prepare several sets of numbers for students to order. Write each set on a different color of paper. Make sure you have one number per child. For beginning learners, create multiple sets so there are fewer numbers in each set. For more advanced learners, create one or two sets so that the ordering is more complex. As class begins, have each student randomly select a number. Explain that students must keep their numbers handy because sometime during the day, you'll call out, "Great Race!" Students will then find others with the same color of paper and line up holding their numbers in order from least to greatest. Whew, what a great race!

Find more student practice on pages 19–20.

Name ______________________ Date ______________________

# Snail Mail

Color each envelope below by the code.
Cut out the envelopes.
Glue them in order from **least** to **greatest.**

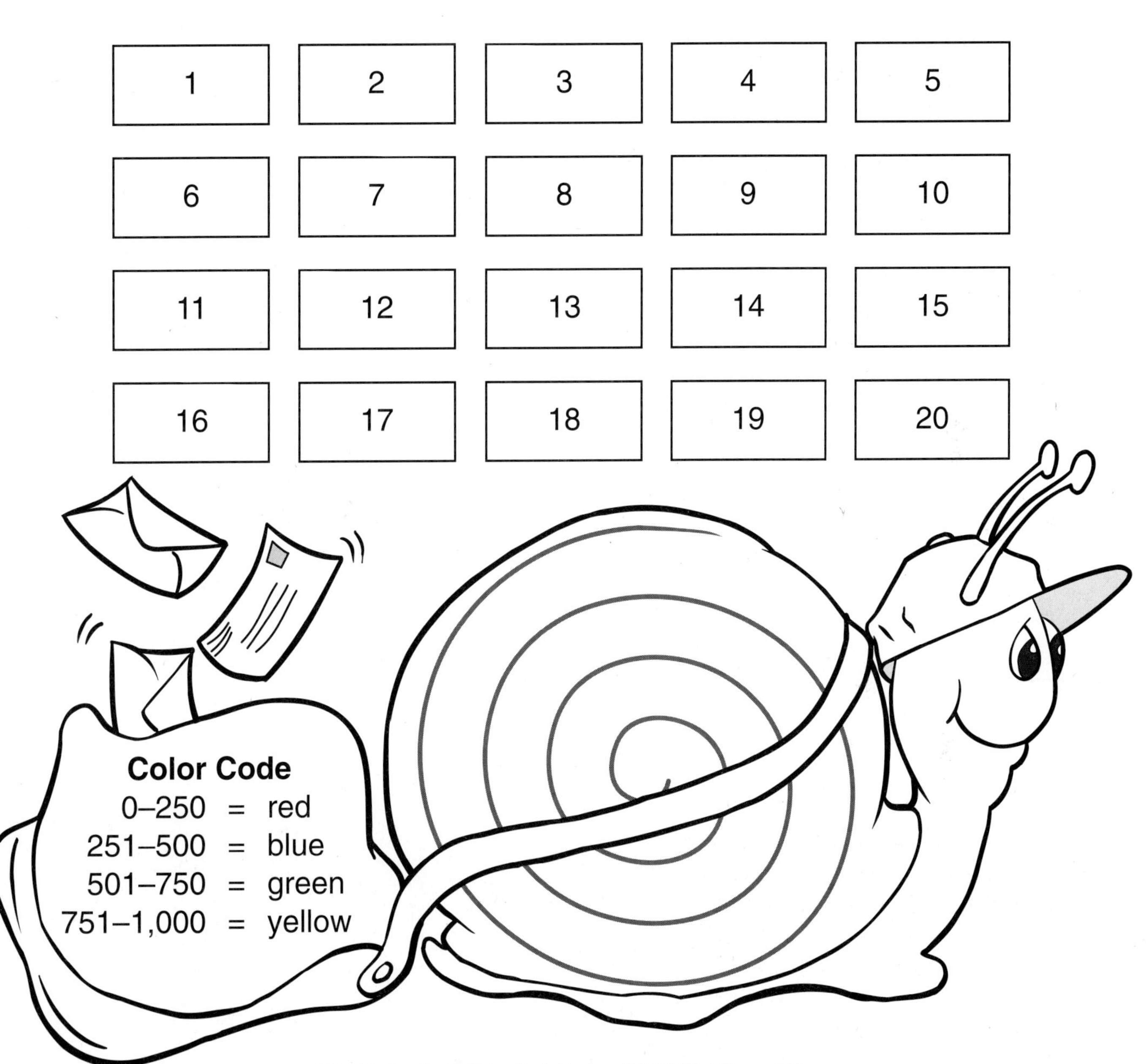




| | | | | | | |
|---|---|---|---|---|---|---|
| 297 | 43 | 703 | 999 | 75 | 184 | 817 |
| 51 | 947 | 607 | 315 | 103 | 537 | 898 |
| 462 | 125 | 498 | 179 | 427 | 217 | |

Name ________________________________ Date ____________________

# Ready, March!

In each box, color the leaf with the greatest number.

Look at the numbers on the leaves you did not color.
Write them in order from least to greatest.

1. ______ 2. ______ 3. ______ 4. ______ 5. ______ 6. ______

7. ______ 8. ______ 9. ______ 10. ______ 11. ______ 12. ______

# <Inequalities>

## Al the Alligator

Munching on this fishy activity boosts students' skills! Give each child 20 fish-shaped crackers and a six-inch paper plate. Show the child how to cut a wedge out of the plate and draw a mouth around the resulting angle. Then have him decorate both sides of the plate to resemble an alligator as shown. Direct each child to place three crackers on the left side of his desk and six crackers on the right side.

Next, tell the story of Al the Alligator, who is so hungry that he wants to eat as much as he can. But when the fish see him coming, they swim away! He only has time to eat one group of fish before the other group is gone. Ask students whether they think Al will eat the pile of six or the pile of three. Then have them place the plate in the center of the two piles so that Al's mouth is open toward the pile he wants to eat. On the board, show them how to write the number sentence $3 < 6$. Guide students through several examples with different numbers of fish. Finally, explain that Al has eaten so many fish that he is full. Then invite them to eat the fish Al has left behind!

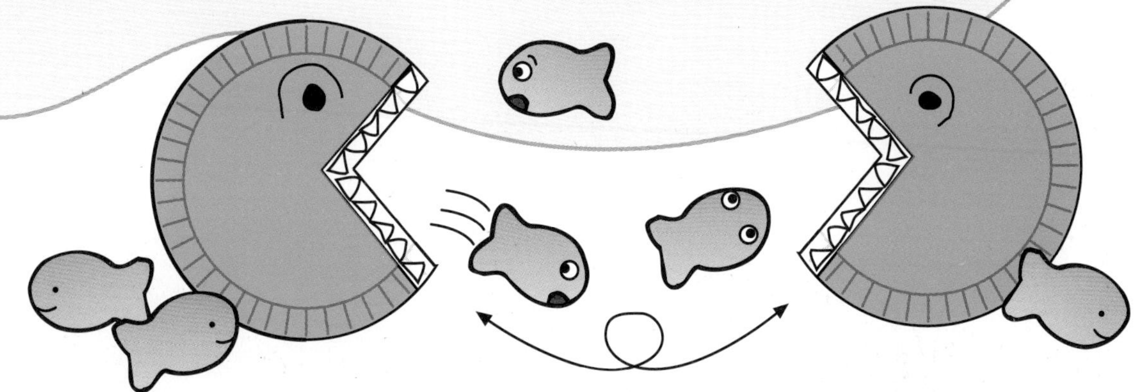

## Grab Bag!

Students scoop up a handful of practice comparing numbers with this whole-class activity! Program three 9" x 12" poster board sheets each with a different comparison symbol (<, >, and =). Place the sheets and two bags of cotton balls on a table. Invite two students to stand by the table. Next, have each student reach into a bag and take one handful of cotton balls. Instruct the child to count her cotton balls and then confer with the other child about the number of cotton balls each is holding. Then direct the pair to choose the appropriate programmed sheet, announce its numbers, and show the sheet it has chosen. Have seated students confirm the pair's choice by giving a thumbs-up or thumbs-down. Scan the seated students for understanding and then invite a new pair of students to repeat the activity.

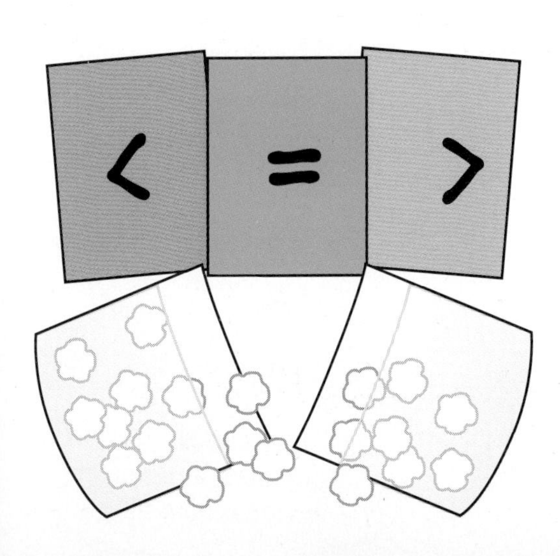

# Acting Out Inequalities

This small-group activity puts students' role-playing skills to work! Divide the class into groups of four and have each group choose one member to play the More Muncher. Next, direct two other group members to each choose a random two-digit number and write it on an index card. Instruct them to stand on either side of the More Muncher, show him their numbers, and have him decide which comparison symbol is appropriate. To show which symbol he has chosen, the More Muncher opens his arms toward the student with the greatest number. Or, if the symbol he chooses is "equal to," he holds both arms straight across his chest as shown. Direct the group to decide whether its equation is correct, and have the fourth member record the group's number sentence on a sheet of paper. Then have students switch roles within their groups and repeat the activity. Ready? Action!

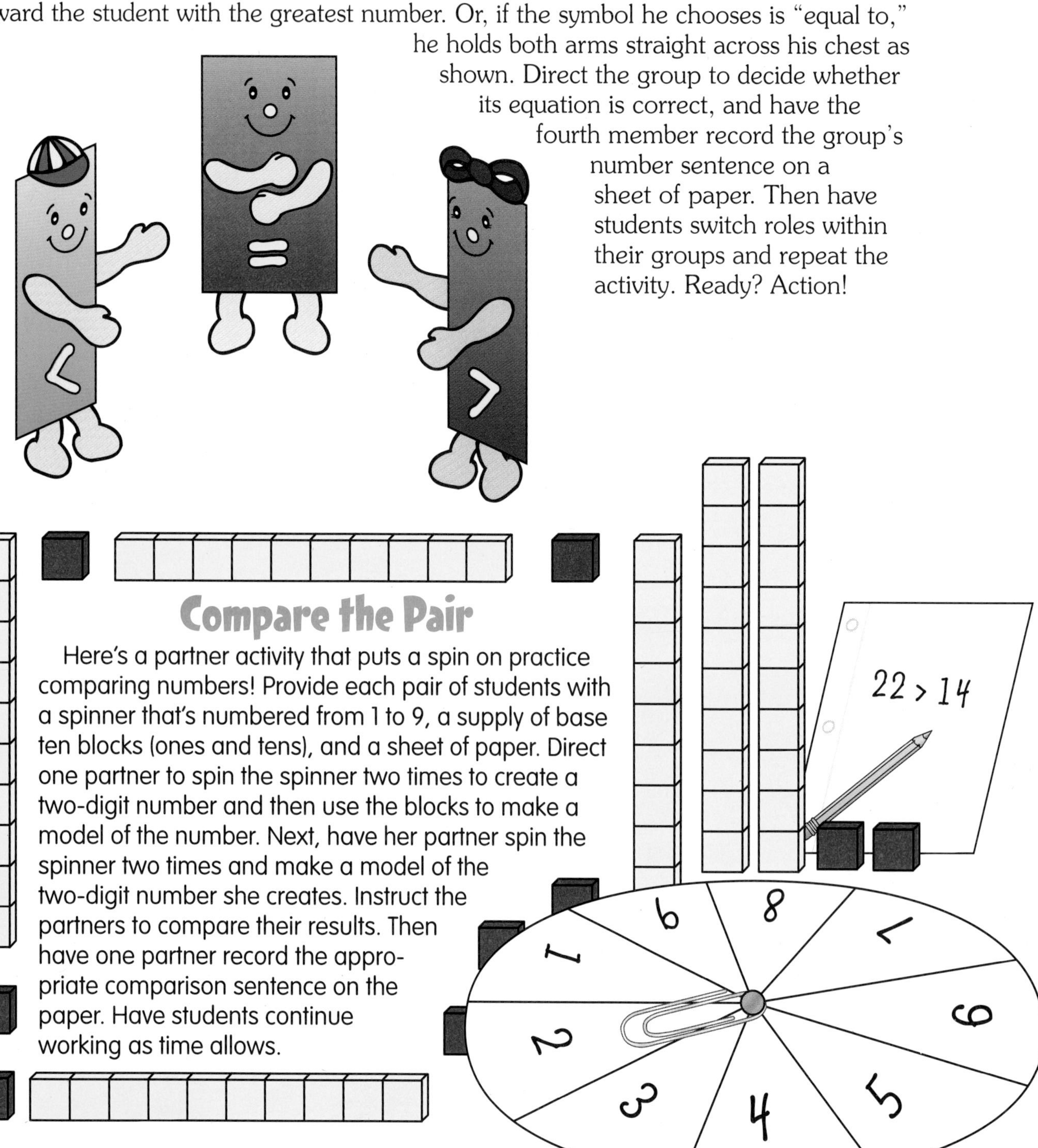


## Compare the Pair

Here's a partner activity that puts a spin on practice comparing numbers! Provide each pair of students with a spinner that's numbered from 1 to 9, a supply of base ten blocks (ones and tens), and a sheet of paper. Direct one partner to spin the spinner two times to create a two-digit number and then use the blocks to make a model of the number. Next, have her partner spin the spinner two times and make a model of the two-digit number she creates. Instruct the partners to compare their results. Then have one partner record the appropriate comparison sentence on the paper. Have students continue working as time allows.

Find more student practice on pages 23–24.

Name ______________________________ Date ____________________

# It's Raining Acorns!

For each acorn, color the part with the greater number green.
Color the part with the lesser number brown.
Use the numbers to complete the number sentence.

A. 
214
241

_____ > _____

B. 
304
340

_____ < _____

C. 
195
591

_____ > _____

D. 
682
826

_____ < _____

E. 
630
603

_____ > _____

F. 
794
947

_____ > _____

G.
289
298

_____ > _____

H.
551
515

_____ < _____

J. 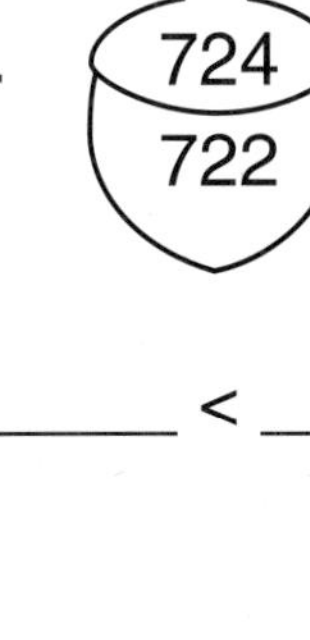
724
722

_____ < _____

K. 
918
891

_____ > _____

I.
125
152

_____ > _____

L.
473
374

_____ < _____

M.
572
275

_____ < _____

N.
547
475

_____ > _____

Name ______________________________ Date ____________________

# Picking Up Pennies

Write a two-digit number that makes each sentence true.

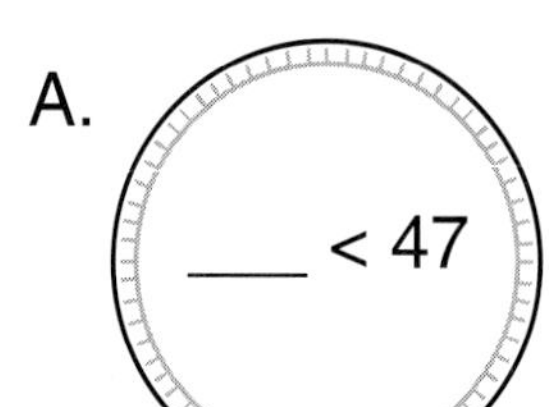

A. ___ < 47

B. 17 > ___

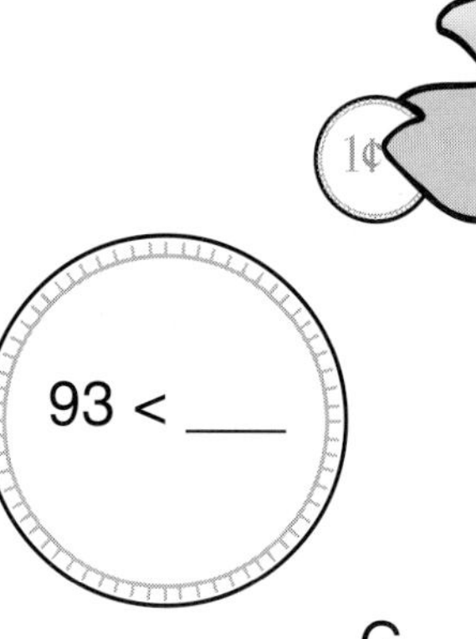

D. 93 < ___

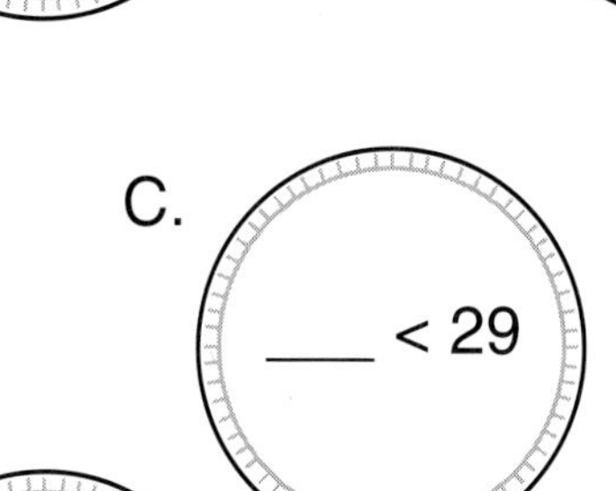

C. ___ < 29

G. ___ < 11

E. ___ < 81

F. 75 > ___

K. 49 < ___

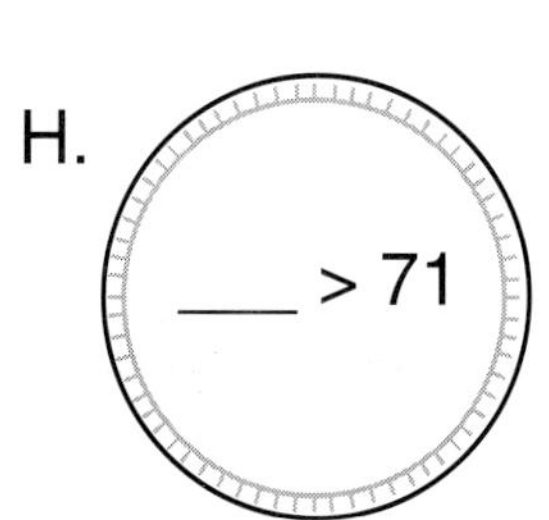

H. ___ > 71

J. ___ > 34

I. 23 > ___

O. 86 < ___

N. ___ > 60

L. ___ < 15

M. 24 > ___

P. ___ > 98

# Odd and Even Numbers

## Small-Group Tic-Tac-Toe

This variation on traditional tic-tac-toe provides small-group practice with even and odd numbers. Program a tic-tac-toe board with the word *even* or *odd* in each space. Make a copy for each student in the group. Then program index cards with the numbers from zero to 20. Place the stack of cards facedown. Explain that to play the game, each player, in turn, selects a card from the top of the stack. She tells the group whether the number is even or odd. If the other players agree that she is correct, she uses a game marker to cover a corresponding space on her card. If she is incorrect, she skips a turn. Players take turns selecting cards and covering spaces until one player covers three spaces in a row horizontally, vertically, or diagonally. Invite the groups to play several rounds until each player has had a chance to go first.

| odd | even | odd |
|---|---|---|
| even | | |
| odd | even | odd |

13

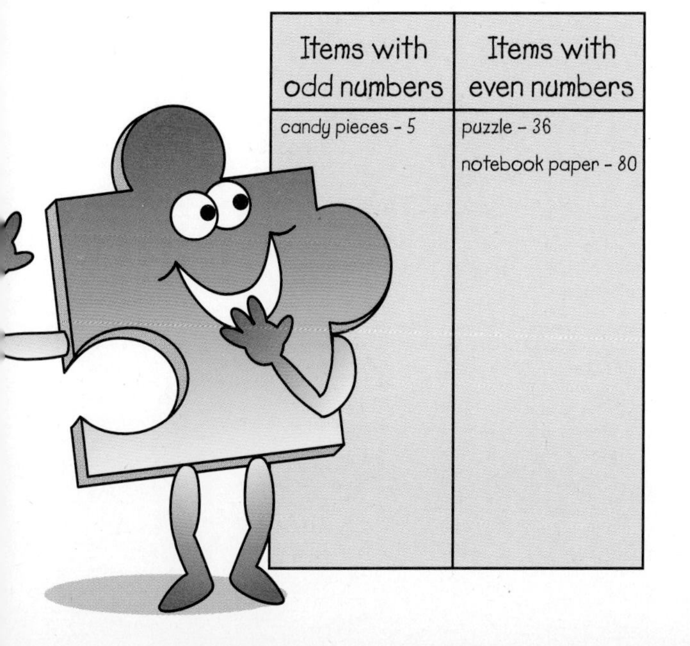


## A Truly Odd (and Even) Scavenger Hunt

Challenge students to see the many odd and even numbers around them with this simple scavenger hunt. To prepare, each child folds a sheet of paper in half. He labels one section "Items with even numbers" and the other section "Items with odd numbers." Next, explain that students will have five minutes to search the classroom to find items that have odd or even numbers on them. For example, if a child finds a puzzle labeled "36 pieces," he writes "Puzzle–36" in the even column. Remind students to record both the number they found and the name of the item containing the number. After the game, collect students' findings on a class chart titled "Even and Odd Numbers Are All Around!"

## Clap, Clap, Clap

Combine the fun and movement of Simon Says with a chance to practice evens and odds by having students play this game. Program two index cards with the word *odd* and two index cards with the word *even*. Place the cards in a bag. Explain to students that you will draw one card from the bag, hold it up for everyone to see, and then give a Simon Says directive, such as "Simon says clap your hands three times." The trick is that students should follow Simon's directions only if the number stated matches the description of the card that was drawn. Mix the cards and repeat with new directions. Simon says that this is a fun way to learn evens and odds!

## Climbing to the Top of the Ladder

Your students will be tops at identifying even and odd numbers through 100 with this partner game. In advance, draw two simple ladders with ten rungs each. Label all of the rungs on one ladder "even" and all of the rungs on the other ladder "odd." Cut apart a hundred chart and place the numbers in a paper lunch bag. Add an identical paper square with a zero on it. Explain to your students that each player places a game marker on the bottom rung of her ladder. In turn, each student draws a number. If the answer matches the label on her ladder, she moves her game piece ahead one rung. If the answer doesn't match, she does not move ahead. The first player to climb to the top of her ladder wins!

## Odds and Evens Are So Easy

*(sung to the tune of "If You're Happy and You Know It")*

To remember odds and evens, sing with me!
Sing with me!
Zero, two, four, six, eight are evens; yes, you see!
Yes, I see!
One, three, five, seven, and nine—they are odd; they are divine.
Odds and evens are so easy; yes, you see!
Yes, I see!

Name ______________________ Date ______________________

# A Balancing Act

Cut out the weight cards below.
Glue each weight on the matching side of the scale.

| 99 | 521 | 178 | 72 | 67 | 351 | 18 | 744 | 139 | 50 |
|---|---|---|---|---|---|---|---|---|---|
| 390 | 46 | 466 | 21 | 837 | 234 | 410 | 35 | 83 | 455 |

Name ______________________________ Date ____________________

# Tasty Tacos

If the number is even, color the taco shell yellow.
If the number is odd, color the taco shell orange.

A.
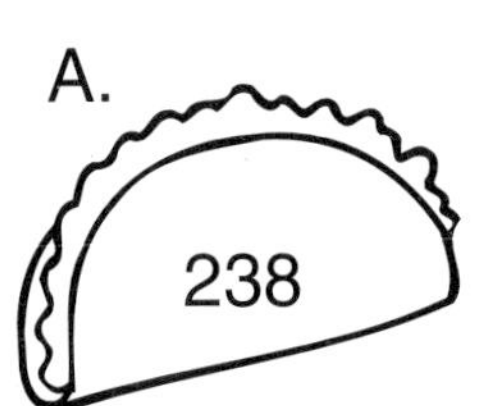


B.
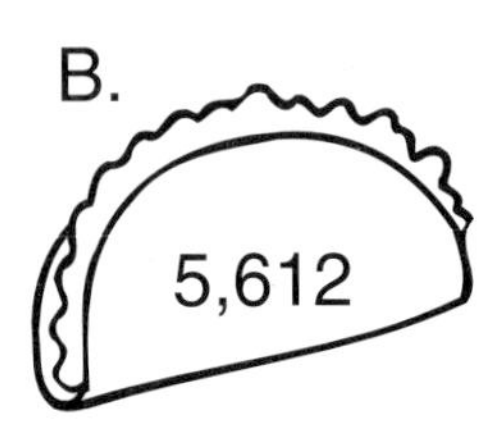


C.
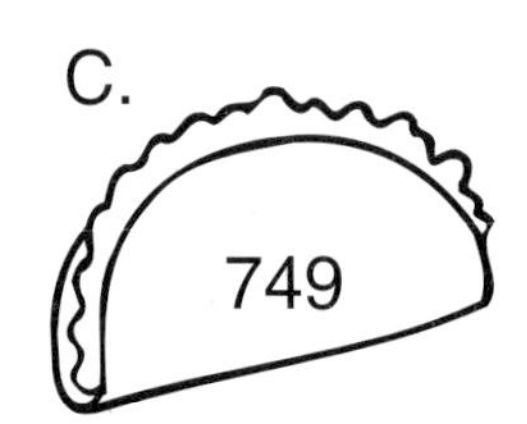


D.
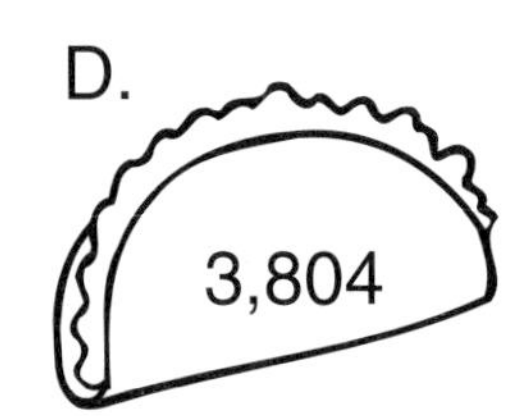


E.
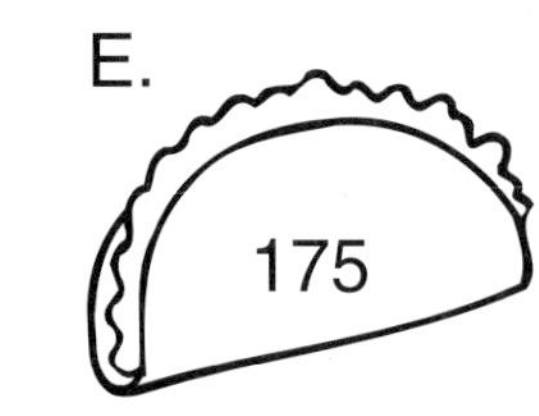


F.
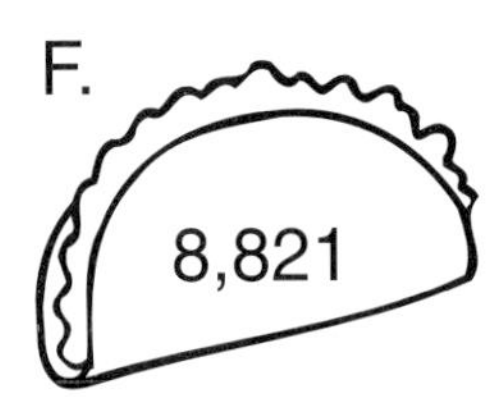


G.
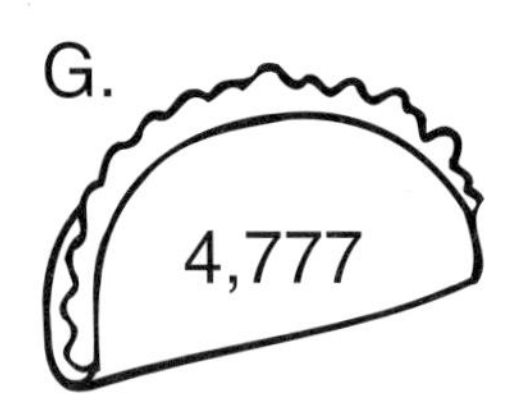


H.
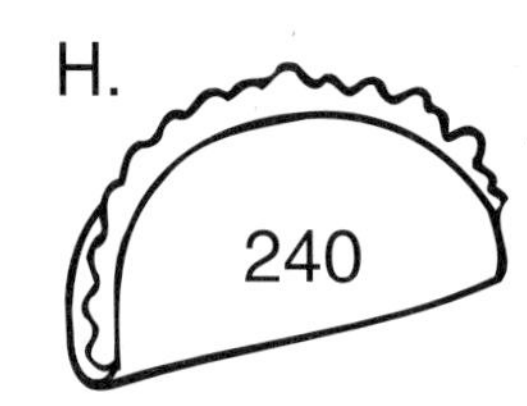


I.
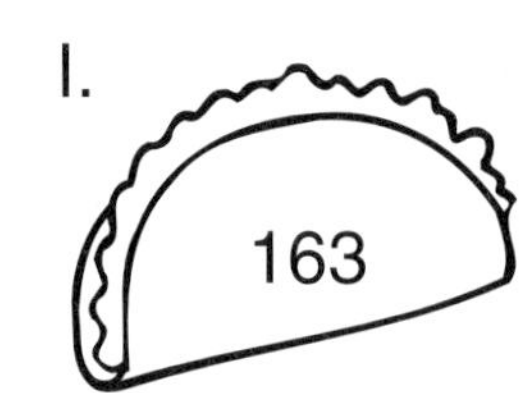


J.
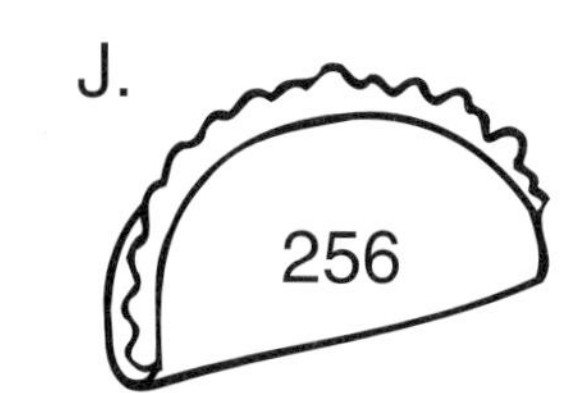


K.
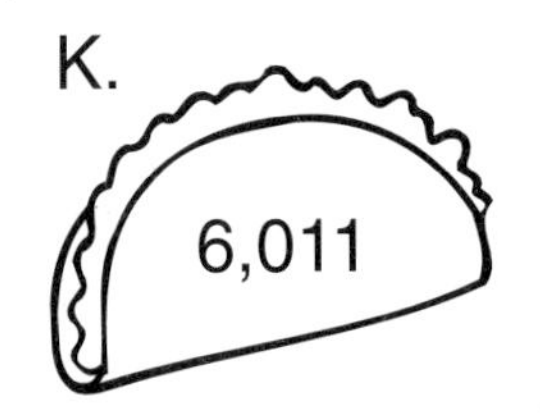


L.
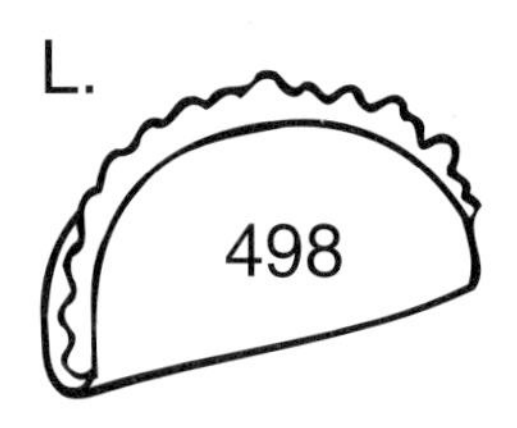


M.
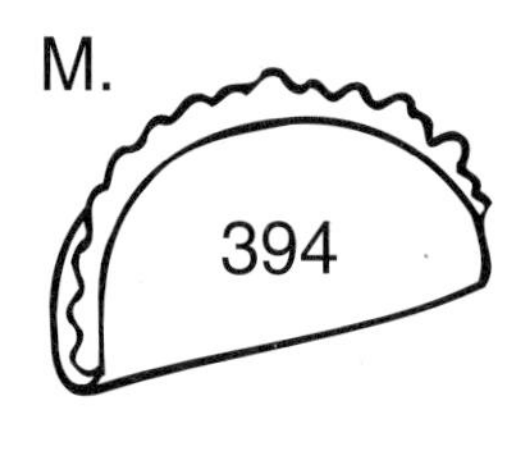

N.
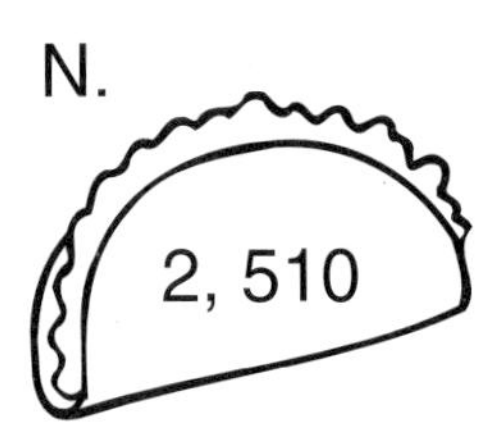


O.
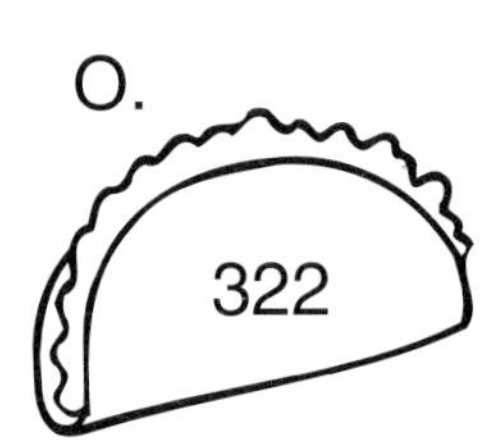


P.
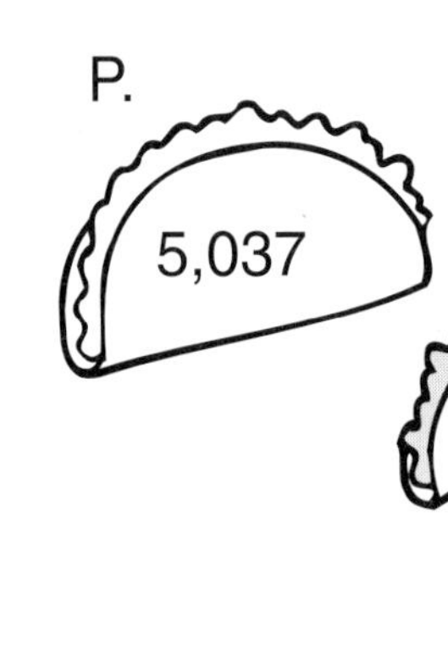


Q.
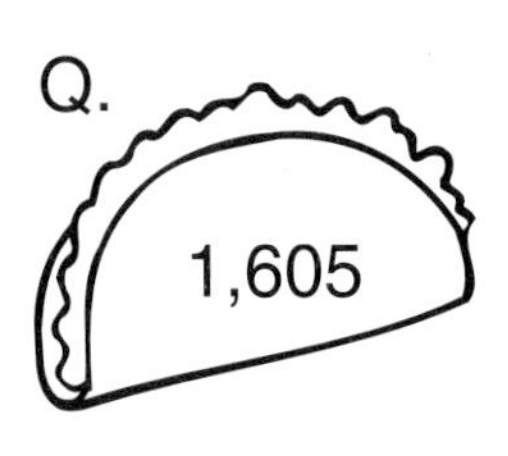


R.
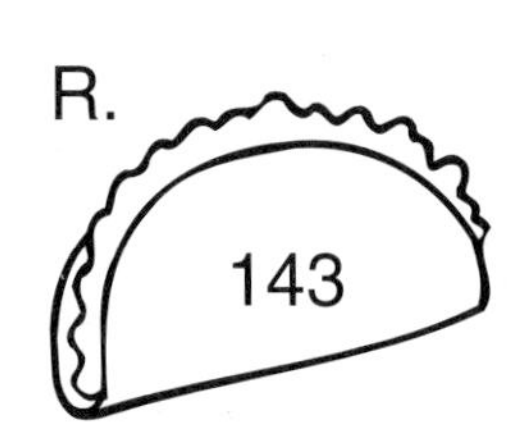

# Addition Facts to 18

## Stick to the Facts!

Looking for a new twist on flash card practice? Add craft sticks to the activity and practice becomes a partner game! To prepare a center, use a green marker to color one end of each of 10 craft sticks. Color one end of each of 10 additional sticks red. Place the sticks in a cup so that the colored ends are not visible. Place the cup and a stack of addition facts flash cards at a center. To play the game, guide the pair through the steps below.

**Steps:**

1. Player 1 draws a stick. If he draws a green stick, Player 2 holds up a flash card with an addition problem for him to solve. If he answers correctly, he keeps the card. If his answer is incorrect, the card is returned to the bottom of the stack.
2. Player 1 continues drawing sticks and solving problems until he draws a red stick. After his turn, he returns the sticks to the cup and gently shakes the cup before passing it to Player 2.
3. Player 2 repeats Steps 1 and 2.
4. Players continue taking turns as time allows. When the game is over, each player counts his cards. The player with more cards is the winner.

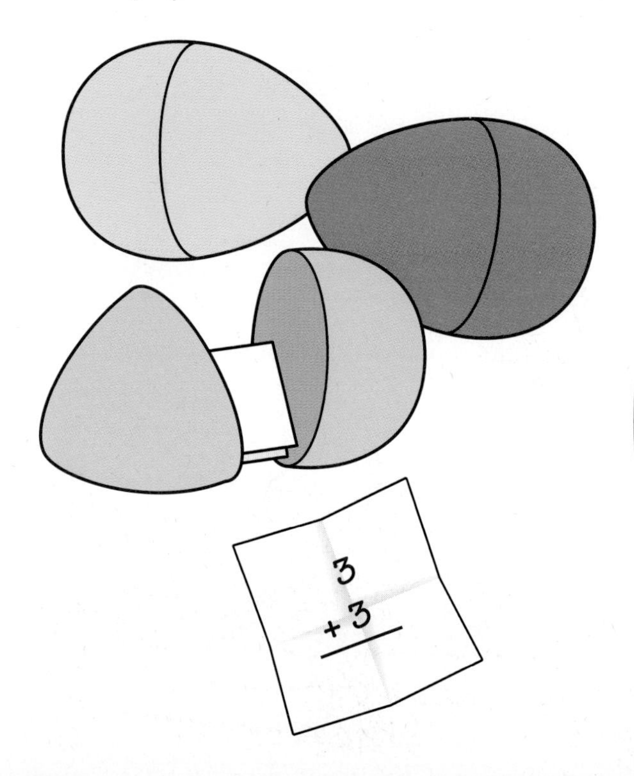


## "Eggs-tra" Special Hunt

Any day is a holiday with this child-pleasing activity! In advance, cut enough three-inch paper squares for each child to have four. Program each square with a different addition fact, fold it, and tuck it into a colorful plastic egg. Next, hide the prepared eggs on the playground. Give each student a sheet of paper and a pencil and then lead students to the playground. Instruct each child to find four eggs. Once she has all four eggs, have her open each egg and copy the problems onto her paper. Then direct her to solve the problems and return the squares to their eggs. Check the student's work and then award her with a sticker or other small treat for each correct answer. If desired, invite students to hide their eggs and repeat the activity.

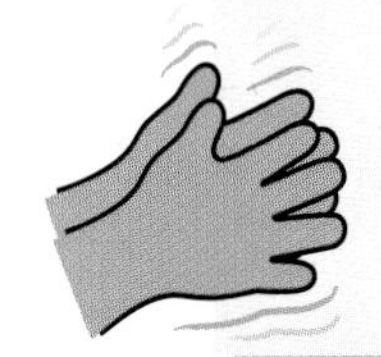

## Clap and Add

Students are sure to applaud this fun practice method! Demonstrate the activity first by having a student volunteer join you in front of the classroom. Show the child how to clap his hand twice and then, on the third beat, hold up his fingers to show a number from 1 to 10. Next, invite the child to join you as you repeat the steps. Announce and solve an equation made from the numbers displayed. For example, if the child is holding up seven fingers and you are holding up three fingers, say, "Seven plus three equals ten." Have the child confirm your answer. (If necessary, help the child count the displayed fingers to find the correct answer.) Then pair students and have each pair practice using the clap-and-add method to review math facts. Clap, clap, add!

## Flying High With Math

This colorful, interactive bulletin board center sends students' addition skills soaring! To make hot-air balloons, cut large circles from colorful construction paper, laminate them, and then mount them on a bulletin board covered with blue paper. To make the balloon baskets, cut the tops off clean, empty juice boxes and then cover each box with construction paper. Mount each box below one of the balloons and then add yarn details as shown. Use a dry-erase marker to program each balloon with a different sum. Next, prepare flash cards by programming several 2" x 4" tagboard strips each with a different addition fact that has a sum matching one of the sums on the balloons. To make the center self-checking, use a crayon that matches the color of the balloon with the correct sum to draw a dot on the back of each strip. Store the strips in a resealable plastic bag near the bulletin board.

To use the center, a student places each strip in the appropriate basket. Then, to check her work, she removes each set of strips and confirms that the color of the dot on the back of each strip matches the color of the balloon. To change the center, simply reprogram the circles and create new strips as described. Off we go into the wild blue yonder!

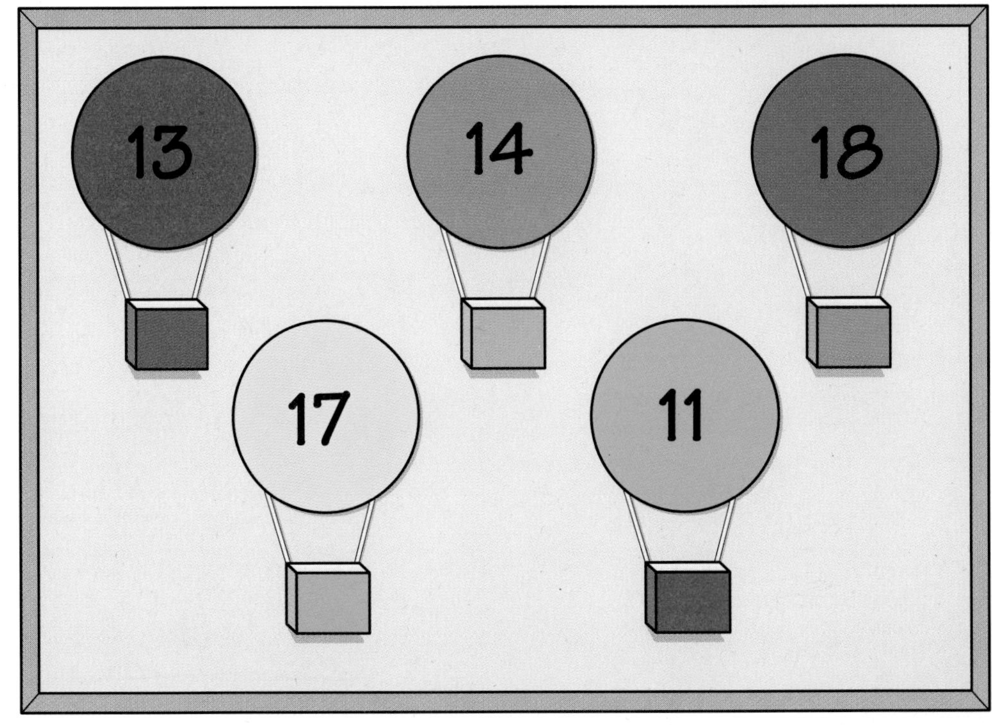

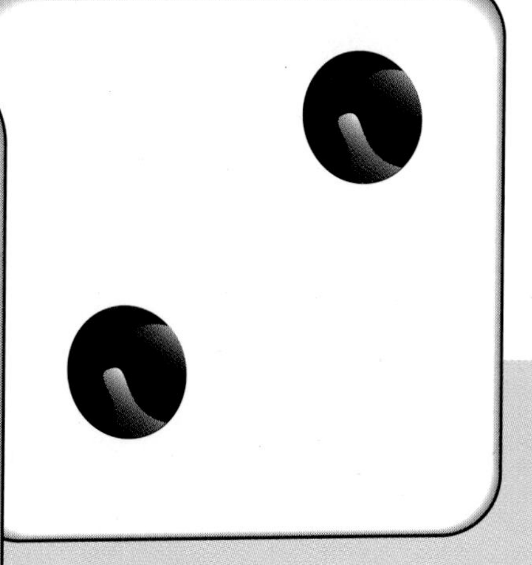

## Roll and Flip

Students will be on a math fact roll with this fast-paced partner game! Pair students and give each partner 11 blank cards. Have the child number one side of each card with a different number from 2 to 12. Next, instruct each partner to arrange her cards number side up in numerical order on the playing surface in front of her. Direct one partner to roll a pair of dice, add the two numbers shown on the dice, and then flip over the card that matches the sum. Then have the second partner take her turn in the same manner. If a child rolls a sum that she has already flipped over, she loses her turn. Play continues until one child has flipped over all of her cards and is declared the winner.

## Deal!

Math fact practice just got easier! Remove the face cards and aces from a deck of playing cards and then give the remaining cards to a pair of students. Direct one partner to deal the cards so that he and his partner each have an equal number of cards. Next, have each partner place his stack of cards facedown in front of him. Direct each one to turn over the top two cards in the stack, add the two numbers shown, and then compare his sum with his partner's. The child with the higher sum takes all four cards and places them facedown on the bottom of his stack. If the sums are identical, the partners both go again until one has a higher sum; that student takes all the upturned cards. Play continues as time allows. At the end of the game, each child counts his cards. The child with more cards wins!

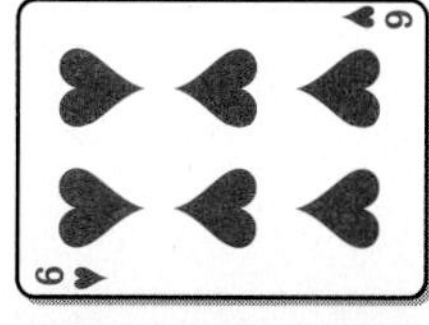

Find more student practice on pages 32–36.

Name________________________ Date______________

# Tulip Totals

Find the two numbers on each tulip with a sum that matches the ladybug's sum. Circle the two numbers.

Name ______________________________ Date ____________________

# Special Delivery

Use the numbers on the boxes to make facts that have the sum shown on the truck. Write the facts below the truck.

A.

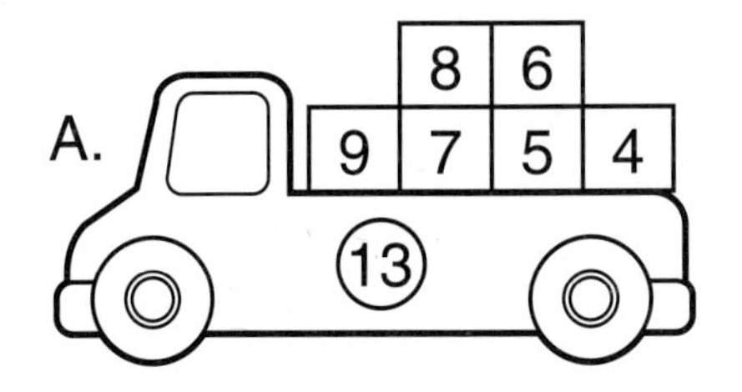


____ + ____ = _____

____ + ____ = _____

____ + ____ = _____

B.

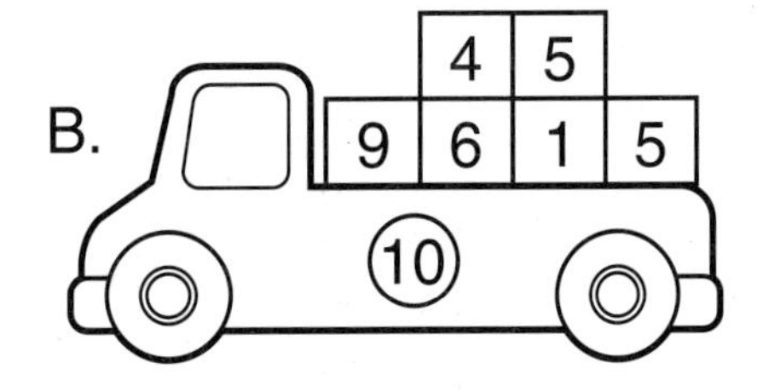


____ + ____ = _____

____ + ____ = _____

____ + ____ = _____

C.

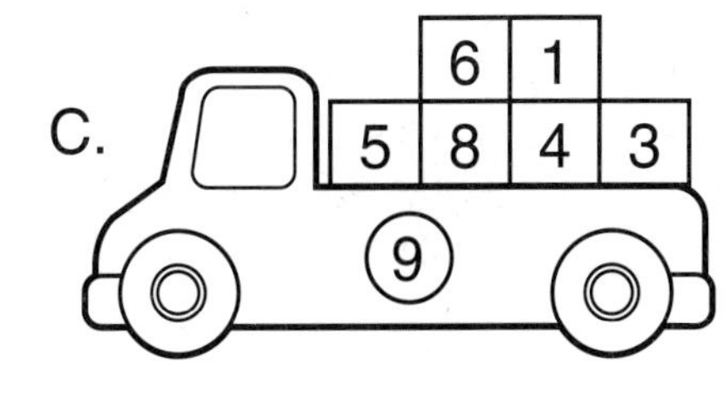


____ + ____ = _____

____ + ____ = _____

____ + ____ = _____

D.

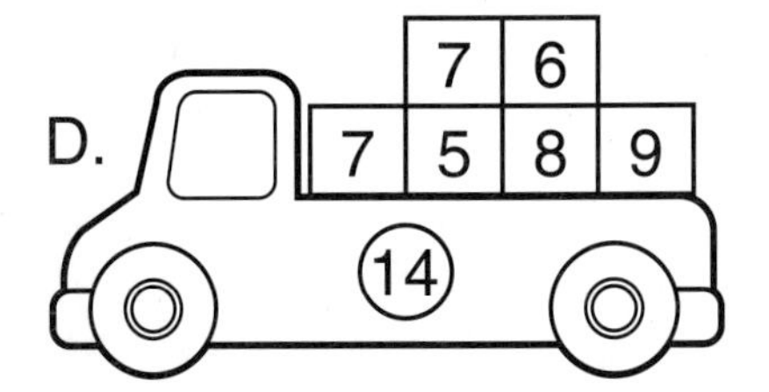


____ + ____ = _____

____ + ____ = _____

____ + ____ = _____

E.

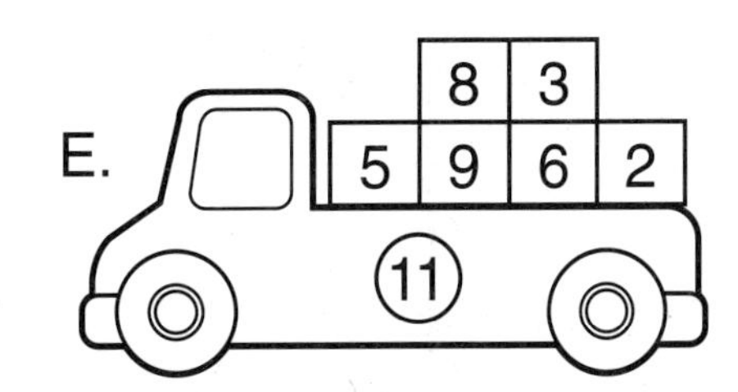


____ + ____ = _____

____ + ____ = _____

____ + ____ = _____

F.

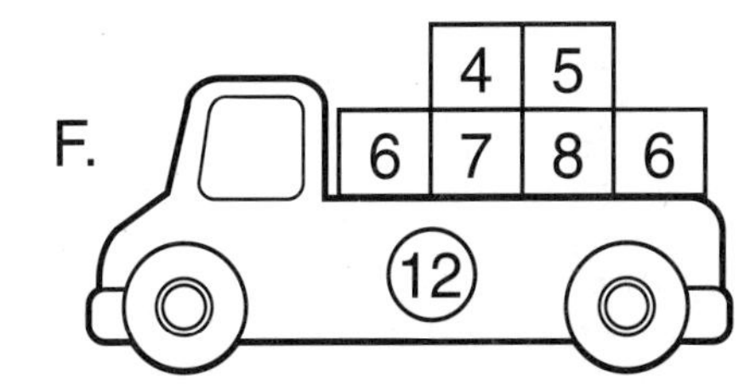


____ + ____ = _____

____ + ____ = _____

____ + ____ = _____

Name ______________________________ Date ____________________

# Hatching the Facts

Solve the facts on each nest.
Color the egg with the matching sum.
Write the leftover sum in the box.
Write an addition fact that equals the sum.

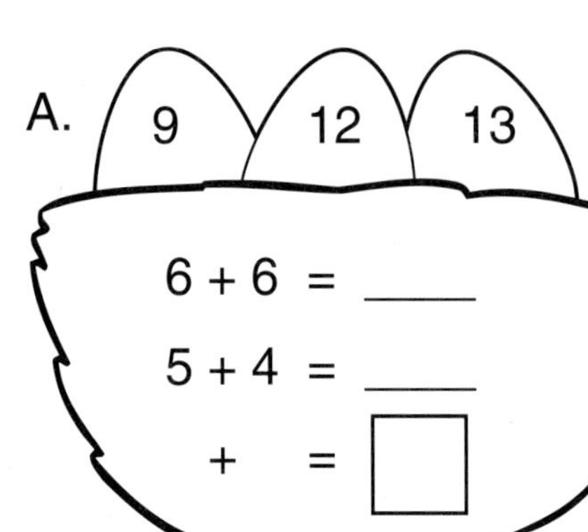

A. Eggs: 9, 12, 13

6 + 6 = ____
5 + 4 = ____
__ + __ = ☐

B. Eggs: 6, 8, 7

5 + 3 = ____
4 + 3 = ____
__ + __ = ☐

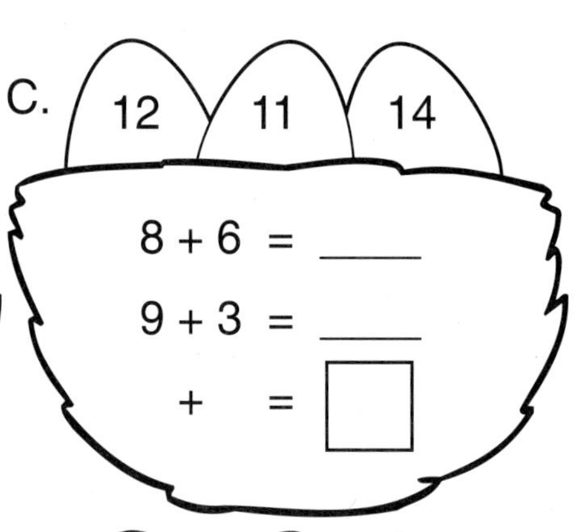

C. Eggs: 12, 11, 14

8 + 6 = ____
9 + 3 = ____
__ + __ = ☐

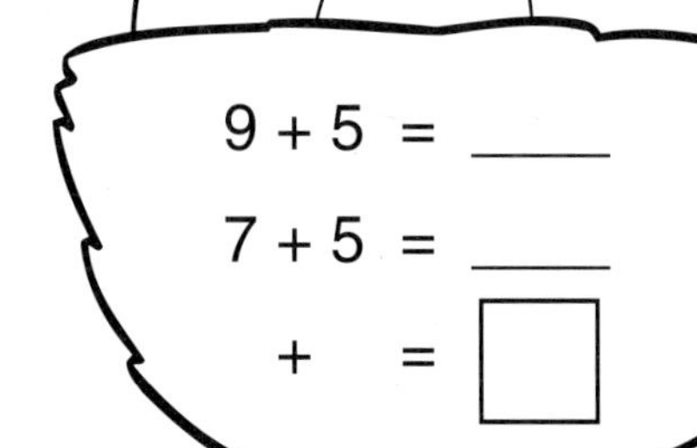

D. Eggs: 12, 14, 15

9 + 5 = ____
7 + 5 = ____
__ + __ = ☐

E. Eggs: 8, 16, 11

8 + 8 = ____
7 + 4 = ____
__ + __ = ☐

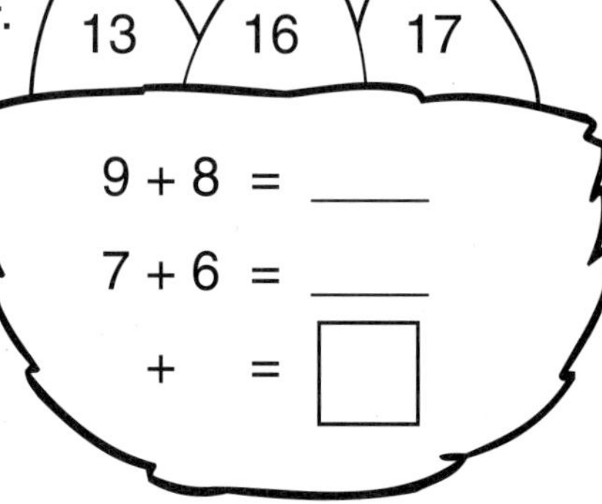

F. Eggs: 13, 16, 17

9 + 8 = ____
7 + 6 = ____
__ + __ = ☐

G. Eggs: 18, 12, 9

6 + 6 = ____
5 + 4 = ____
__ + __ = ☐

H. Eggs: 11, 13, 8

9 + 2 = ____
6 + 7 = ____
__ + __ = ☐

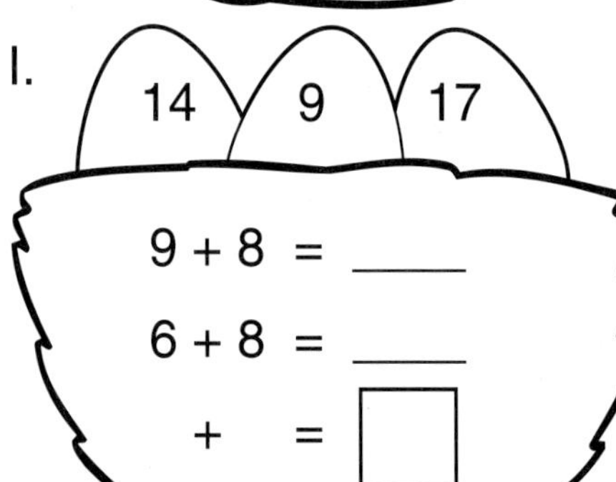

I. Eggs: 14, 9, 17

9 + 8 = ____
6 + 8 = ____
__ + __ = ☐

J. Eggs: 16, 8, 15

3 + 5 = ____
6 + 9 = ____
__ + __ = ☐

Name________________________________ Date____________________

# Froggy Facts

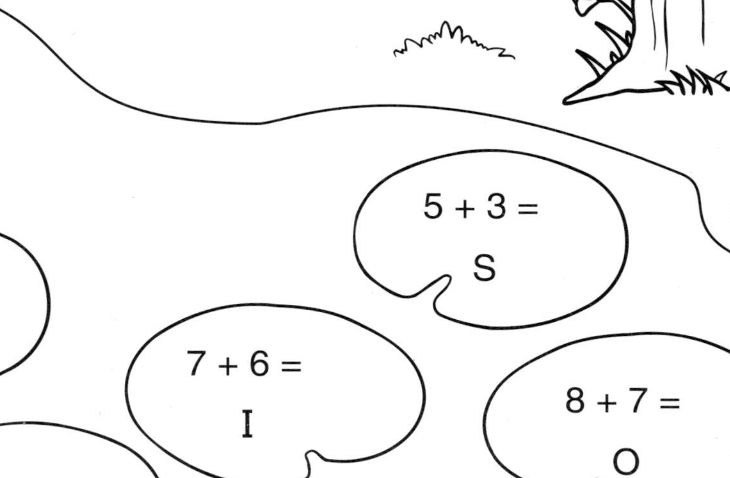

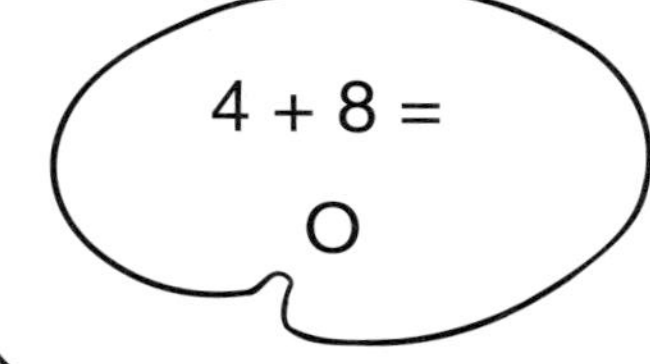

Add.

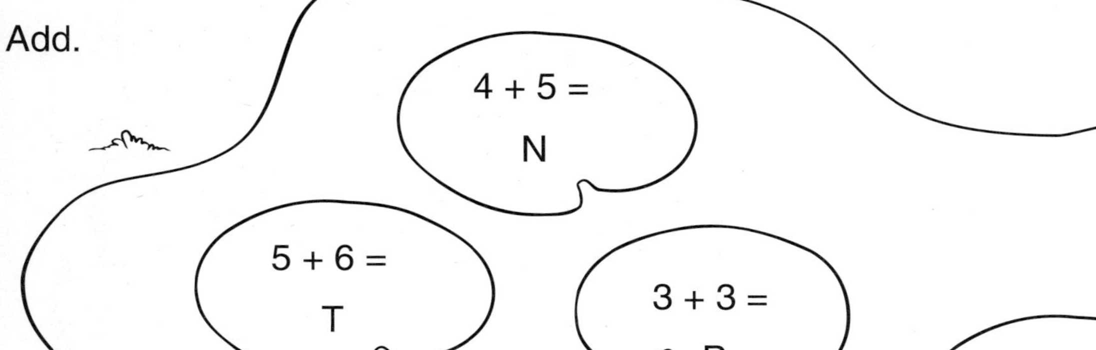

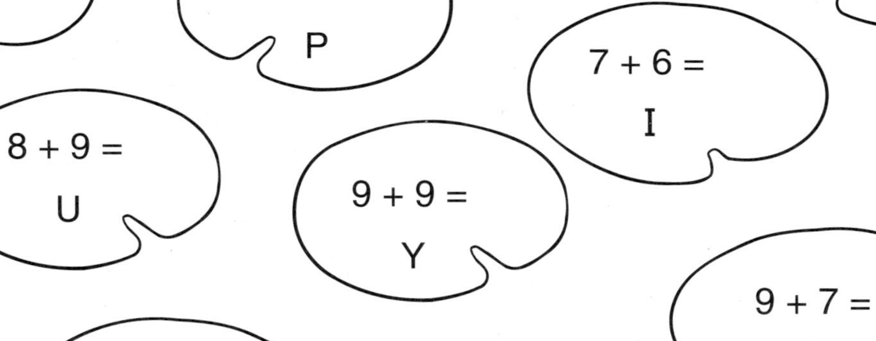

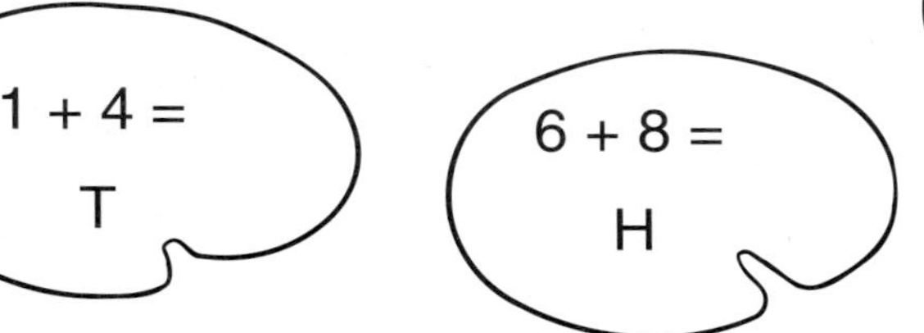

4 + 5 =
N

5 + 6 =
T

3 + 3 =
P

8 + 9 =
U

9 + 9 =
Y

5 + 5 =
P

1 + 4 =
T

6 + 8 =
H

7 + 6 =
I

5 + 3 =
S

8 + 7 =
O

9 + 7 =
O

4 + 3 =
I

4 + 8 =
O

## Why doesn't a frog jump when it's sad?

To solve the riddle, match the letters to the numbered lines below.

__ __   __ __   __ __ __   "__ __ - __ __ __ __ __"!

7 11   13 8   5 15 12   17 9   14 16 10 6 18

Name ______________________ Date ______________________

# Too Puzzled to Sleep

Circle each set of numbers that makes an addition fact in the puzzle below.
Seven facts go across, and five go up and down.
The first one has been done for you.

6 + 8 = ?
4 + 5 = ?

| | | | | | |
|---|---|---|---|---|---|
| 6 | 6 | 12 | 7 | 8 | 15 |
| 8 | 5 | 5 | 10 | 3 | 2 |
| 14 | 7 | 6 | 13 | 11 | 2 |
| 9 | 9 | 18 | 5 | 2 | 7 |
| 1 | 16 | 4 | 14 | 6 | 9 |
| 10 | 3 | 2 | 5 | 15 | 12 |

 Key p. 215

# Subtraction Facts to 18

## Stick to It!

Students will have no problem sticking with this fun subtraction center. To prepare, label nine clean containers with the numbers 1–9. Also program a supply of craft sticks with basic subtraction facts to 18 that result in differences of nine or less. To make the activity self-checking, use a permanent marker to write the corresponding subtraction facts on the back of each container. Then store the sticks in a separate plastic container at a center.

To begin, a student chooses a stick and reads the subtraction fact. Then he determines the difference and places the stick in the corresponding container. He continues in this manner with the remaining sticks and then checks his work. Subtraction? No problem!

## Take Away Cookies

This special batch of subtraction cookies is difficult to resist! Gather two small cookie sheets and cut a supply of tan construction paper circles. Use a brown crayon to add several dots to each circle to resemble a chocolate chip cookie. Next, program the blank side of each cookie with a different subtraction fact. Cut each cookie to separate the subtraction fact from its answer as shown. Then store the pieces in a cookie tin. Place the tin and cookie sheets at a center. To use this partner center, each player takes a cookie sheet and then places the cookie pieces problem side down on the playing surface. Player 1 turns over two cookie pieces and reads each one. If the two pieces make a correct subtraction sentence, she puts them together to form a cookie on her cookie sheet. If the two pieces do not make a true subtraction sentence, she turns them over again. Then Player 2 takes a turn. Play continues in this manner until all the cookies have been taken away. Yum, yum!

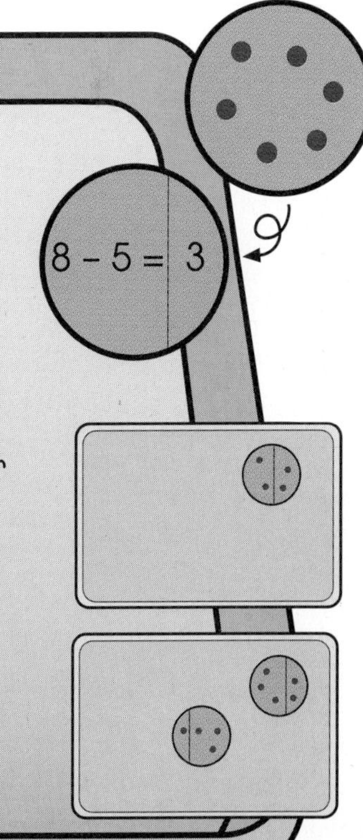

## Same Difference

This small-group game is just right for providing practice with writing subtraction facts. Gather a dry-erase board and marker for each child in a small group. Call out a number and challenge each student to write a basic subtraction fact in which the difference matches the announced number. (For example, if you call out, "Six," students may write "18 – 12 = 6," "9 – 3 = 6," or any other correct subtraction fact.) At your signal, have students flip their boards for their group members to see. Then lead students to realize that there can be many different subtraction facts for each difference. Ditto that difference!

18 – 12 = 6

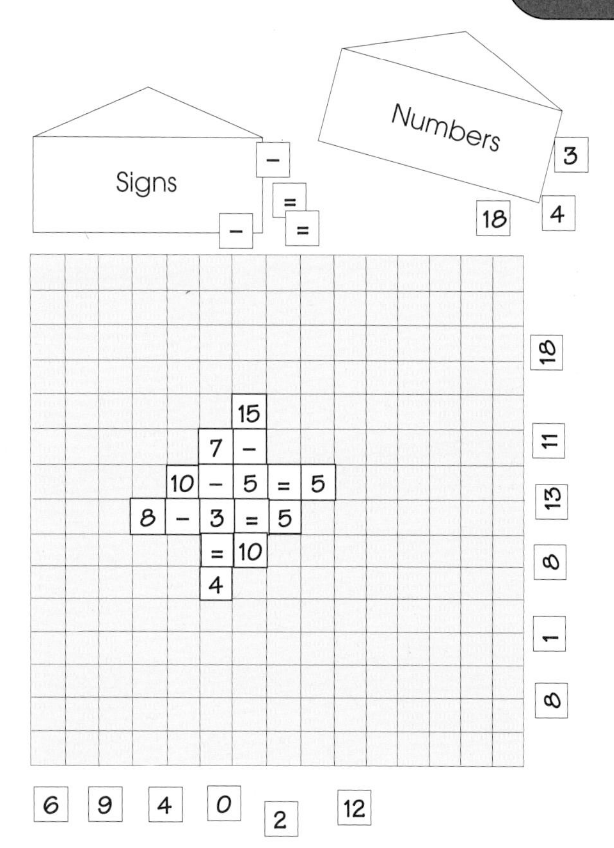


## Subtraction Scramble

Students will scramble to play this partner game, which reinforces subtraction skills. Prepare a 15 x 15 gameboard with one-inch squares. Also program one-inch grid paper with several sets of numbers from 0 to 18 and subtraction and equal signs. Copy the programmed paper several times and then cut it apart to make game cards. Place the cards in labeled envelopes.

To play, each child draws six number cards and places them faceup in front of her. Player 1 uses her number cards and the sign cards to make a subtraction fact on the gameboard as shown. If the subtraction fact is correct, she earns two points and takes from the envelope the same number of cards as she used. If the subtraction fact is not correct, she gathers her cards. If she cannot make a subtraction fact with the cards she has, she may exchange her cards with some from the envelope. Player 2 builds upon the cards on the board to make a different subtraction fact. Play continues until one player accumulates ten points.

## Bloomin' Flowers

Flowers and subtraction skills will blossom with this whole-group activity. In advance, cut a petal pattern and a three-inch circle for each child. Program each circle with a number from 0 to 17. To begin, each child glues the circle onto a 9" x 12" sheet of light-colored construction paper and signs his name. Next, he traces the petal pattern near the circle and writes a subtraction fact on the petal that matches the difference on the circle. At your signal, he passes his sheet to the next student along a predetermined route. On his new paper, he traces a petal and writes a subtraction fact that is different from the one already on the paper. Students continue passing flowers in this same manner until the flowers are full of petals. Then each child returns his paper to the original owner, who checks the subtraction facts and then draws a stem and leaves to complete the flower. Now that's a full bloom!

## Magic Numbers

Three in a row! That's the key to this subtraction version of tic-tac-toe. Pair students and give each pair a tic-tac-toe grid. Before beginning, each child chooses a magic number for her partner. To begin, Player 1 writes in one square on the gameboard a subtraction fact that equals her magic number. Player 2 checks her partner's subtraction fact. If the answer is correct, Player 1 draws the assigned symbol (X or O) atop it and her turn is over. If the answer is not correct, she erases the fact and her turn is over. Play alternates in this manner until the game is completed.

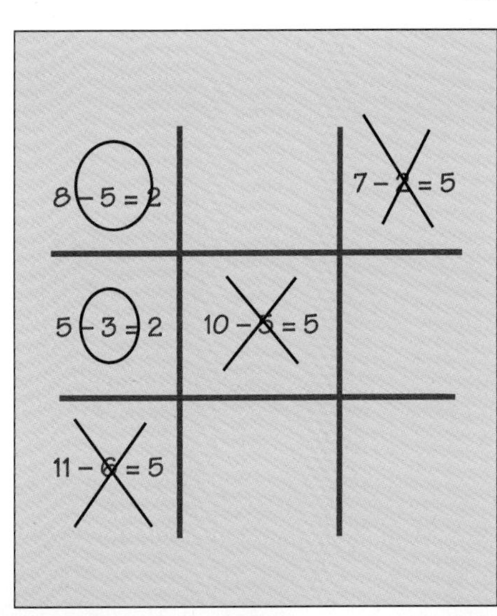

Find more student practice on pages 40–44.

 Name ________________ Date ________

# Munch! Munch!

Cut.
Glue to match differences.

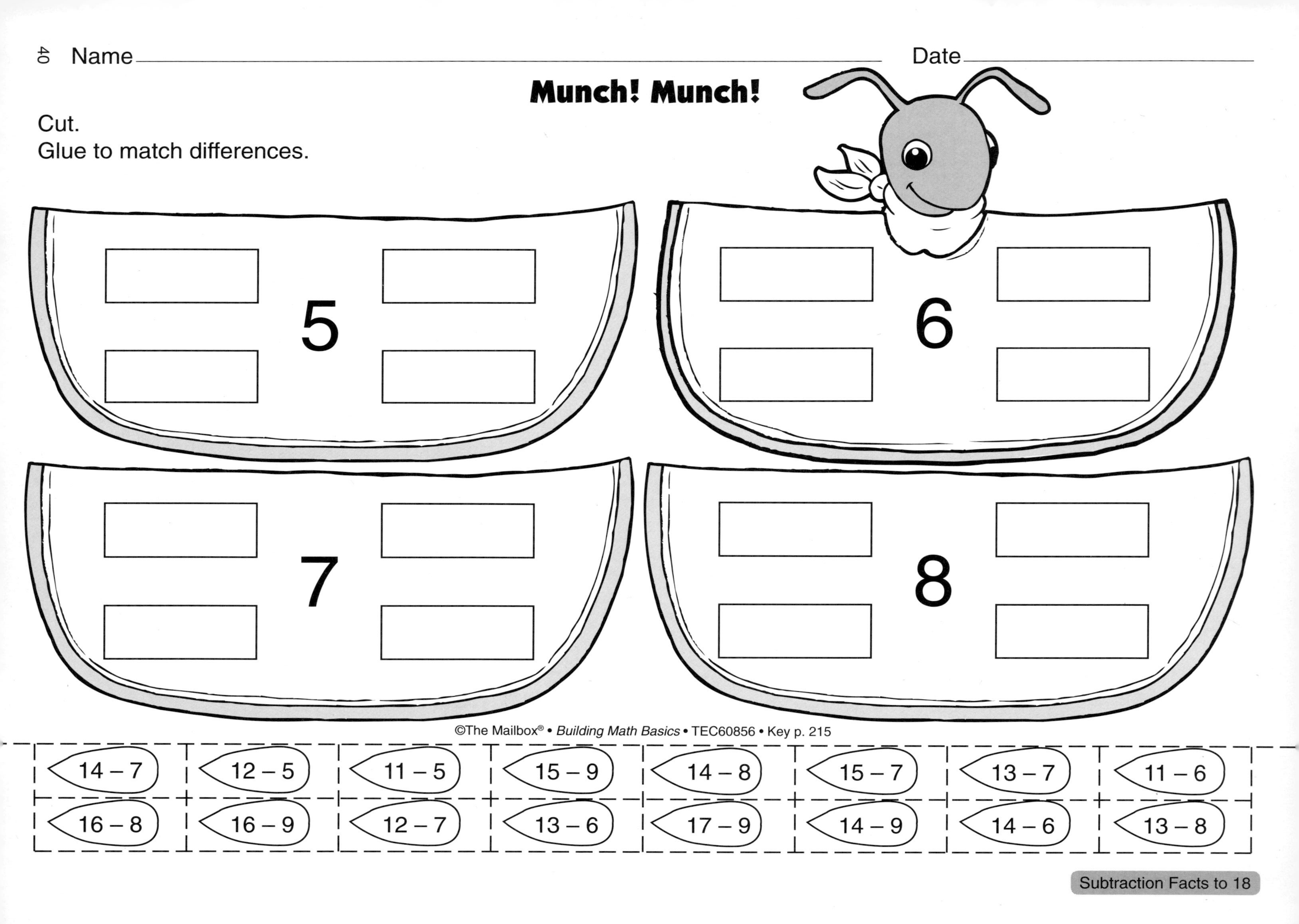

Name ____________________ Date ____________

# Lots of Popping!

Subtract.
Cross out a matching answer.

10 − 5 = ___

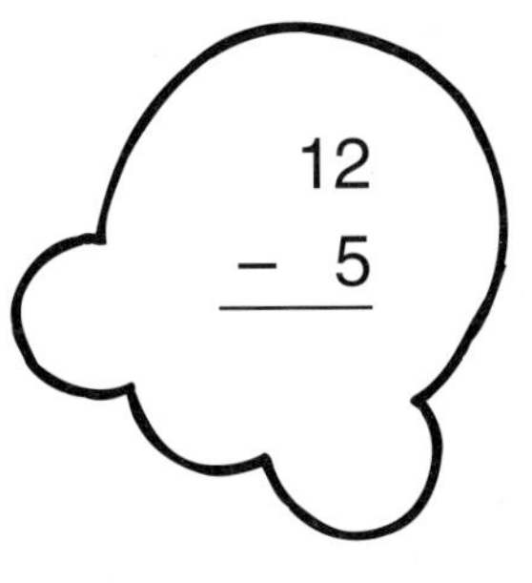

12 − 5 = ___

14 − 6 = ___

11 − 2 = ___

17 − 9 = ___

13 − 7 = ___

12 − 8 = ___

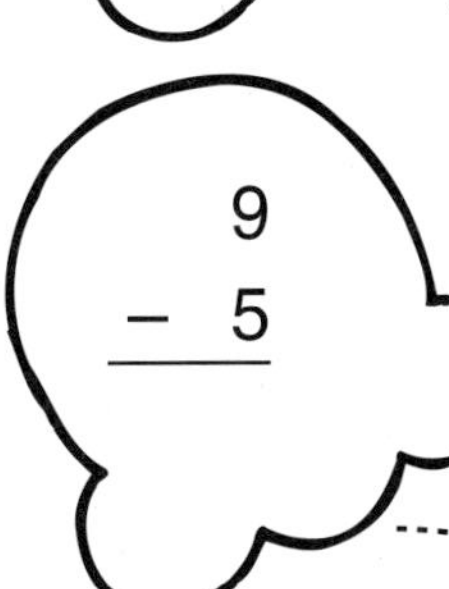

9 − 5 = ___

18 − 9 = ___

12 − 6 = ___

15 − 8 = ___

14 − 7 = ___

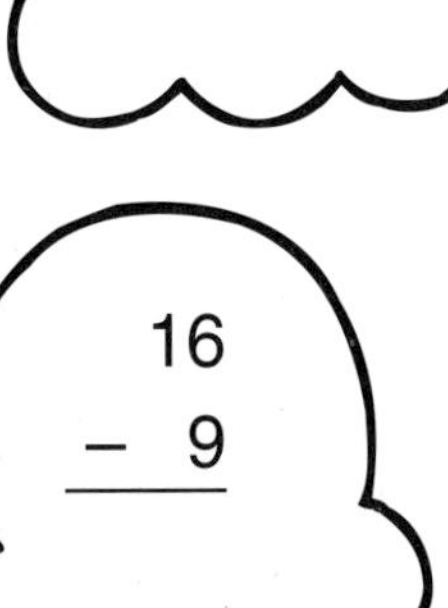

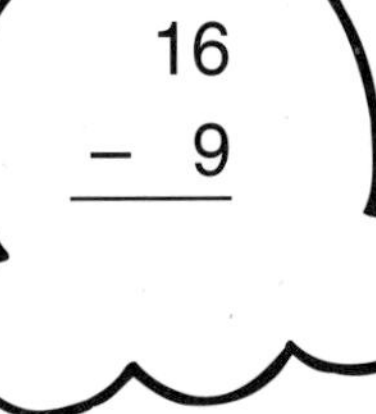

16 − 9 = ___

14 − 8 = ___

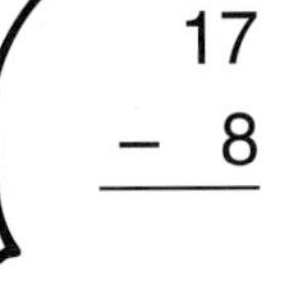

17 − 8 = ___

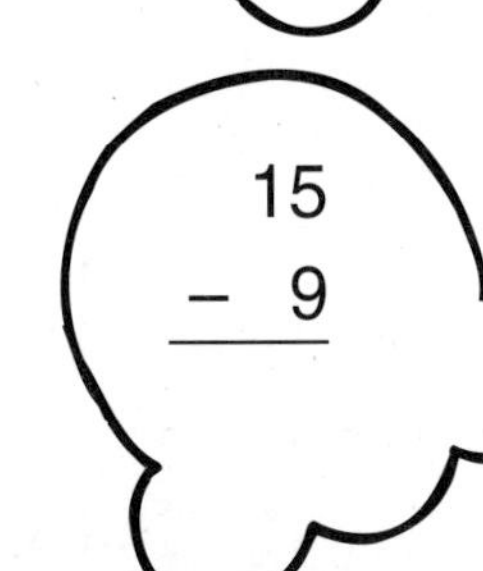

15 − 9 = ___

| | | | |
|---|---|---|---|
| 7 | 4 | 9 | 6 |
| 9 | 8 | 7 | 6 |
| 4 | 5 | 9 | 7 |
| 6 | 7 | 6 | 8 |

Name ______________________ Date ______________

# Toys for Max!

Subtract.
Color the bone with the correct answer to reveal which toy Max fetches.

1. 12 – 7 = ____
2. 11 – 6 = ____
3. 13 – 9 = ____
4. 10 – 8 = ____
5. 16 – 8 = ____
6. 14 – 6 = ____
7. 18 – 9 = ____
8. 17 – 9 = ____
9. 11 – 8 = ____
10. 16 – 7 = ____
11. 15 – 9 = ____
12. 14 – 7 = ____

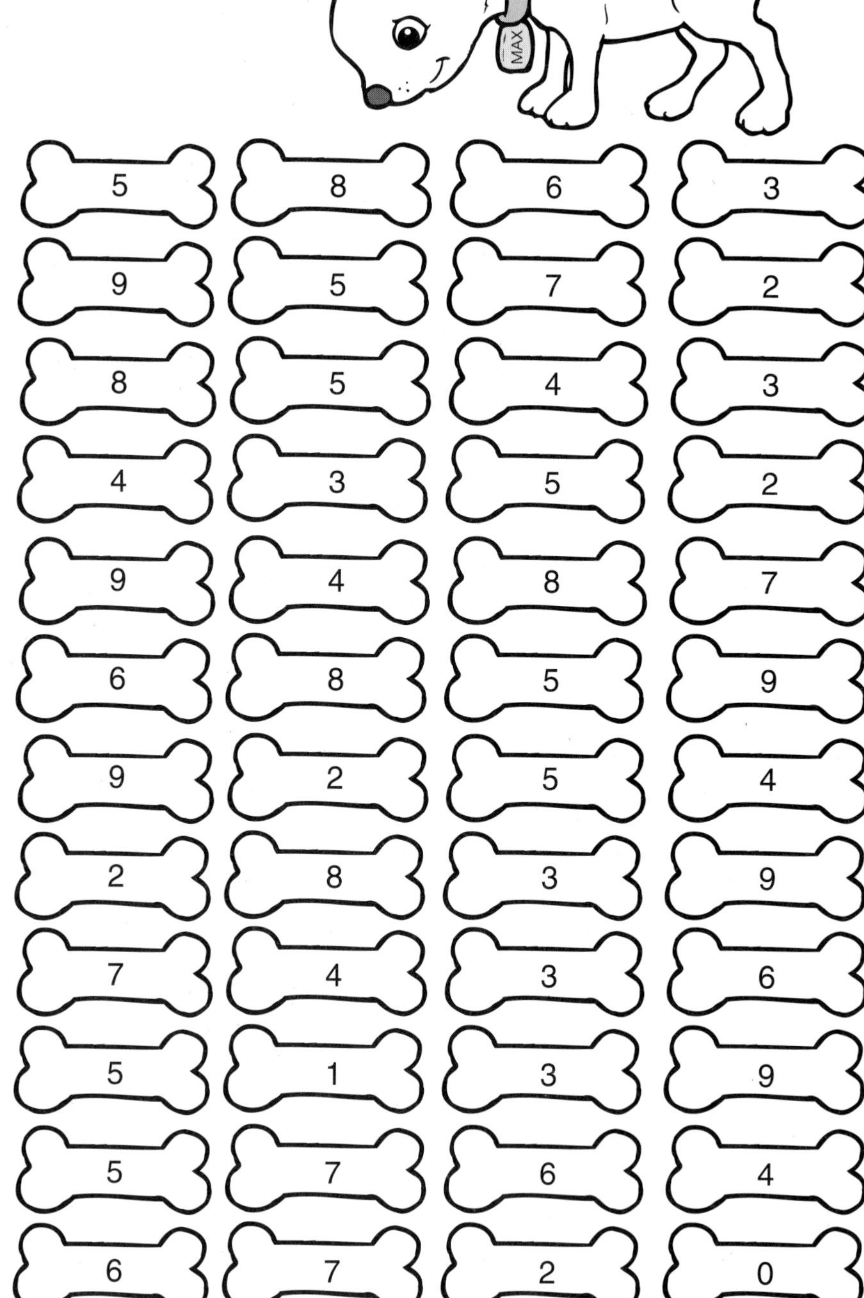

Name ____________________ Date ____________________

# Mmm! Peanuts!

Solve each problem.
Color the peanut with the matching answer.

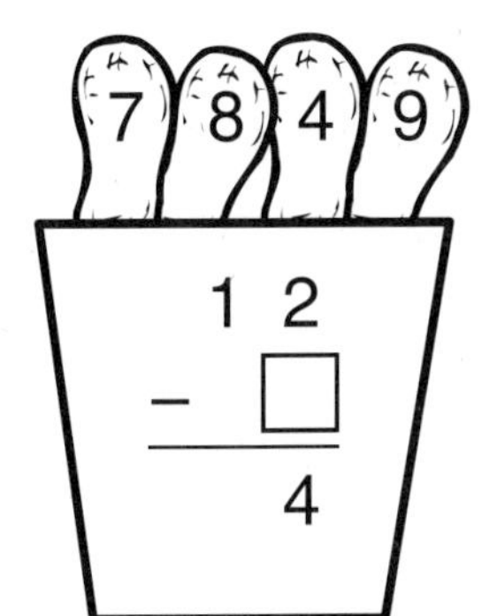


7 4 3 8

1 7
– □
9

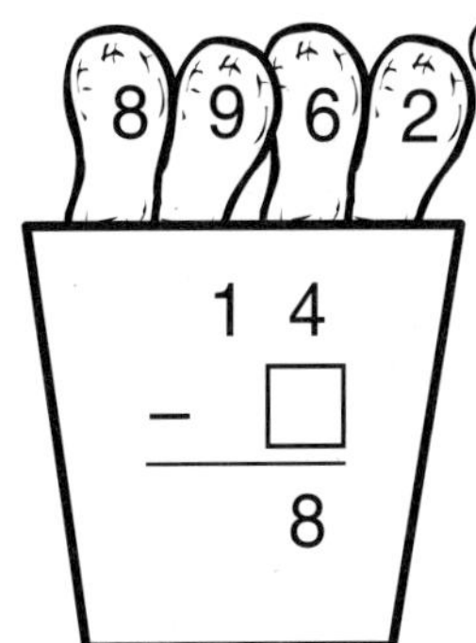


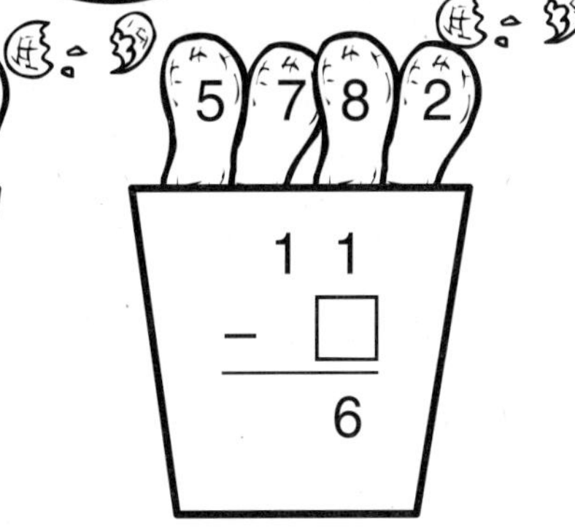


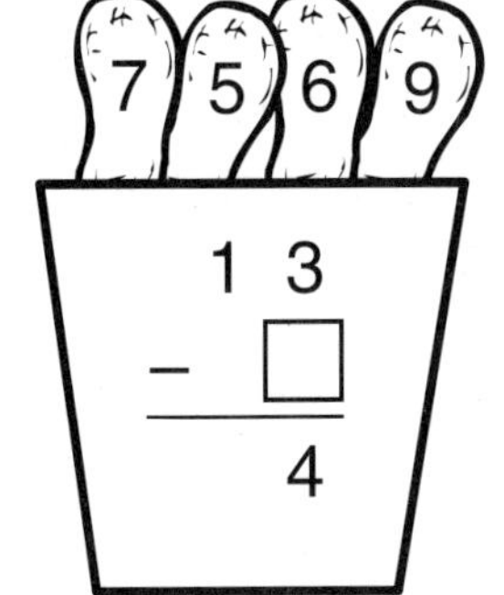


6 9 3 7

1 5
– □
9

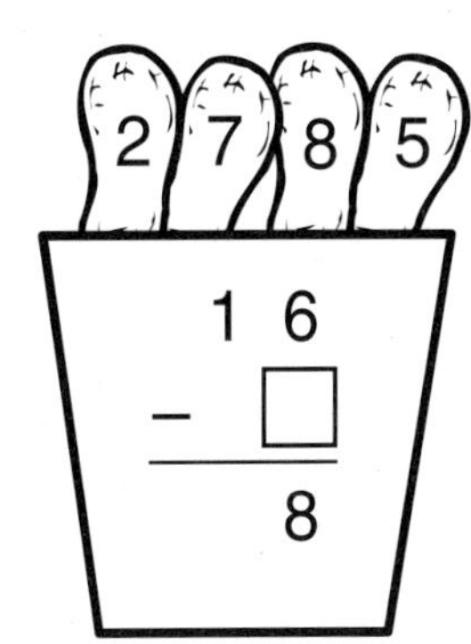


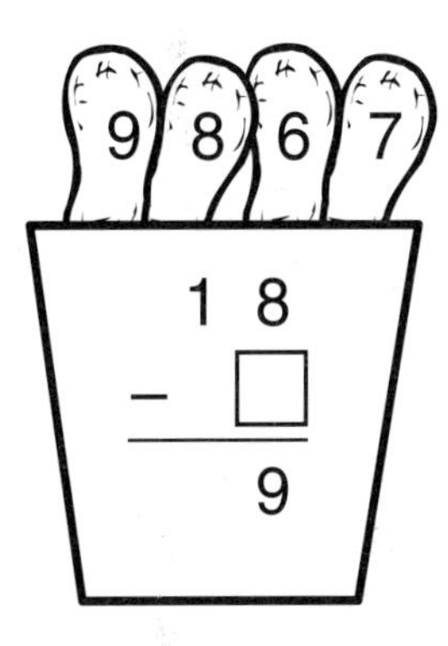


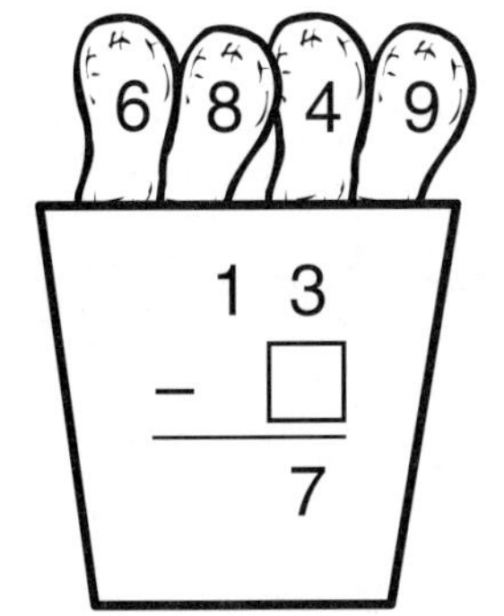


6 9 5 7

1 5
– □
8

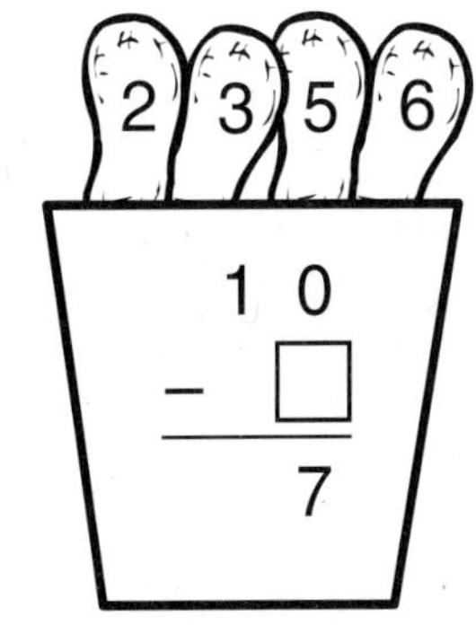


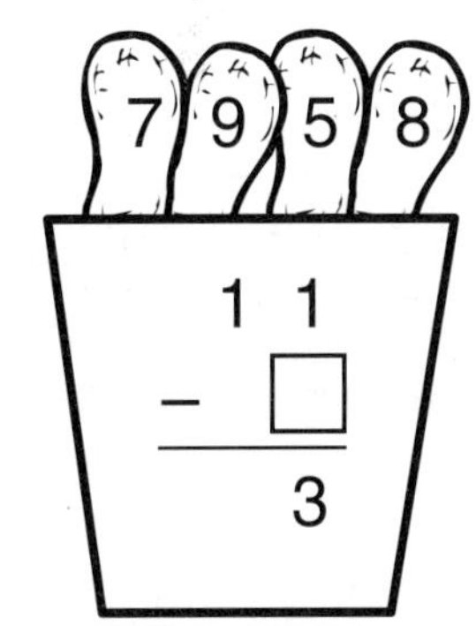


7 5 9 6

1 6
– □
7

6 5 7 4

1 4
– □
9

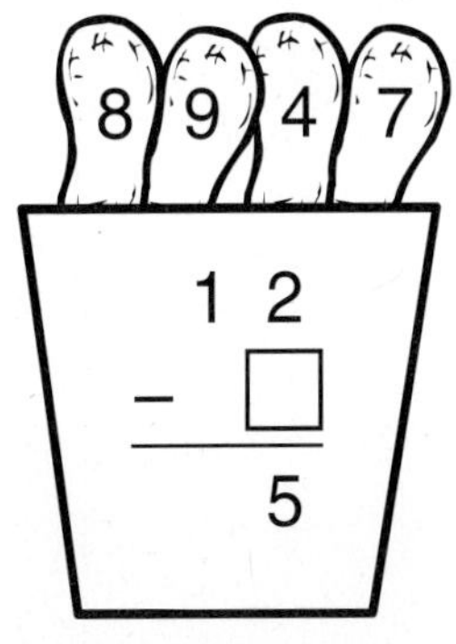


5 8 6 7

1 3
– □
8

Name______________________ Date______________

# Out of This World

Use the numbers on the stars to write two subtraction problems.
Color each star as you use the number.

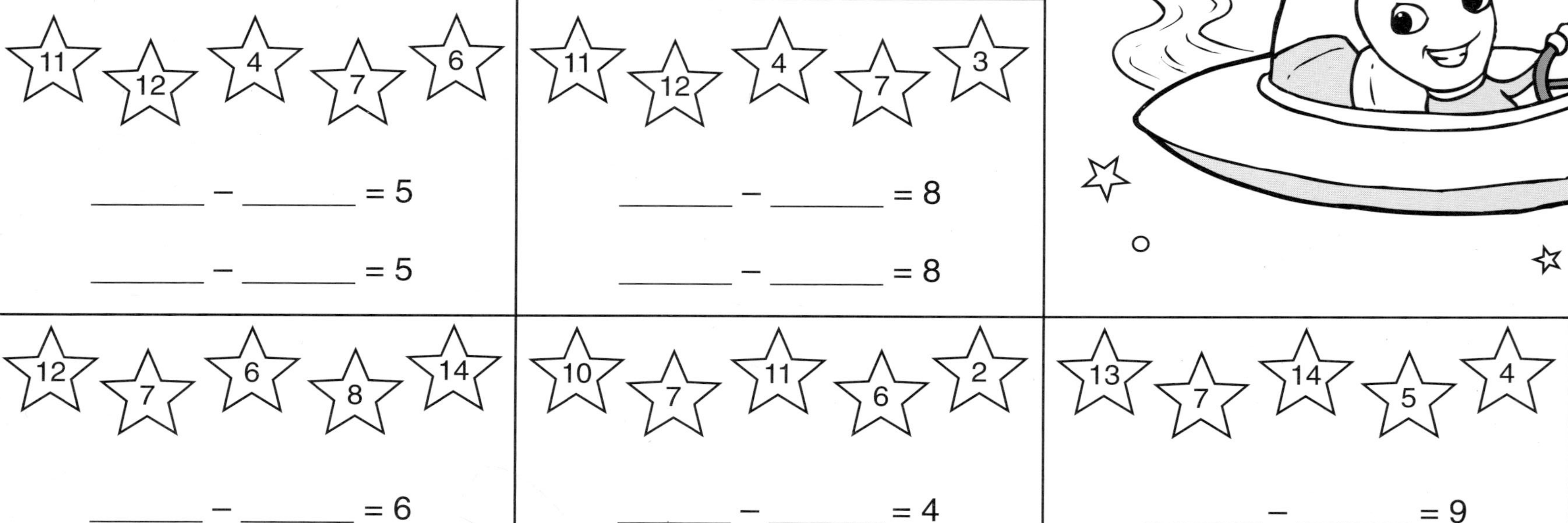

11 12 4 7 6

_____ – _____ = 5

_____ – _____ = 5

11 12 4 7 3

_____ – _____ = 8

_____ – _____ = 8

12 7 6 8 14

_____ – _____ = 6

_____ – _____ = 6

10 7 11 6 2

_____ – _____ = 4

_____ – _____ = 4

13 7 14 5 4

_____ – _____ = 9

_____ – _____ = 9

11 9 2 4 18

_____ – _____ = 9

_____ – _____ = 9

14 15 7 8 3

_____ – _____ = 7

_____ – _____ = 7

1 11 9 8 12

_____ – _____ = 3

_____ – _____ = 3

# Addition and Subtraction Fact Families

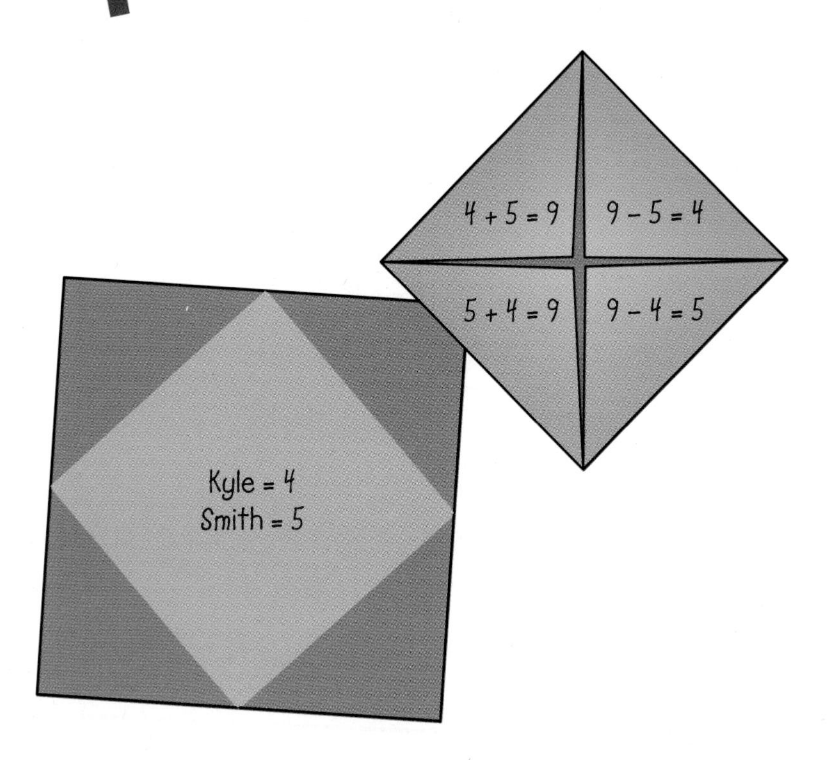


## Name Numbers

What's the result of this fact family activity? A display that features the number of letters in students' names! Give each child a nine-inch square of construction paper. Have him use a marker to write his name in the center of his paper. Then instruct him to count the number of letters in each part of his name and record the numerals as shown. Next, have him fold each corner of his paper into the center. Direct each child to write a different fact family fact on each corner flap, using the numbers in his name. Display the projects on a bulletin board and encourage students to fold in the flaps to view the facts of each classmate's fact family.

## Diggin' Fact Families

Youngsters get the dirt on fact family success with this cute worm project. Cut a supply of three-inch circles from construction paper so that each child has five circles. For each child, program one circle with sticky dots and a different fact family, as shown, to make a worm head. Cut a 3" x 12" brown construction paper strip (dirt) for each student. Give each child a head, four circles, and a brown strip. Have her use the numbers on her worm head to write four fact family facts, one on each circle. Then instruct her to glue her fact family worm to her dirt. Attach the projects to a bulletin board titled "Diggin' Fact Families!"

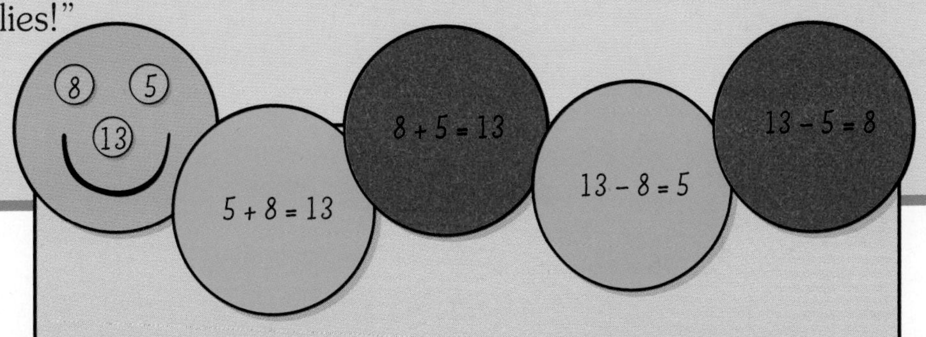

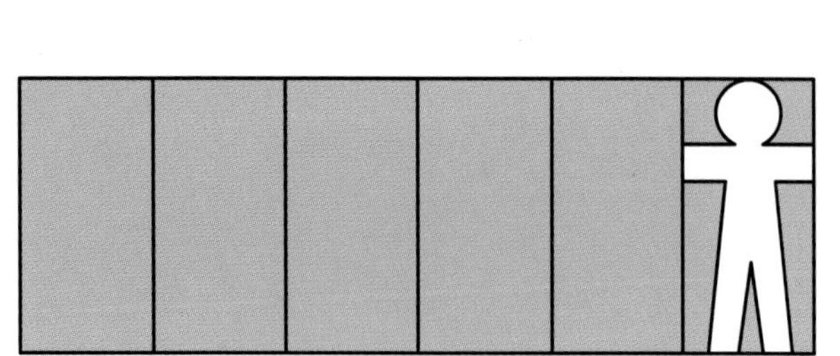

## All in the Family

Students will have fun creating this closely connected fact family. In advance, make a paper doll pattern, as shown, that fits a 3" x 6" piece of paper. Give each student a 6" x 18" strip of construction paper. Have him use a pencil and ruler to mark the strip at three-inch increments. Then direct him to accordion-fold the strip using the marks as a guide. Next, have him trace the paper doll pattern on his folded paper and cut out the resulting string of dolls.

Assign a set of fact family numbers to each child. Have him write the numerals on the first doll and then write a different fact on each of the next four dolls. Finally, instruct him to write his name on the last doll. If desired, assign another set of fact family numbers and have him repeat the process on the back. String the dolls around the room to create a fact family reunion!

## Draw Two!

This fast-action card game will have students mastering fact family facts in a flash! To prepare, write each number from 1 to 9 on a different blank card twice to create a set of 18 cards. Shuffle the cards and place them facedown at a center along with paper and pencils. Invite three students to visit the center at a time. Designate two children to play the game and the third child to be the monitor. To play, each player draws two cards, keeping them facedown. When both students are ready, instruct the monitor to say, "Go." Each player turns her cards over and creates an addition sentence using the numbers on the cards. Next, she writes the other three facts to complete her fact family. Then the monitor checks each player's work and returns the cards to the bottom of the stack before switching roles with one of the players. Play continues in the same manner for a predetermined number of rounds.

| | |
|---|---|
| $5 + 3 = 8$ | $4 + 1 = 5$ |
| $3 + 5 = 8$ | $1 + 4 = 5$ |
| $8 - 3 = 5$ | $5 - 1 = 4$ |
| $8 - 5 = 3$ | $5 - 4 = 1$ |
| $6 + 9 = 15$ | $7 + 2 = 9$ |
| $9 + 6 = 15$ | $2 + 7 = 9$ |
| $15 - 9 = 6$ | $9 - 7 = 2$ |
| $15 - 6 = 9$ | $9 - 2 = 7$ |

# Loopy Fact Families

Students are in the loop while completing this fun fact family activity! Give each child ⅓ cup of Froot Loops cereal. Have the student sort the cereal pieces by color and then pair the cereal colors to make three sets. Instruct her to draw a set of cereal pieces on her paper, as shown, and then write and solve an addition sentence using the corresponding numbers. Have her use the fact family numbers to complete three more facts. Instruct her to repeat the process with her two remaining sets of cereal pieces. If desired, reward good work with a fact family cereal snack!

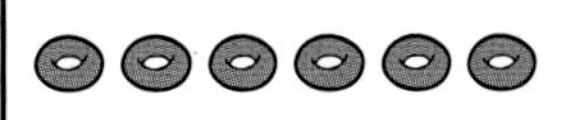
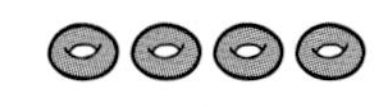

8 + 3 = 11
3 + 8 = 11
11 − 8 = 3
11 − 3 = 8

6 + 4 = 10
4 + 6 = 10
10 − 4 = 6
10 − 6 = 4

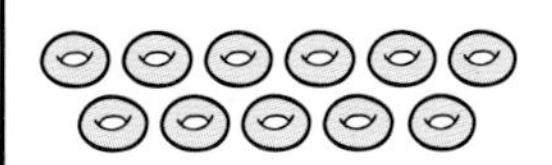
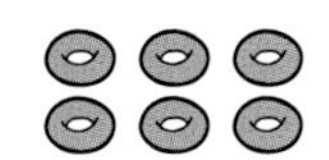

11 + 6 = 17
6 + 11 = 17
17 − 11 = 6
17 − 6 = 11

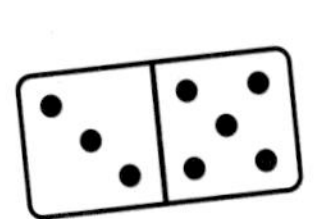
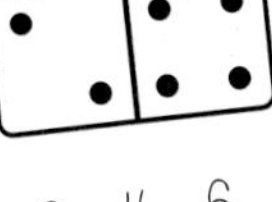

3 + 5 = 8
5 + 3 = 8
8 − 3 = 5
8 − 5 = 3

2 + 4 = 6
4 + 2 = 6
6 − 4 = 2
6 − 2 = 4

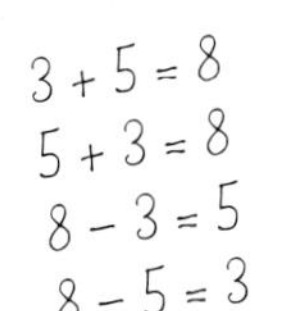

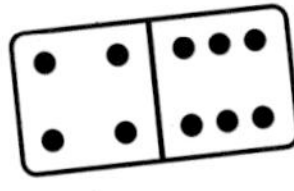

6 + 5 = 11
5 + 6 = 11
11 − 6 = 5
11 − 5 = 6

4 + 6 = 10
6 + 4 = 10
10 − 4 = 6
10 − 6 = 4

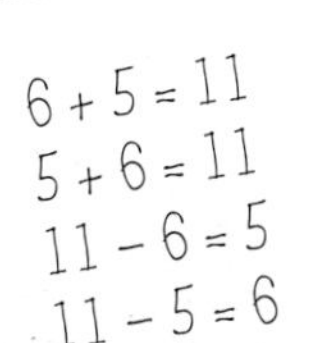

## Domino Fact Families

Grab your dominoes for simple fact family practice. In advance, fold a sheet of paper into fourths. In each section, trace a domino and draw a line through the center of each outline as shown. Make a copy for each child. Give each student a copy and four dominoes. Have him draw the dots for each domino on a different outline on his paper. For each domino, instruct him to write and solve an addition sentence using the numbers on the domino. Have him write three more facts using the fact family numbers. Then have each child share his results with a classmate. "Dots" all folks!

Find more student practice on pages 48–49.

Name____________________ Date__________

# Digging for Treasure

Use the numbers on the coconuts to write fact families on each pair of lines.
The first one has been started for you.

Name______________________________ Date________________

# A School to Hoot About

Solve each problem.
Cut out the boxes and sort them into fact families.
Glue each fact family in a separate window.

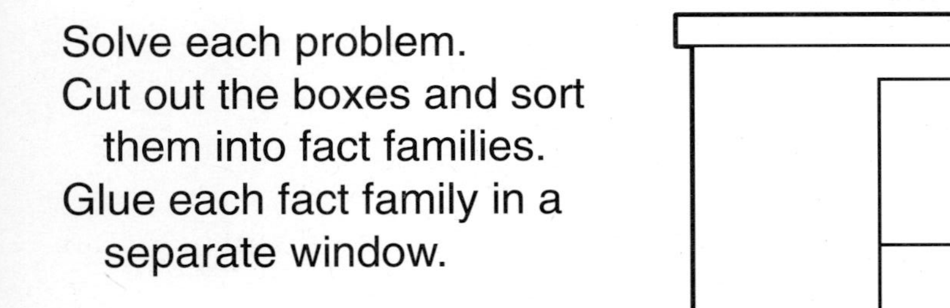

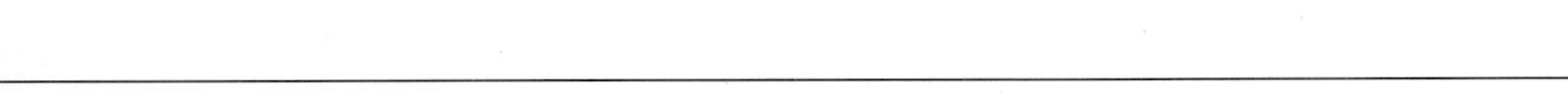



| | | | | | | | | | | | |
|---|---|---|---|---|---|---|---|---|---|---|---|
| 8 + 5 | 9 + 5 | 15 − 7 | 12 − 8 | 8 + 7 | 14 − 9 | 4 + 8 | 11 − 6 | 13 − 5 | 9 + 7 | 14 − 5 | 16 − 7 |
| 5 + 6 | 7 + 9 | 13 − 8 | 6 + 5 | 5 + 8 | 8 + 4 | 15 − 8 | 5 + 9 | 16 − 9 | 11 − 5 | 7 + 8 | 12 − 4 |

# Commutative Property

## Hop Along!

Leap into the commutative property with this "toad-ally" fun math activity! Cut out a large lily pad shape from green bulletin board paper. Program the lily pad with two parallel number lines and then laminate it for durability. Next, label each of several blank cards with a pair of addends. Then attach a frog sticker or a frog cutout to the end of a dry-erase marker. Post the lily pad on a wall within students' reach and place the cards, a marker, and a cloth nearby.

To begin, ask a student volunteer to select a card, read the numbers, and then use the marker to show the frog making two jumps on the first number line to represent the numbers on the card. For example, if the card says 6 and 5, the child jumps 6 spaces and then 5 more. Have him repeat this process on the second number line, but this time instruct him to switch the order of the numbers shown on the card. Direct students' attention to the number under the ending jump on each number line. Lead students to realize that the order of the addends doesn't matter—the sums stay the same. Then use the cloth to wipe the lily pad clean and invite additional students to take turns repeating the activity. Ribbit!

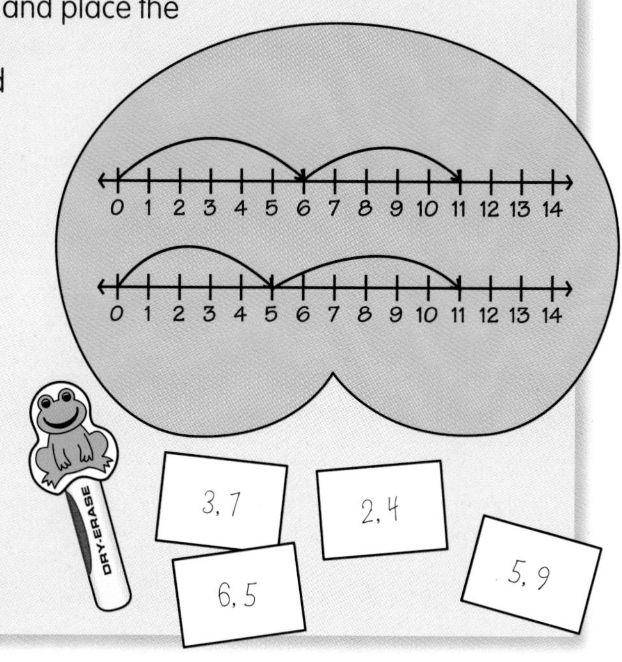


## Beanbag Bonanza!

Everyone's a winner with this beanbag game! In advance, program a sheet of poster board with a 4 x 4 grid and label each section with a different number. Place the grid on the floor near your chalkboard. Divide the class into two teams. Instruct each team to line up several feet from the grid. Provide each team with a different-colored beanbag; then have the first player on each team toss her beanbag on the grid. At your signal, instruct each player to form an addition equation on the chalkboard using the number on which her beanbag landed as the first addend and the number on which her opponent's beanbag landed as the second addend. Each player to correctly solve her equation earns a point for her team. After each round, have students compare the two sums to see that no matter the order of the addends, the sums are equal. Continue play until each child has had a turn or until a predetermined number of points has been earned.

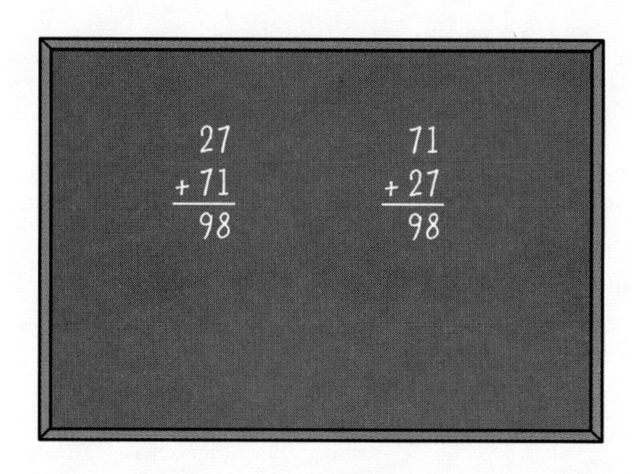


| 5 | 27 | 8 | 18 |
|---|---|---|---|
| 46 | 12 | 9 | 6 |
| 34 | 71 | 25 | 60 |
| 9 | 52 | 3 | 17 |

## Flip-Flop Addition

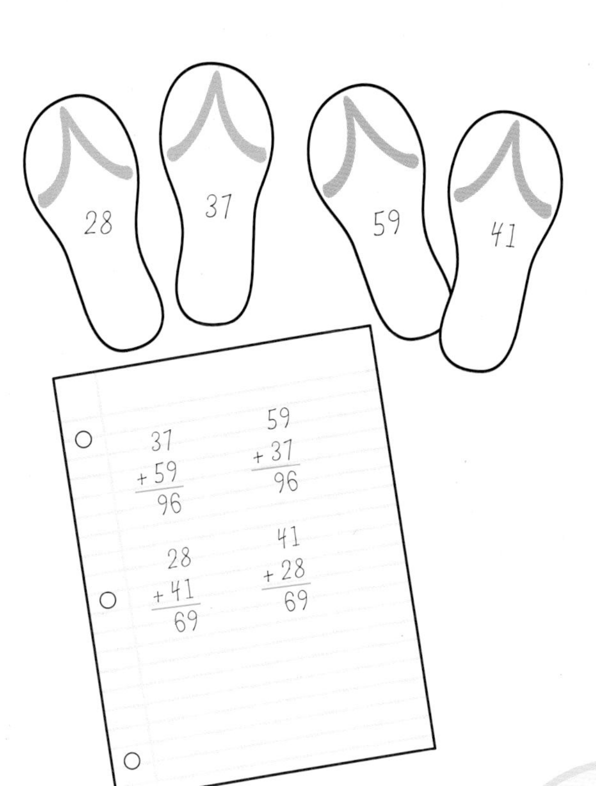


When addends flip-flop, will they produce the same sum? Sure! Make several pairs of construction paper flip-flops, each pair having different-colored straps. Program each flip-flop with a different two-digit number; then laminate them for durability. Store the flip-flops in a sand pail and place the pail at a center along with paper and pencils. Have each child who visits the center pair the flip-flops. Then instruct him to use the numbers on them as addends to write two different addition equations on his paper. Have him repeat the process for the remaining pairs. Wow! The sum is always the same for each pair of equations.

## Crazy Daisies!

Watch students' understanding of the commutative property bloom with this colorful project! Make a white construction paper copy of page 52 for each child. Have each child cut out the flower and wheel. Then have her carefully cut out the two squares on the flower's center. Help each student use a brad to fasten the number wheel behind the flower.

Have each child turn the number wheel to show an addition sentence. Instruct her to record the equation on a petal and then solve it. Have her repeat this process until she has recorded all eight equations on the petals. Direct her to choose four different crayons and color each pair of petals that have matching sums the same color. Lead students to understand that changing the order of the addends does not change the sum.

If desired, have students embellish their flowers with green construction paper leaves and stems. Then attach the projects to a bulletin board titled "Crazy Daisies!"

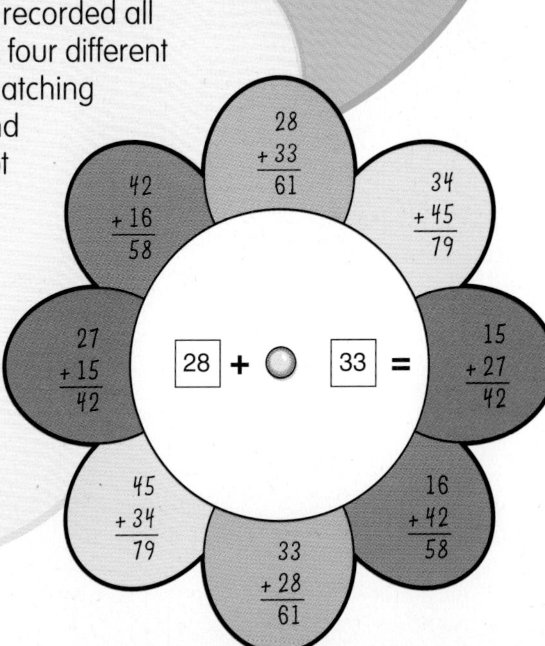

## Flower and Wheel Patterns

Use with "Crazy Daisies!" on page 51.

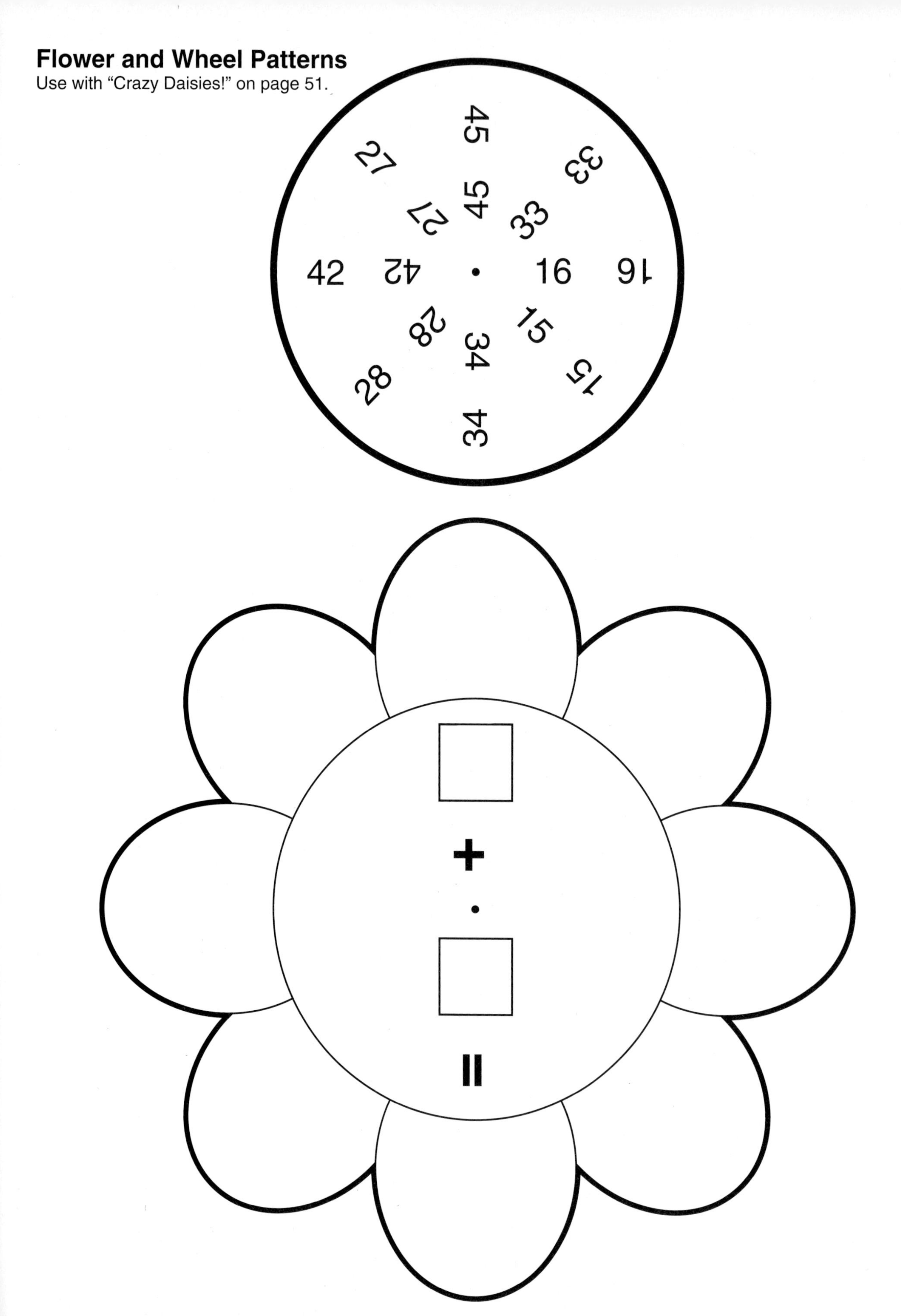

# Addition of Larger Numbers

## Addition Shake Up

Shake up students' addition practice with this easy partner activity! Give each pair of students a copy of a hundred chart, and have the students cut the numbers apart and place them in a plastic bag. Also provide the pair with paper, pencils, and a calculator. Have one child shake the bag and draw out two number cards. Instruct him to use the numbers to create an addition equation on his paper and then solve it. Instruct the other child to check his partner's answer with the calculator. Then direct students to switch roles and repeat the activity. Have students continue play for a predetermined number of rounds. For an added challenge, have students choose three numbers from the bag and add them together. What a great way to shake up addition practice!

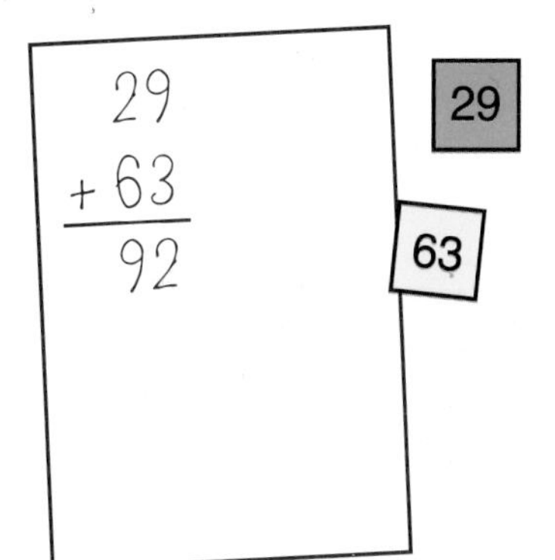

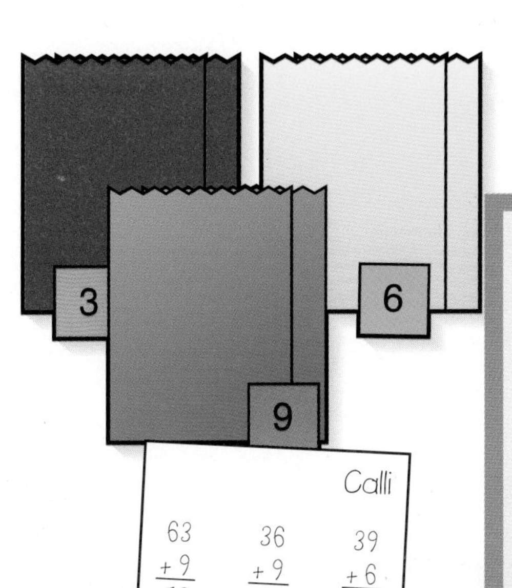


## It's in the Bag!

Two-digit addition practice is in the bag with this easy to prepare game! In advance, write a single-digit number on each of a supply of small blank cards so that each student will have three cards. Next, divide the cards among three different bags. Have each child draw one card from each bag. Then challenge her to use the number cards to make as many two-digit plus one-digit addition equations as she can and then solve them. (There will be six possibilities for each combination of numbers.) After a desired amount of time, pair students and have them check each other's addition.

## Walk Across Your State

Addition practice is a step away with this class project, which requires a map of your state, a pedometer, and a piece of chalk. Post a map of your state in an easily accessible location. Help students calculate the number of miles it would take to walk from one part of the state to another. Then, each day, have a different child wear the pedometer. Before dismissal, record on the chalkboard the number of steps the child took (rounding as necessary). At the end of the week, add the daily step totals to determine the distance students walked that week. (If desired, have students work independently to solve the equation and then rework the problem as a group to check answers.) Then highlight this distance on the map. Continue in this manner each week until students reach the desired goal. Put on your walking shoes!

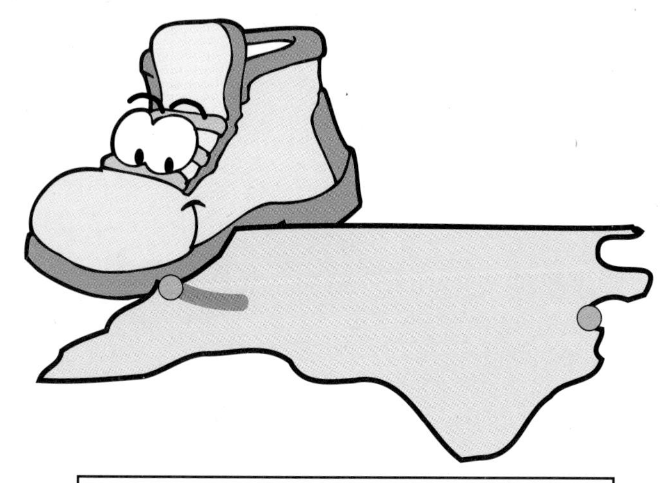

| Week of Nov. 29 - 7 miles | | Total: |
|---|---|---|
| M | 1 | 63 miles |
| T | 2 | + 7 miles |
| W | 1 | 70 miles |
| Th | 1. | |
| F | 2 | |

## House of Digits

tens house | ones house
2 | 4
3 | 8

This idea provides a neighborly way to practice addition of larger numbers. Pair students and give each twosome a 6" x 9" piece of construction paper. Have one child in the pair fold the paper in half and cut one end into a point to represent a rooftop. Instruct the other child to unfold the paper, label the resulting houses as shown, and then position it between her partner and herself. Provide each child in a pair with a supply of small sticky notes, a sheet of paper, and a pencil. To play, have each player write a one-digit numeral on each of two sticky notes and position them on the house in the column closest to her. After the four numerals are positioned, have each child write and solve the addition problem on her paper before comparing answers with her partner. Continue play for a desired number of rounds.

## Newspaper Numbers

This idea for practicing addition of larger numbers is certainly newsworthy! Pair students and provide each twosome with a page from the sports section of a newspaper, paper, pencils, a highlighter, and a calculator. Have one child highlight two scores on the sports page and then use them to make an addition equation on her paper. After she solves the equation, have her partner check her answer with a calculator. For each correct answer, she earns a point. Have students switch roles and continue play until a predetermined number of points have been earned by each player. Score one for addition practice!

542
+ 31
573

1, 2, 3, 4, & 5

## Build the Biggest Number

No hard hat is required to construct the largest number with this quick activity! Write a group of numerals such as "1," "2," "3," "4," and "5" on the chalkboard. Challenge each student to group the numbers as desired to create a three-digit number and a two-digit number. Then have her add the two numbers together to try to make the largest sum. For example, a student might choose to write "123 + 45," or "543 + 12." After providing time for calculations, enlist students' help in determining who created the greatest sum. Then invite the winner to choose the next five numbers for another round of play.

Find more student practice on pages 56–60.

 Name______________________ Date______________

# A Perfect Catch

Add.
Color the fish with the matching answer.

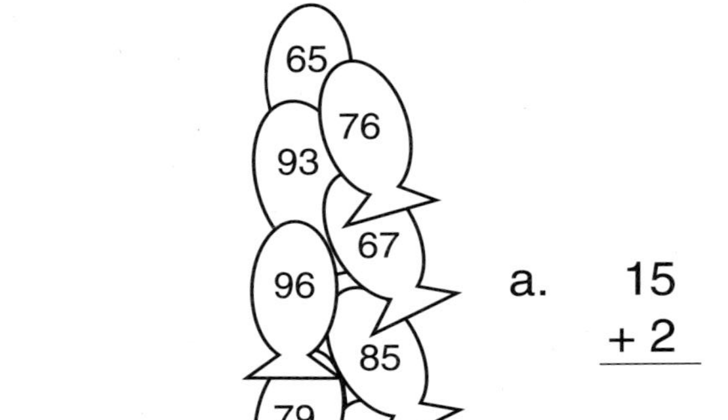


a. 15 + 2 = ____

b. 50 + 19 = ____

c. 73 + 6 = ____

d. 16 + 51 = ____

e. 40 + 34 = ____

f. 62 + 3 = ____

g. 91 + 2 = ____

h. 36 + 42 = ____

i. 71 + 5 = ____

j. 43 + 25 = ____

k. 28 + 71 = ____

l. 54 + 42 = ____

m. 83 + 2 = ____

n. 32 + 12 = ____

o. 21 + 32 = ____

Name ______________________ Date ______________________

# Picture Perfect

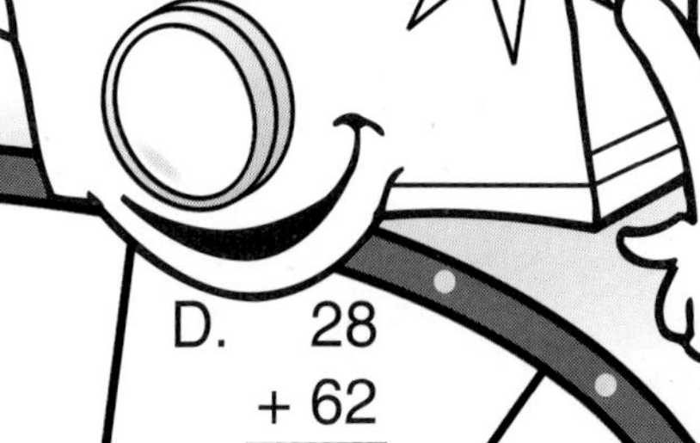

Add.
Show your work.

A. 89 + 53

B. 15 + 35

C. 91 + 59

D. 28 + 62

E. 58 + 17

F. 35 + 97

G. 63 + 48

H. 79 + 34

I. 42 + 89

J. 88 + 36

K. 46 + 18

L. 13 + 89

M. 67 + 28

N. 94 + 97

O. 56 + 27

P. 37 + 49

Q. 24 + 39

R. 72 + 58

Name________________ Date__________

# Dig In!

Add.
Show your work.
Use your answers to complete each puzzle.

## 1

| A | B | C |
|---|---|---|
| D | | |
| E | | |

**Across**

A. 176 + 613

D. 521 + 323

E. 864 + 121

**Down**

A. 104 + 685

B. 215 + 633

C. 543 + 402

## 2

| A | B | C |
|---|---|---|
| D | | |
| E | | |

**Across**

A. 153 + 401

D. 237 + 752

E. 405 + 242

**Down**

A. 436 + 160

B. 312 + 272

C. 336 + 161

## 3

| A | B | C |
|---|---|---|
| D | | |
| E | | |

**Across**

A. 388 + 301

D. 742 + 135

E. 690 + 106

**Down**

A. 523 + 164

B. 450 + 429

C. 172 + 804

Name ______________________________ Date ____________________

# Target Practice

Where will the arrows land?
Add.
Color by the code.

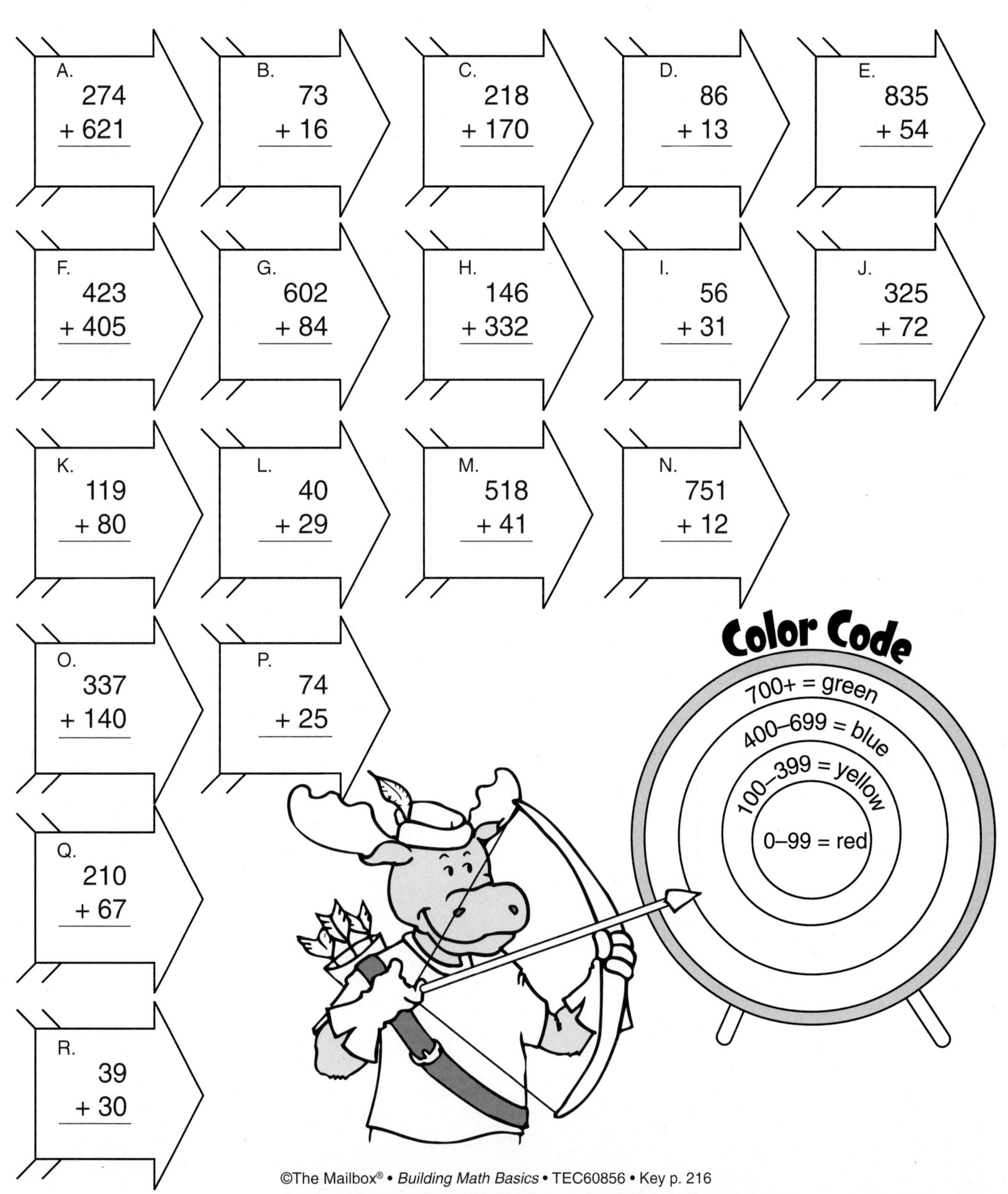

Name ______________________ Date ______________________

# A Clean Sweep

Add.
Show your work.

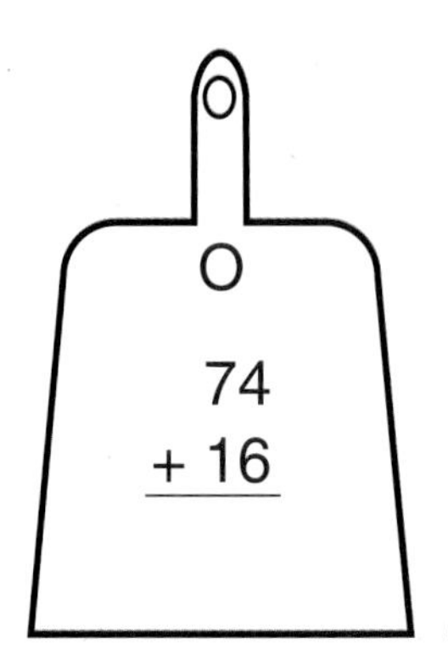


D
830
+ 295

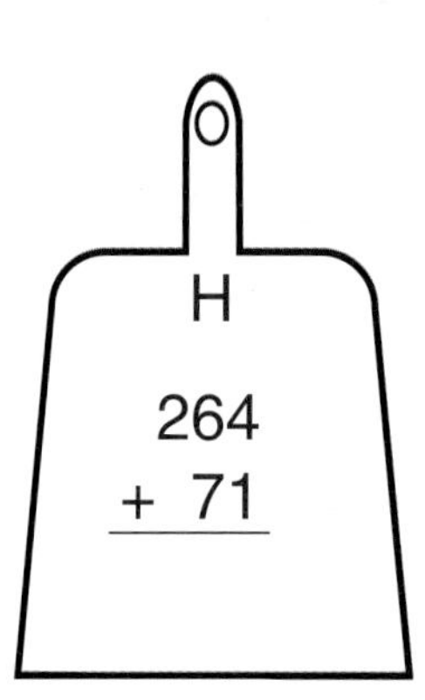


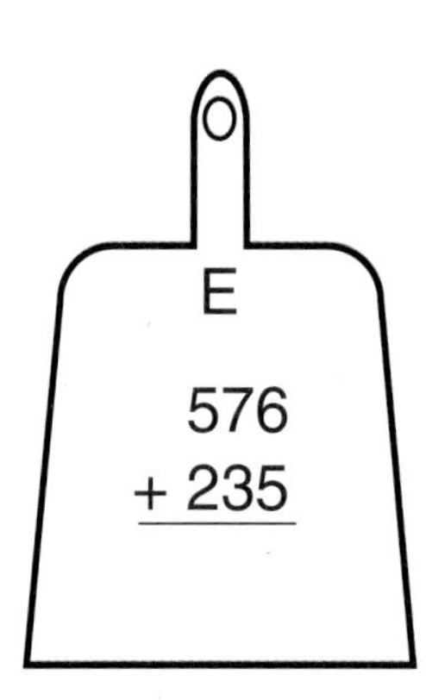


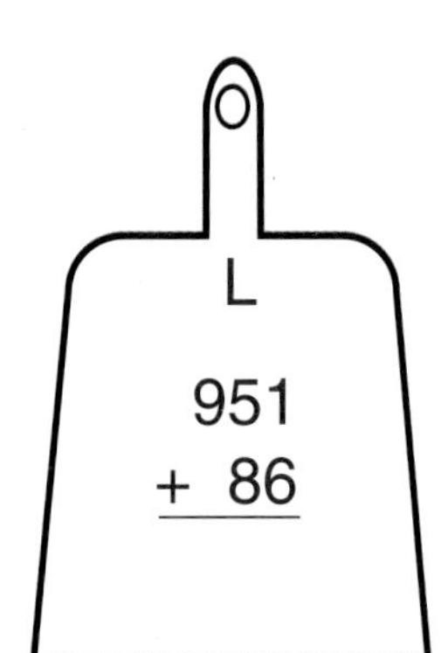


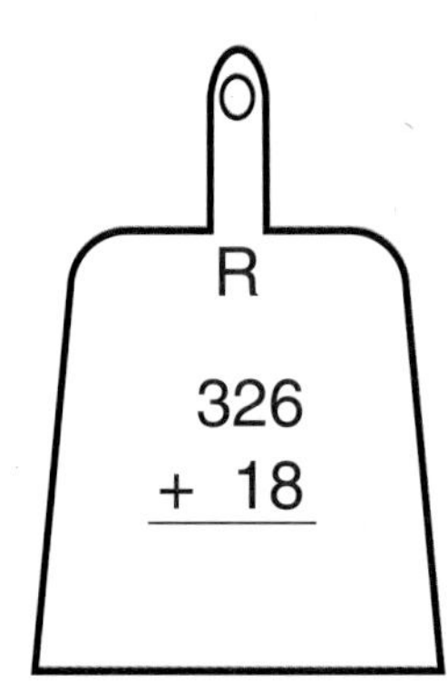


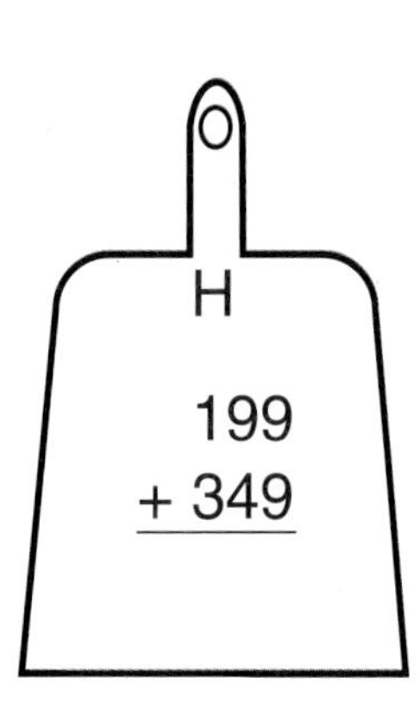


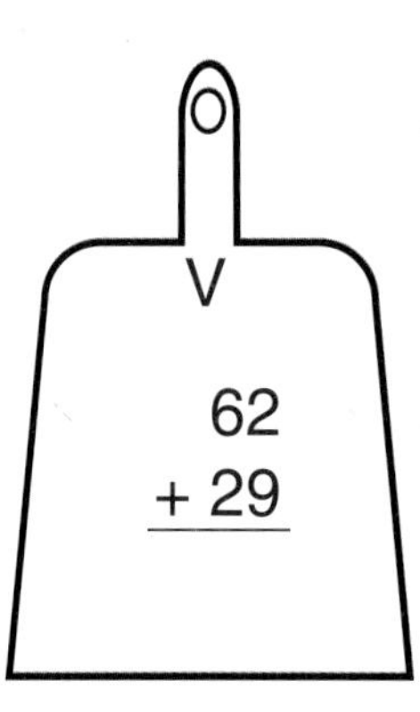


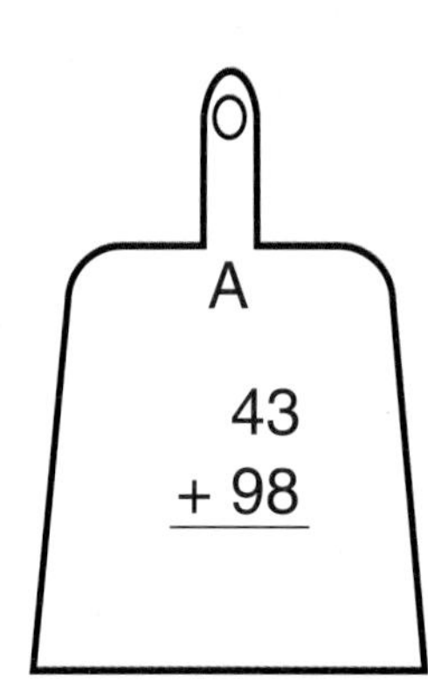


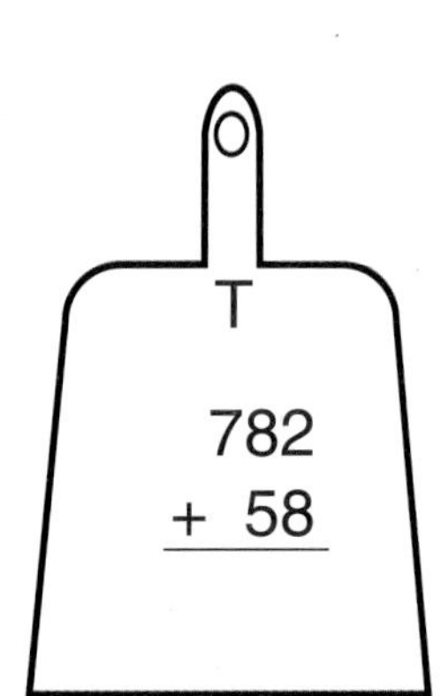


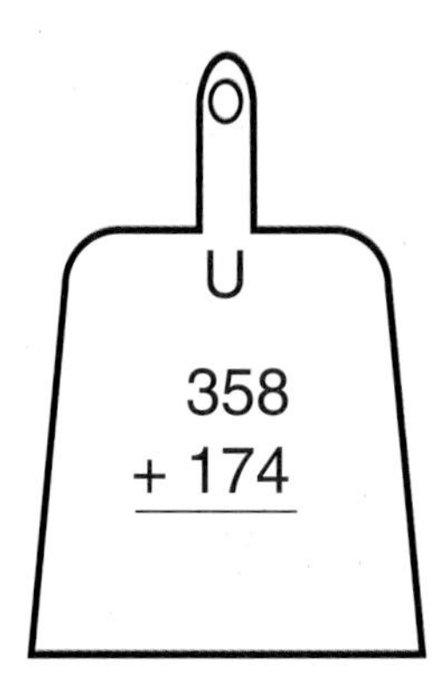


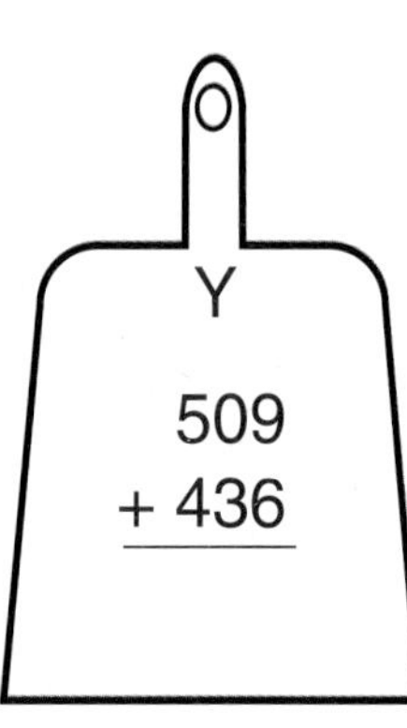


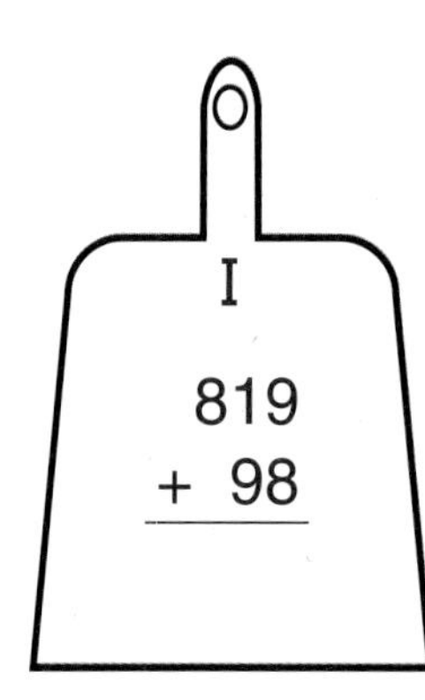


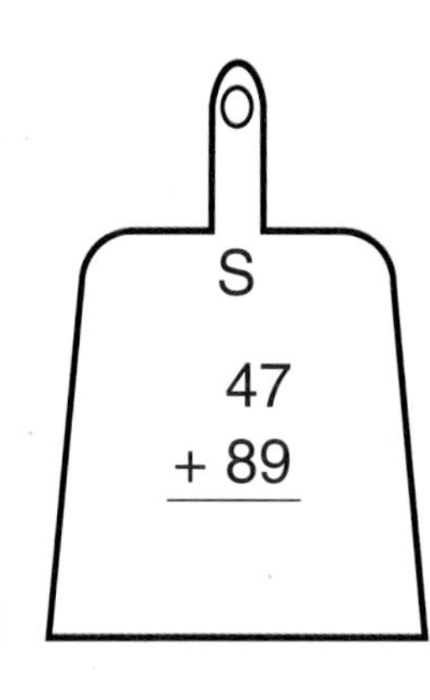


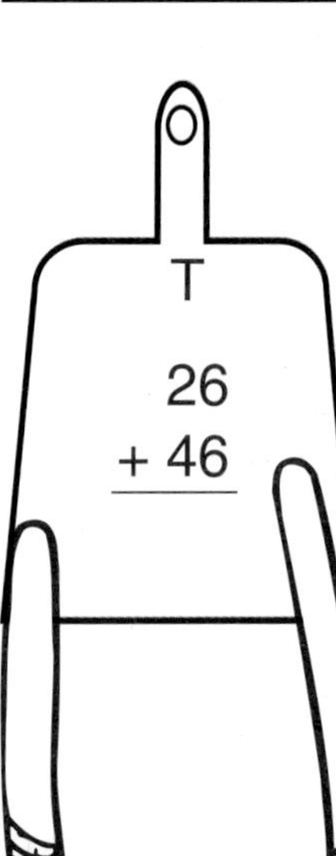


**What did one broom say to the other broom?**

To solve the riddle, match each letter to a numbered line below.

___ ___ ___ ___    ___ ___ ___    ___ ___ ___ ___ ___
335 141 91 811    945 90 532    335 811 141 344 1,125

___ ___ ___    ___ ___ ___ ___ ___ ___    ___ ___ ___ ___?
840 548 811    1,037 141 72 811 136 840    1,125 917 344 72

# Subtraction of Larger Numbers

## May I See Your Number?

Boost students' subtraction skills with this fast-paced activity! In advance, cut out a small slip of paper for each student. Program each slip with a different three-digit number and then place the slips in a bowl. Have each student draw a strip and direct him to not reveal his number. In addition, give each student a sheet of paper.

Have students pair themselves on your signal. Instruct each child to reveal his number to his partner. Next, have the child use the two numbers to write a subtraction problem on his paper. (Remind him that the largest number should be on top.) Direct each child to solve his problem and then check his answer against his partner's. Then give another signal and have the child find a different partner. Continue in this manner until each child has solved a desired number of problems. Now that's partner work that results in great practice!

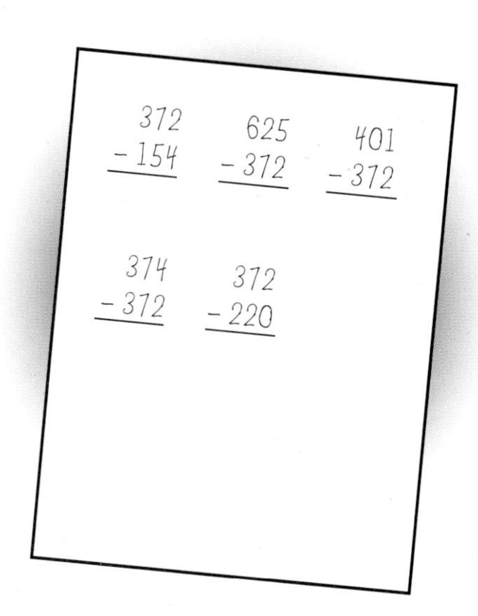


## On Target

This engaging center is right on target when it comes to reinforcing subtraction skills!

To prepare, make a list of several pairs of subtraction problems that each have the same answer. Use two colors of construction paper to make one target for each pair of problems as shown. Next, program the center of each target with one of the answers to the subtraction problems. Around the outside of the target, write the four numbers that were used in the subtraction problems. Store the targets in a center along with a class supply of paper.

Direct each visitor to the center to choose a target and use the four numbers around the target to write the six possible subtraction problems. Then have her solve each problem and circle the two that have the answer that is in the center of the target. Have her continue choosing targets and solving problems as time allows. Bull's-eye!

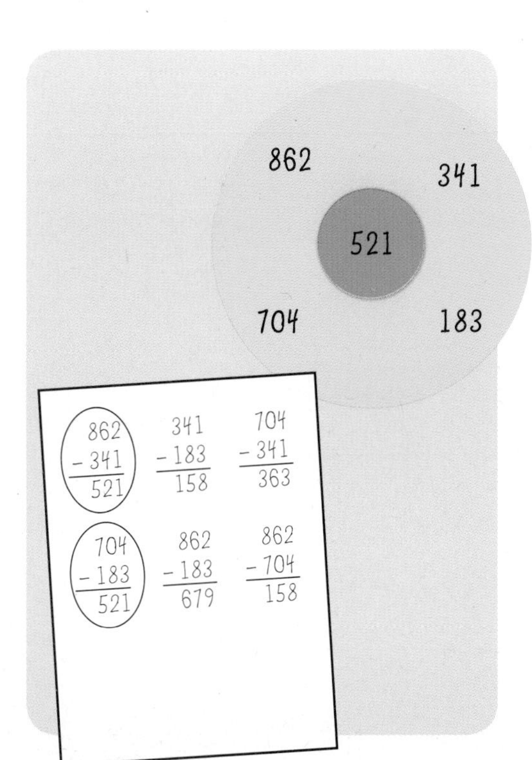

## Pile on the Points!

This small-group game of chance keeps students on their toes! To play the game, prepare the listed materials and then guide the group through the directions on this page.

**Materials needed for a group of five students:**

- 4 laminated gameboards programmed as shown
- 4 dry-erase markers
- wet paper towels
- a spinner programmed with the digits from 0 to 9

**Steps:**

1. Assign one student to be the caller and have him use the spinner. Have each of the remaining players take a gameboard.
2. The caller spins the spinner and calls out the number on which it stops.
3. Each player writes the called number in one of his boxes. He cannot change the number's position once it has been written.
4. Players repeat steps 2 and 3 until each box is filled in.
5. Each player studies his boxes to determine whether he has created a math problem that can be solved with a positive answer. If he has not, he doesn't earn any points this round. If he has, he solves his problem and records the answer in his points column.
6. After a predetermined amount of time, have each child add his points. The child with the highest number of points wins!

## Countdown to the 100th Day!

Looking for a quick, daily review of subtraction across zeros? Try this! Program a large sheet of paper with the sentences shown (but without the answers) and then laminate the paper. Each morning, have a different student volunteer use a dry-erase marker to fill in the first two blanks. Next, have the child write and solve a subtraction problem that shows how many days are left until the 100th day of school. Then have her fill in the third blank. Confirm the child's work and then guide the class to read the paragraph aloud. At the end of the day, wipe off the words so the page is ready for the next day. Wow! Time flies!

## Nothing Left

Your students use strategy, teamwork, and subtraction skills to play this whole-class game! Divide the class into two teams and have each team sit together. At the top of the chalkboard, write a three-digit number between 500 and 999. To play, the first team collectively chooses a number between ten and 100 to subtract from the number on the board. Next, one team member writes the number on the board and then solves the problem. The second team chooses a number between ten and 100 and subtracts it from the resulting number. Teams continue taking turns until one team gets zero for the answer and is declared the winner.

588
- 88
500
- 90
410
- 58
352
- 95
257
- 65
192
- 90
102
- 12
90
- 90
0

## Wacky Word Problems

The word is out—this activity is sure to tickle your students' funny bones! Place a large dictionary, index cards, and a pencil in a center. Direct each student who visits the center to think of a pair of words that sound silly together. Have him write the words on a card and look up the first one in the dictionary. Next, have him record the page number that the word appears on as shown. Instruct him to look up the second word and record its page number too. Then have him use the two numbers to write a subtraction problem. Finally, invite him to illustrate his comical word pairing. If desired, post the completed cards on a bulletin board titled "Silly Subtraction Pairs." Who can come up with the funniest combination?

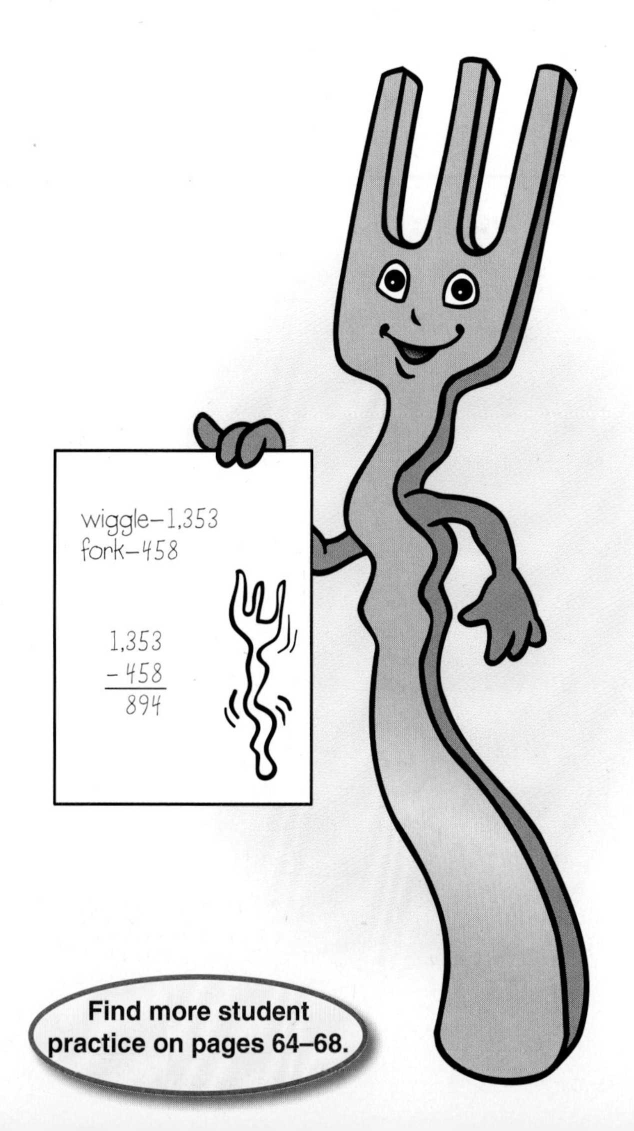


Find more student practice on pages 64–68.

Name ____________________ Date ____________________

# A Gem of a Find

Subtract.
Color the gem with the matching answer.

A. $\begin{array}{r} 18 \\ -\ 6 \\ \hline \end{array}$

B. $\begin{array}{r} 56 \\ -\ 10 \\ \hline \end{array}$

C. $\begin{array}{r} 77 \\ -\ 35 \\ \hline \end{array}$

D. $\begin{array}{r} 38 \\ -\ 5 \\ \hline \end{array}$

E. $\begin{array}{r} 67 \\ -\ 44 \\ \hline \end{array}$

F. $\begin{array}{r} 81 \\ -\ 10 \\ \hline \end{array}$

G. $\begin{array}{r} 29 \\ -\ 23 \\ \hline \end{array}$

H. $\begin{array}{r} 62 \\ -\ 1 \\ \hline \end{array}$

I. $\begin{array}{r} 36 \\ -\ 2 \\ \hline \end{array}$

J. $\begin{array}{r} 95 \\ -\ 72 \\ \hline \end{array}$

K. $\begin{array}{r} 44 \\ -\ 4 \\ \hline \end{array}$

L. $\begin{array}{r} 81 \\ -\ 31 \\ \hline \end{array}$

M. $\begin{array}{r} 56 \\ -\ 3 \\ \hline \end{array}$

N. $\begin{array}{r} 23 \\ -\ 2 \\ \hline \end{array}$

O. $\begin{array}{r} 54 \\ -\ 30 \\ \hline \end{array}$

P. $\begin{array}{r} 92 \\ -\ 40 \\ \hline \end{array}$

Q. $\begin{array}{r} 43 \\ -\ 13 \\ \hline \end{array}$

R. $\begin{array}{r} 19 \\ -\ 12 \\ \hline \end{array}$

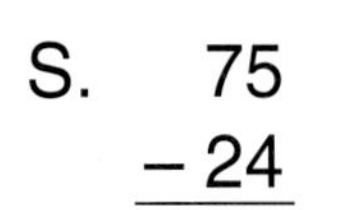

S. $\begin{array}{r} 75 \\ -\ 24 \\ \hline \end{array}$

T. $\begin{array}{r} 39 \\ -\ 11 \\ \hline \end{array}$

Name ____________________________ Date ____________________

# Munching Max

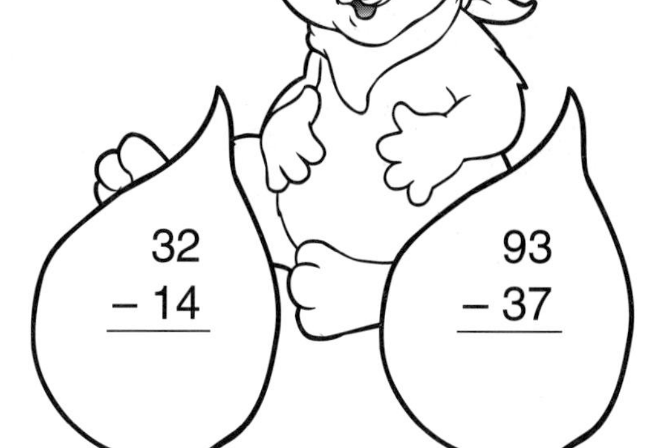

Subtract.
Show your work.
Help Max find his dinner.
If the answer is even, color the leaf green.

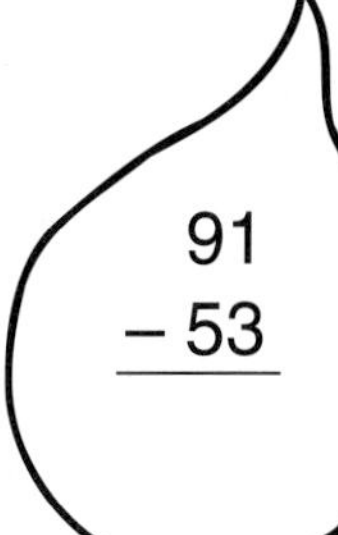

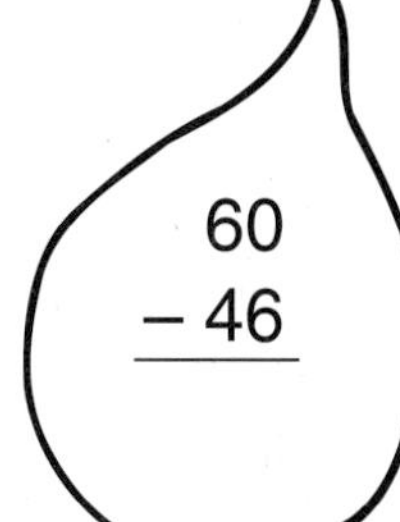

$$\begin{array}{r} 91 \\ -\ 53 \\ \hline \end{array} \quad \begin{array}{r} 11 \\ -\ 9 \\ \hline \end{array} \quad \begin{array}{r} 60 \\ -\ 46 \\ \hline \end{array} \quad \begin{array}{r} 32 \\ -\ 14 \\ \hline \end{array} \quad \begin{array}{r} 93 \\ -\ 37 \\ \hline \end{array}$$

$$\begin{array}{r} 71 \\ -\ 9 \\ \hline \end{array} \quad \begin{array}{r} 62 \\ -\ 13 \\ \hline \end{array} \quad \begin{array}{r} 53 \\ -\ 6 \\ \hline \end{array} \quad \begin{array}{r} 84 \\ -\ 25 \\ \hline \end{array} \quad \begin{array}{r} 41 \\ -\ 18 \\ \hline \end{array}$$

$$\begin{array}{r} 57 \\ -\ 9 \\ \hline \end{array} \quad \begin{array}{r} 74 \\ -\ 8 \\ \hline \end{array} \quad \begin{array}{r} 50 \\ -\ 14 \\ \hline \end{array} \quad \begin{array}{r} 12 \\ -\ 8 \\ \hline \end{array} \quad \begin{array}{r} 26 \\ -\ 9 \\ \hline \end{array}$$

$$\begin{array}{r} 20 \\ -\ 9 \\ \hline \end{array} \quad \begin{array}{r} 75 \\ -\ 27 \\ \hline \end{array} \quad \begin{array}{r} 40 \\ -\ 27 \\ \hline \end{array}$$

$$\begin{array}{r} 24 \\ -\ 6 \\ \hline \end{array} \quad \begin{array}{r} 36 \\ -\ 9 \\ \hline \end{array} \quad \begin{array}{r} 62 \\ -\ 27 \\ \hline \end{array}$$

Name ______________________ Date ______________

# A Gummy Situation

Subtract.
Color by the code.

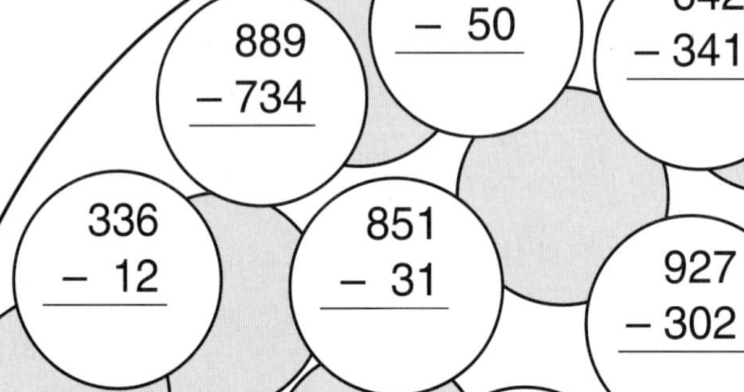

| | | | |
|---|---|---|---|
| 889 − 734 | 184 − 50 | 642 − 341 | 162 − 42 |
| 336 − 12 | 851 − 31 | 927 − 302 | 478 − 16 |
| 678 − 514 | 215 − 114 | 559 − 305 | 397 − 17 |
| 451 − 40 | 986 − 625 | 593 − 172 | 715 − 602 |
| 684 − 481 | 140 − 30 | 693 − 120 | 370 − 50 |

**Color Code**

0–150 = red
151–300 = yellow
301–450 = blue
451–850 = green

Name________________________ Date______________

# Ink's High Jinks

Subtract
Show your work.

| | | | | | | |
|---|---|---|---|---|---|---|
| 173 − 95 = A | 764 − 298 = S | 304 − 19 = K | 285 − 79 = O | 480 − 297 = E | | |
| 693 − 164 = T | 212 − 57 = P | 650 − 325 = H | 705 − 617 = I | 531 − 79 = N | 763 − 388 = E | 820 − 53 = G |
| 318 − 172 = U | 842 − 438 = T | 571 − 86 = B | 144 − 57 = R | | | |
| 436 − 192 = F | 927 − 64 = C | 902 − 549 = N | 451 − 63 = O | | | |

**Why did the farmer name his pig Ink?**

To solve the riddle, match the letters to the numbered lines below.

___ ___ ___ ___ ___ ___ ___   ___ ___
485 375 863 78 146 466 183   88 529

___ ___ ___ ___   ___ ___ ___ ___ ___ ___ ___
285 183 155 404   87 146 353 452 88 353 767

___ ___ ___   ___ ___   ___ ___ ___   ___ ___ ___!
206 146 404   388 244   529 325 375   155 183 452

Name ______________________ Date ______________

# Having a Ball!

Subtract.
Use the answers to complete each puzzle.

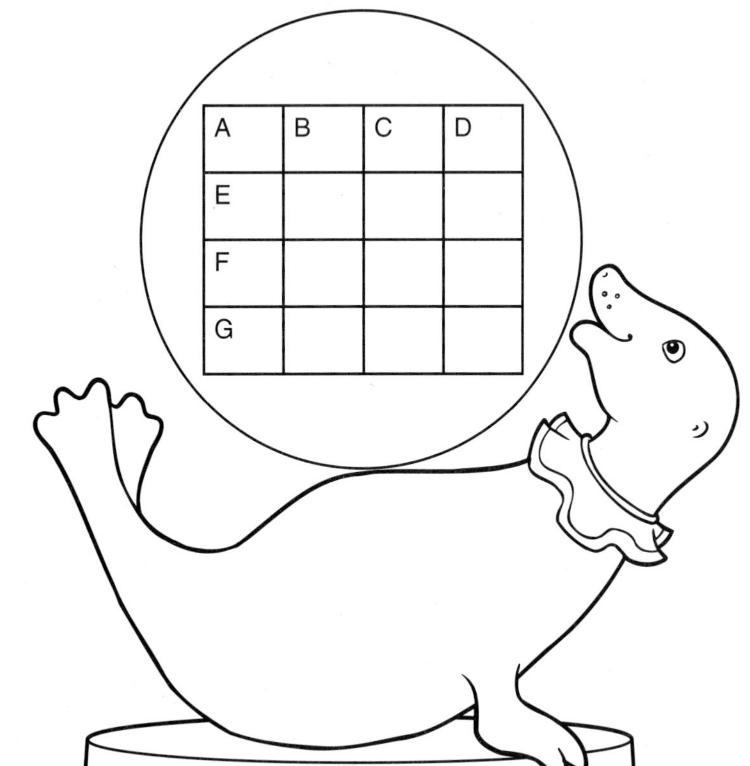


**Across**

A. 5,996 − 153

E. 4,758 − 446

F. 2,793 − 663

G. 1,589 − 347

**Down**

A. 5,563 − 142

B. 8,592 − 280

C. 4,287 − 153

D. 3,714 − 512

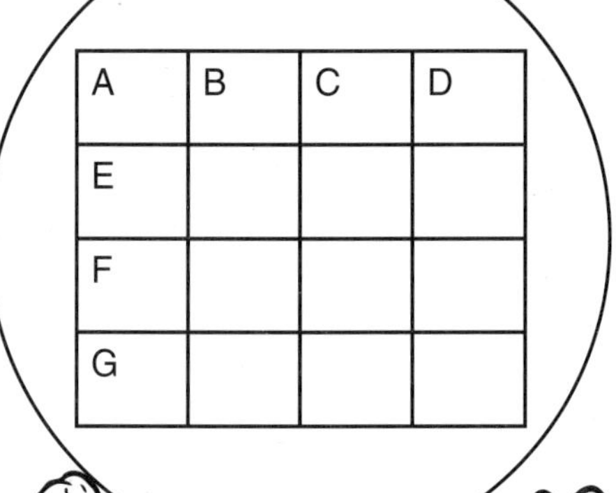

**Across**

A. 7,867 − 505

E. 4,987 − 371

F. 3,874 − 610

G. 5,492 − 241

**Down**

A. 7,648 − 213

B. 3,932 − 310

C. 6,489 − 324

D. 2,894 − 253

# Understanding Multiplication

## Chilly Computation

Provide solid multiplication practice with this ice cube tray activity. Here's how! Place at a center the tray, a supply of multiplication flash cards (without the answers), plastic counters, paper, and pencils. To use the center, a student draws a flash card and then copies and solves the multiplication problem on his paper. Next, he places the appropriate number of manipulatives in the corresponding number of tray sections to represent the problem. He counts the manipulatives and compares the total to his answer, making any corrections as shown. He repeats this process to solve a predetermined number of problems. Then he submits his work for teacher approval. Cool!

## Fabulous Fabric Arrays

Use scraps of fabric to piece together multiplication practice! Gather pieces of fabric that have checkered patterns. Cut out a class supply of pieces so that they show various multiplication arrays. Next, use a strip of masking tape and a permanent marker to label the back of each fabric piece with the corresponding multiplication fact.

To begin, give a piece of fabric to each student. Have her study the array and write the multiplication fact. Then tell her to look at the back to check her work. At your signal, have students switch fabric pieces. Repeat this process as desired. Then place the fabric pieces at a center for additional practice. What a fabulous idea!

## Taking a Stand!

Students are sure to enjoy standing up to be counted for this multiplication activity! To begin, invite eight students to the front of the room. Write a multiplication fact for the number of students on a card (such as 4 x 2 = 8). Secretly show the fact to the volunteers, shielding it from the seated students' view. Then have the youngsters quietly arrange themselves into groups to represent the multiplication fact. (Remind them that there are two possible arrangements.) Challenge the seated students to study their classmates' arrangement and state the multiplication fact. Confirm the correct answer by revealing the card to the seated students. Repeat this process with additional students and multiplication facts. Bravo!

## Fine-Feathered Facts

It's time to hatch a new understanding of multiplication! Give each pair of students a disposable cup containing about 30 dried beans (eggs) and six mini muffin cups (nests). Tell students to imagine that there are three bird nests, each containing two eggs. Have each pair set out three nests and then place two eggs inside each one. Ask youngsters to tell how many eggs there are in all *(six)*. Write the corresponding multiplication problem on the board (3 x 2 = 6), explaining that three nests containing two eggs each equals six eggs in all. Pose additional problems in this manner, having students write a multiplication problem for each one. Then have partners work together to model their own problems. Now that's giving students a bird's-eye view of multiplication!

Find more student practice on pages 71–72.

Name ______________________________ Date ______________________

# In Full Bloom

How many petals are in each set?
On each flower, write the number of petals.
Write an addition sentence.
Then write a multiplication sentence.
The first one has been done for you.

A.

4 + 4 = 8

2 x 4 = 8

B.

___ + ___ + ___ = ___

___ x ___ = ___

C.

___ + ___ = ___

___ x ___ = ___

D.

___ + ___ + ___ = ___

___ x ___ = ___

E.

___ + ___ + ___ = ___

___ x ___ = ___

F.

___ + ___ + ___ + ___ = ___

___ x ___ = ___

G.

___ + ___ + ___ + ___ = ___

___ x ___ = ___

H.

___ + ___ + ___ + ___ + ___ = ___

___ x ___ = ___

Name________________________ Date______________

# Baking a Fresh Batch

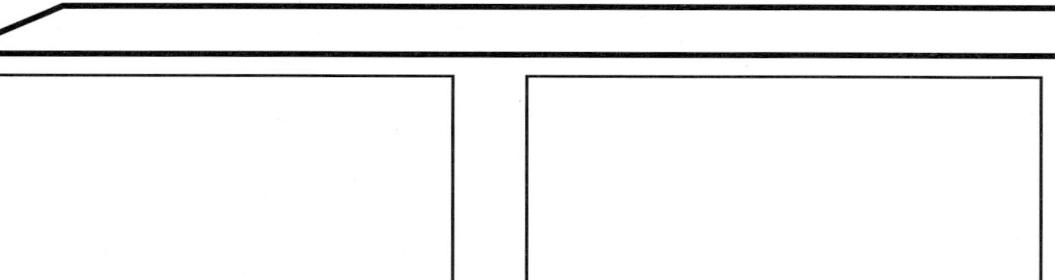

How many chocolate chips are in each set?
Draw the number of cookies.
On each cookie, draw the number of chips.
Write the multiplication sentence.
The first one has been started for you.

A. 2 cookies
3 chips each

_2_ x _3_ = ___

B. 4 cookies
2 chips each

___ x ___ = ___

C. 3 cookies
4 chips each

___ x ___ = ___

D. 2 cookies
6 chips each

___ x ___ = ___

E. 5 cookies
2 chips each

___ x ___ = ___

F. 4 cookies
4 chips each

___ x ___ = ___

G. 6 cookies
3 chips each

___ x ___ = ___

H. 3 cookies
5 chips each

___ x ___ = ___

# Basic Multiplication Facts (0–9)

## On Board With Multiplication Facts!

Give a new purpose to a hands-on tool that's readily available in your classroom—a Geoboard! Provide each student with a Geoboard, a rubber band, a pencil, and paper. Start the activity by directing each student to write a multiplication problem, such as 1 x 4, on her paper. Then explain that the student can solve the problem by looping her rubber band around one column of four pegs on the Geoboard. Have her record the answer. Repeat with different facts. Finally, challenge students to write a problem, display it on the Geoboard, and then have a partner write the equivalent number sentence. Students' multiplication skills will really stretch!

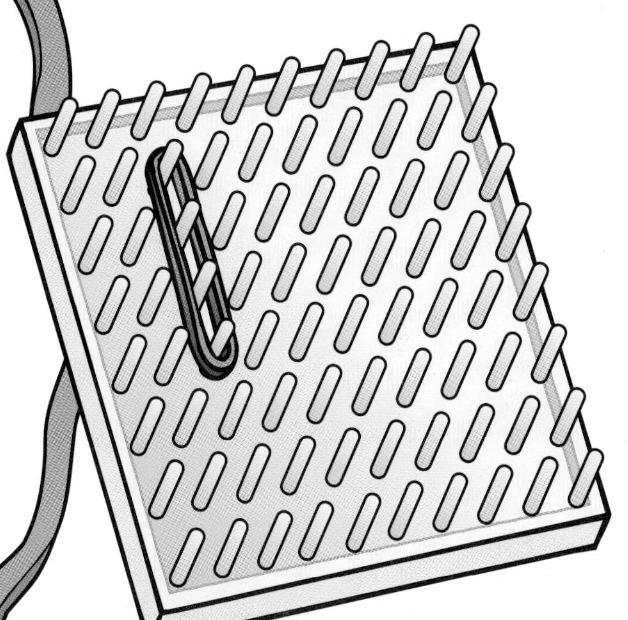

## Flipping Over Multiplication Facts

Provide mouthwatering multiplication practice with this clever small-group game called Multiplication Flip. Choose ten or more pairs of multiplication facts that have the same products. Cut circles from brown construction paper to make pancakes, one for each fact. Then write each fact (but not the answer) on a different pancake. Place the pancakes facedown in rows on a cookie sheet. Give a group of two to four students a spatula. Direct each student to use the spatula to flip over two pancakes and determine their products. If their products are the same, the student takes those pancakes. If the products aren't equal, he flips the pancakes back over. Players take turns until all of the pancakes are gone. The player with the most pancakes wins.

# Unifix Tricks

Fact practice is a snap when you combine flash cards, skip counting, and Unifix cubes. Give each student a supply of Unifix cubes in two colors. Choose one factor, such as five, and have students use one color of Unifix cubes at a time to snap together sets with five cubes each. Then direct students to connect a specified number of sets, alternating colors to make one long row. Show students how to skip-count to figure out the total number of cubes; then show them how that equates to a math sentence. Finally, introduce multiplication fact flash cards and have students use the cubes to skip-count their way to each answer. Multiplication facts won't be tricky anymore!

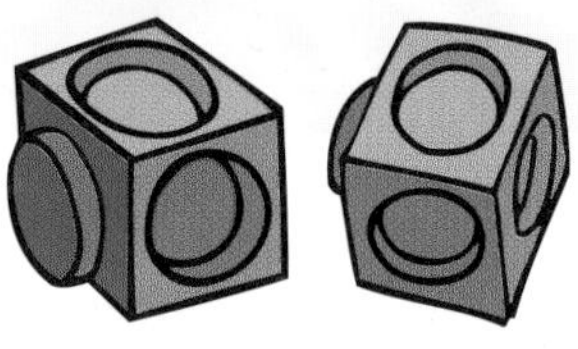
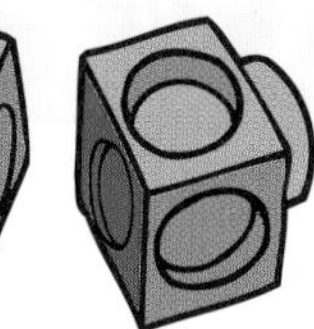

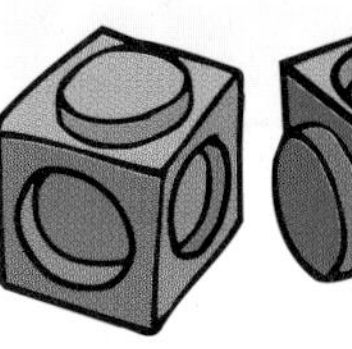

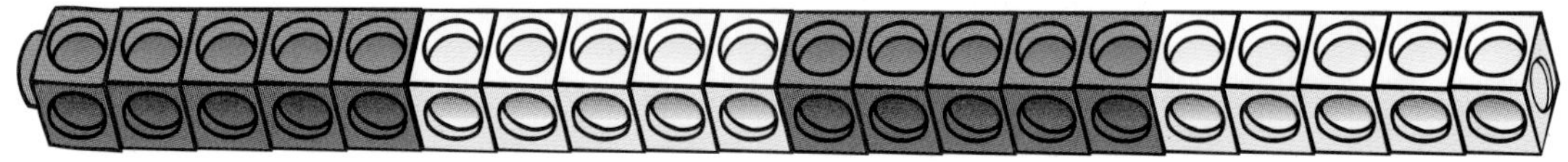

# Double the Facts, Double the Fun

Learning doubles is twice the fun with this artsy activity! For each student, gather assorted colors of paint, cotton swabs, ten strips of 2" x 4" white paper, two sheets of 12" x 18" construction paper, glue, and a marker.

On the board list the twos facts from 2 x 0 up to 2 x 9. Have students provide the answers. Next, distribute the supplies. Have each student fold each paper strip in half to make a 2" x 2" square. Then have him unfold each one. Demonstrate how to dip a cotton swab in paint and lightly dab a dot of paint on one side of the paper strip. Then refold the strip and gently press the sides together. Unfold to reveal that one dab of paint has been doubled to two. Repeat using different colors to show how two dabs, three dabs, and so forth can be doubled. One strip will remain, which represents 2 x 0. Allow the paint to dry. Then glue the two sheets of construction paper together to make one long sheet. Glue the strips, in order, to the construction paper and label each doubles fact. Doubles are no trouble!

## It's a Date!

Put outdated monthly calendars to a timely new use! Give one old calendar page to each student. Choose a multiplication fact, such as 3 x 3, and challenge each student to block off her calendar squares in pencil to represent that equation and determine the answer. Once she has checked her answer, the student traces over the blocks in colored marker and writes the number sentence in the box. Repeat with other facts and colors. Very timely!

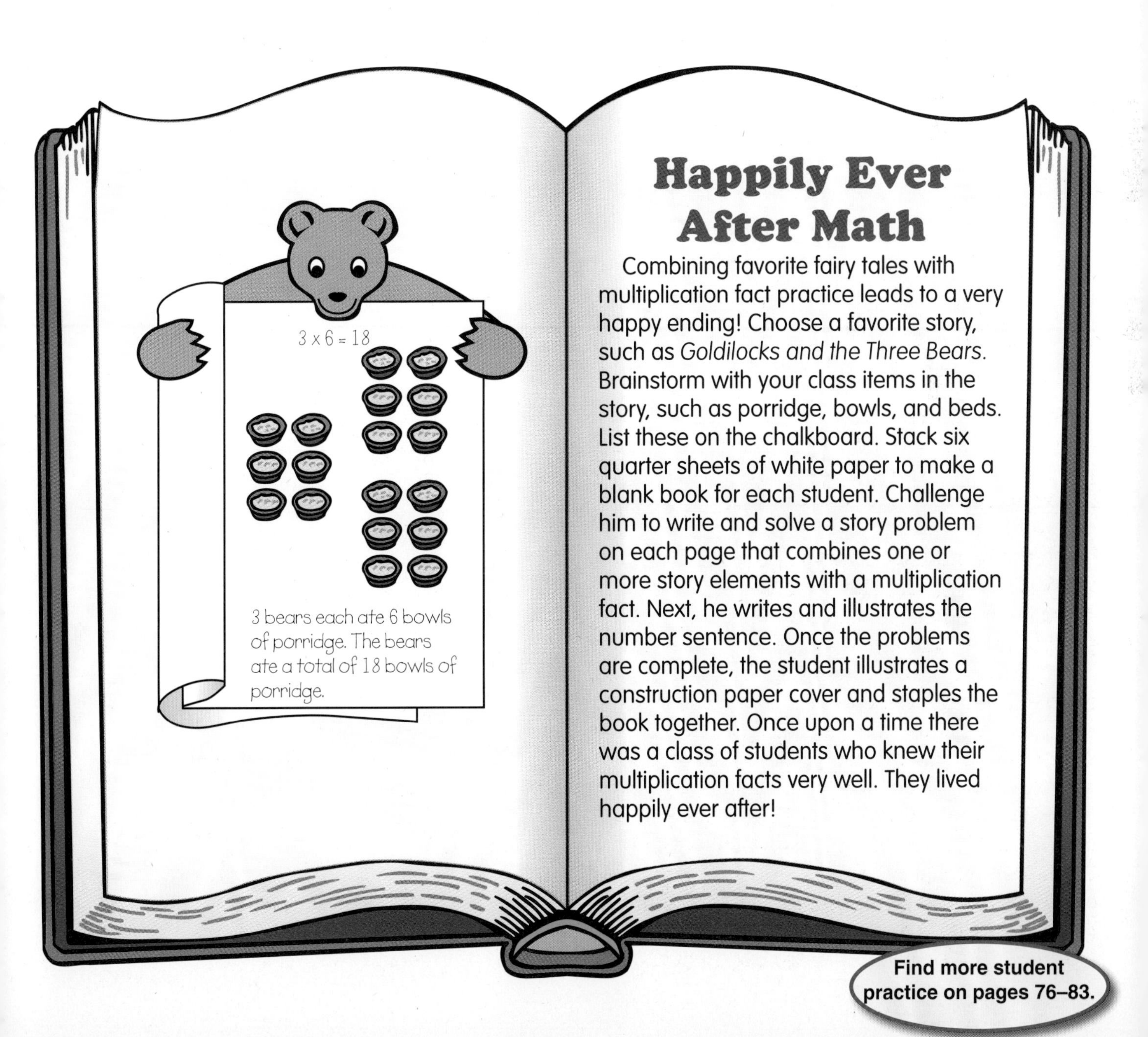


## Happily Ever After Math

Combining favorite fairy tales with multiplication fact practice leads to a very happy ending! Choose a favorite story, such as *Goldilocks and the Three Bears.* Brainstorm with your class items in the story, such as porridge, bowls, and beds. List these on the chalkboard. Stack six quarter sheets of white paper to make a blank book for each student. Challenge him to write and solve a story problem on each page that combines one or more story elements with a multiplication fact. Next, he writes and illustrates the number sentence. Once the problems are complete, the student illustrates a construction paper cover and staples the book together. Once upon a time there was a class of students who knew their multiplication facts very well. They lived happily ever after!

Find more student practice on pages 76–83.

Name ______________________________ Date ____________________

# Smooth Sailing

Use the code.
Write the missing factors.
Multiply.

**Code**

△ = 0 □ = 1 ○ = 2

a. 3 x ○ = ____

b. □ x 7 = ____

c. □ x 6 = ____

d. 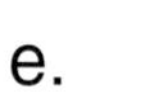 △ x 2 = ____

e. 1 x △ = ____ 

f. 5 x □ = ____

g. 4 x △ = ____

h. ○ x 9 = ____

i. 4 x ○ = ____

j. 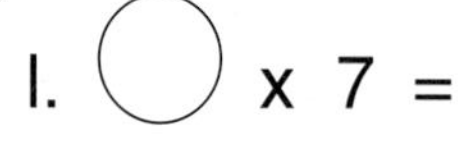 △ x 8 = ____

k. 2 x □ = ____

l. ○ x 7 = ____

m. 5 x △ = ____ 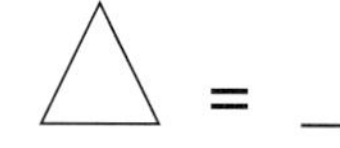

n. □ x 9 = ____

o. 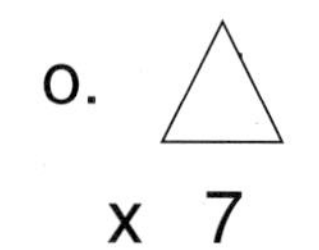 △ x 7 = ____

p. 1 x ○ = ____

q. ○ x 6 = ____

r. 3 x △ = ____

s. □ x 8 = ____

Name ______________________ Date ______________________

# A Cowboy Contest

Who will win the horseshoe game?
Write 0, 1, or 2 to complete each fact.
Make a tally mark on the matching cowboy's scoreboard.
Color the winner.

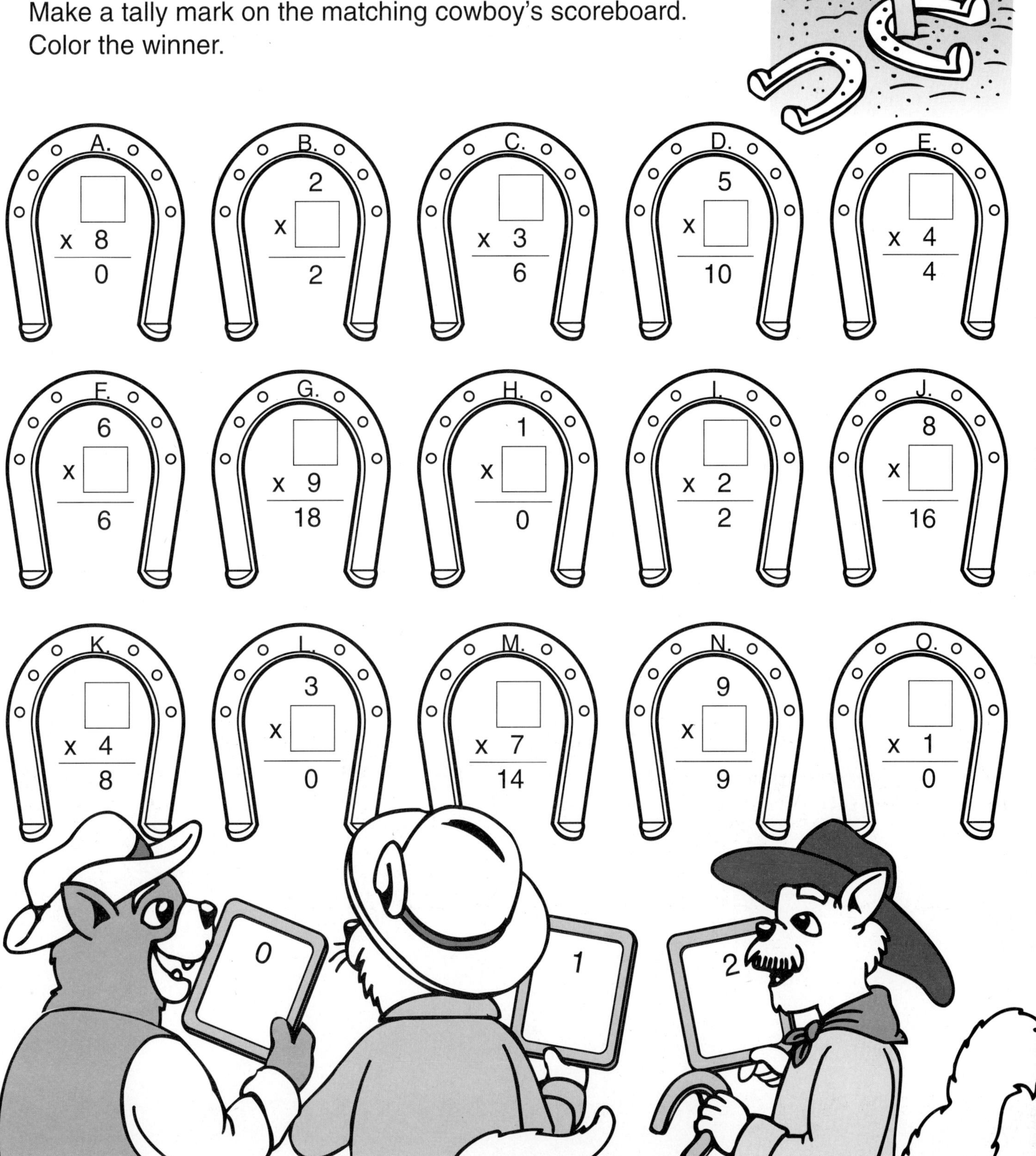

Name____________________ Date____________

# Goodies to Go

Read the baker's sign.
Write a multiplication problem for each order.
Solve the problem.
The first one has been done for you.

One bakery box holds
- 1 cake
- 2 doughnuts
- 3 muffins **or**
- 4 cookies

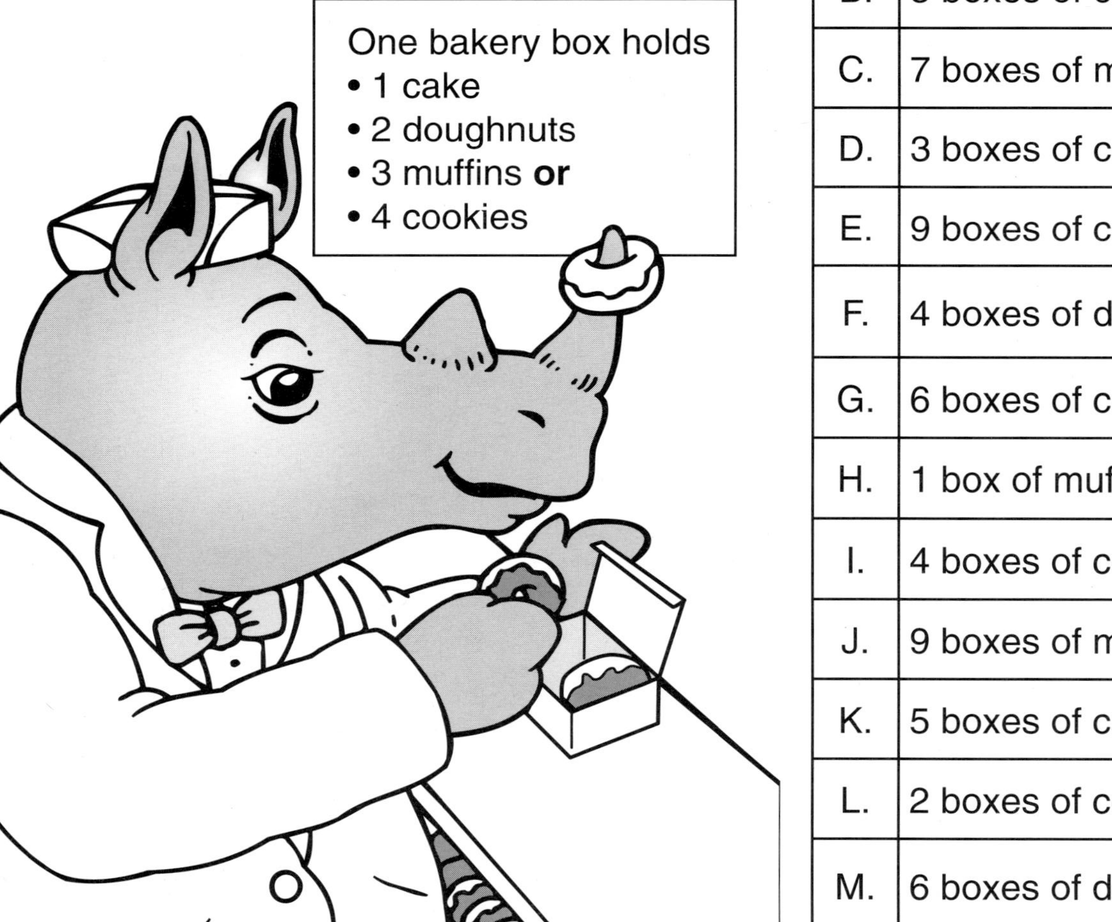

| | Orders | Total Number of Treats |
|---|---|---|
| A. | 2 boxes of muffins | 2 x 3 = 6 muffins |
| B. | 5 boxes of cookies | ___ x ___ = ___ cookies |
| C. | 7 boxes of muffins | ___ x ___ = ___ muffins |
| D. | 3 boxes of cakes | ___ x ___ = ___ cakes |
| E. | 9 boxes of cookies | ___ x ___ = ___ cookies |
| F. | 4 boxes of doughnuts | ___ x ___ = ___ doughnuts |
| G. | 6 boxes of cookies | ___ x ___ = ___ cookies |
| H. | 1 box of muffins | ___ x ___ = ___ muffins |
| I. | 4 boxes of cookies | ___ x ___ = ___ cookies |
| J. | 9 boxes of muffins | ___ x ___ = ___ muffins |
| K. | 5 boxes of cakes | ___ x ___ = ___ cakes |
| L. | 2 boxes of cookies | ___ x ___ = ___ cookies |
| M. | 6 boxes of doughnuts | ___ x ___ = ___ doughnuts |

Name ______________________________ Date ______________________

# Road Race

Multiply each number on the wheel by the center number.
The first one has been done for you.

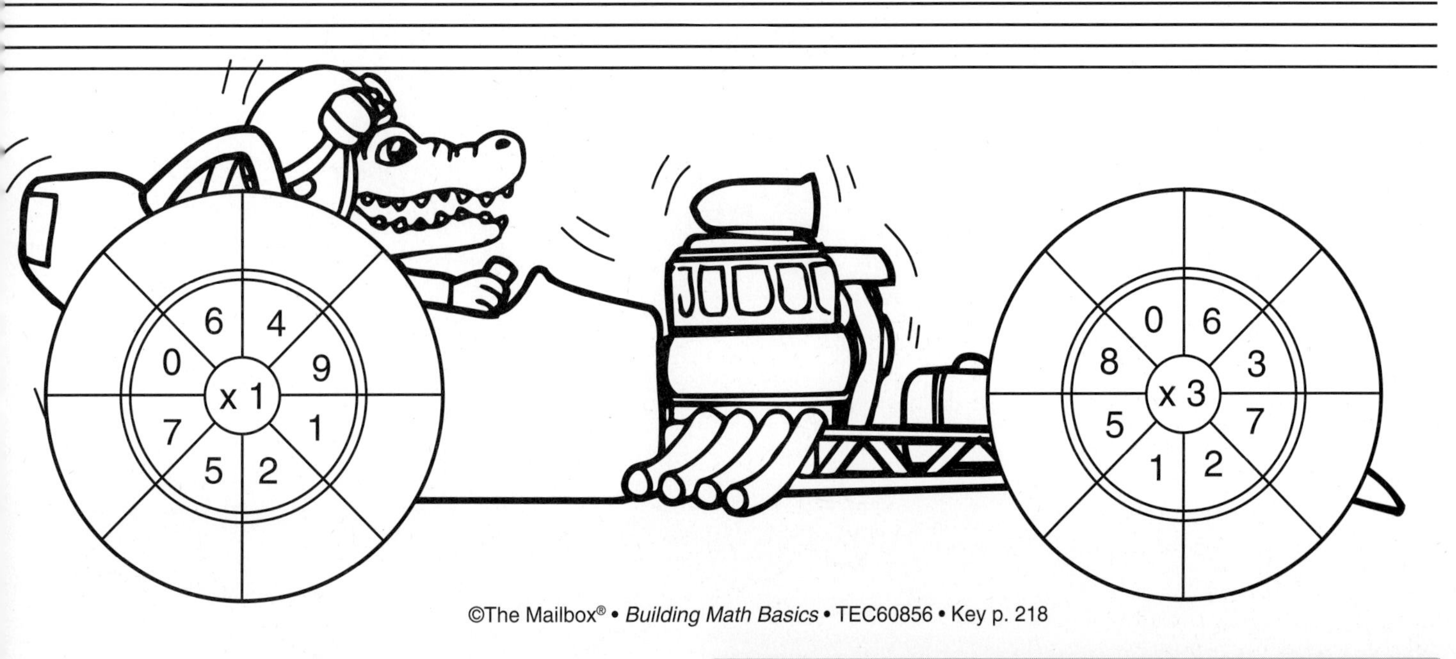

Name ______________________ Date ______________________

# High-Flying Penguin!

Multiply.
Color by the code.

$5 \times 7 =$ ___

$6 \times 4 =$ ___

$9 \times 4 =$ ___

$3 \times 5 =$ ___

$6 \times 7 =$ ___

$5 \times 4 =$ ___

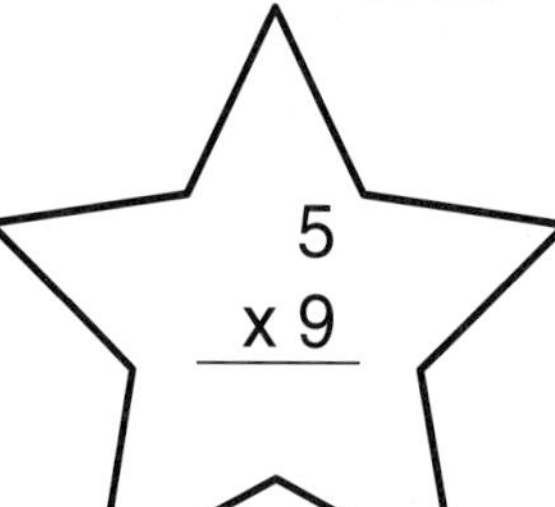


$2 \times 7 =$ ___

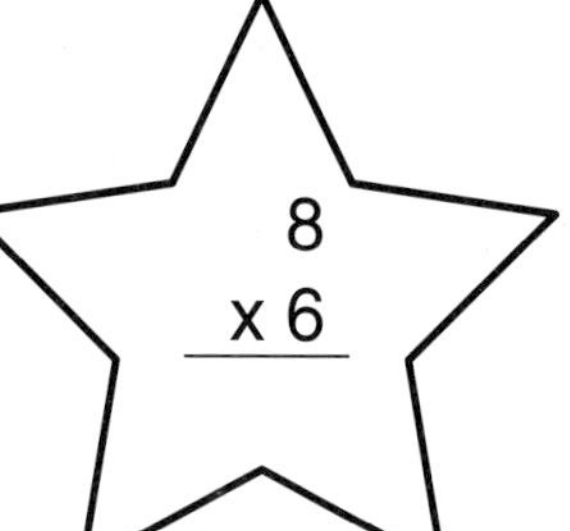


$9 \times 6 =$ ___

$9 \times 3 =$ ___

$8 \times 0 =$ ___

$7 \times 6 =$ ___

$1 \times 6 =$ ___

$6 \times 8 =$ ___

$5 \times 2 =$ ___

$4 \times 8 =$ ___

$5 \times 5 =$ ___

$6 \times 6 =$ ___

**Color Code**

| | | |
|---|---|---|
| 0–10 | = | red |
| 11–20 | = | blue |
| 21–30 | = | green |
| 31–40 | = | orange |
| 41–50 | = | yellow |
| 51–60 | = | purple |

Name ______________________________ Date____________________

# Two Scoops!

Write the missing factor or product.

a.

x 3

18

b.

c.

d.

e.

f.

g.

h.

i.

j.

5

x 5

k.

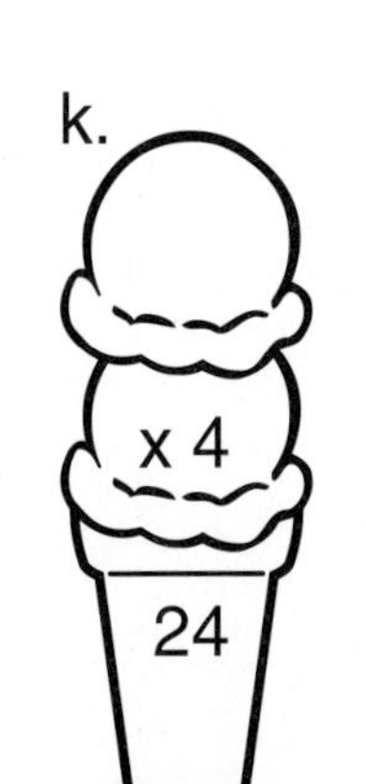


l.

m.

n.

o.

p.

Name ______________________ Date ______________

# What's Cooking?

Cut out the vegetable cards below.
Place the cards on the pots so that each pair of factors equals the product on the pot's lid.
Glue the cards in place.
Write a multiplication sentence below the pot.

A. 63 ___ x ___ = ___

B. 40 ___ x ___ = ___

C. 56 ___ x ___ = ___

D. 54 ___ x ___ = ___

E. 45 ___ x ___ = ___

F. 28 ___ x ___ = ___

G. 72 ___ x ___ = ___

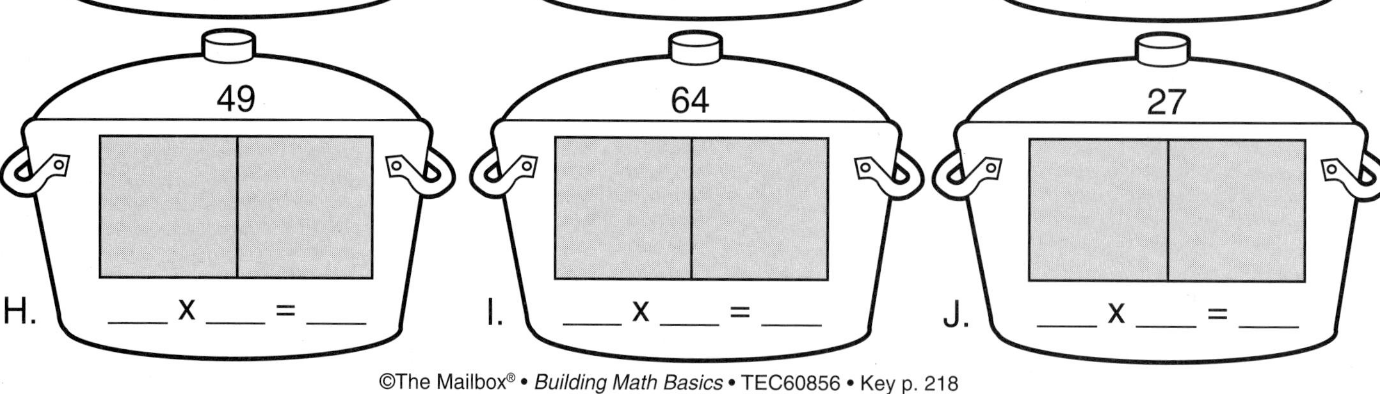




3 4 5 5 6 7 7 7 7 7

8 8 8 8 8 9 9 9 9 9

Name____________________ Date____________

# Pleasant Dreamssss

For each letter, write a problem that equals the matching product on the snake.
Each problem must use 7, 8, or 9 as a factor.
Multiply.

a. ☐ x ☐

b. ☐ x ☐

c. ☐ x ☐

d. ☐ x ☐

e. ☐ x ☐

f. ☐ x ☐

g. ☐ x ☐

h. ☐ x ☐

i. ☐ x ☐

j. ☐ x ☐

k. ☐ x ☐

l. ☐ x ☐

a. 72 b. 64 c. 8 d. 35 e. 81 f. 36 g. 42 h. 16 i. 54 j. 40 k. 18 l. 21 m. 48 n. 49 o. 7 p. 45 q. 64 r. 28 s. 27 t. 14 u. 24 v. 9 w. 56 x. 32

m. ☐ x ☐

n. ☐ x ☐

o. ☐ x ☐

p. ☐ x ☐

q. ☐ x ☐

r. ☐ x ☐

s. ☐ x ☐

t. ☐ x ☐

u. ☐ x ☐

v. ☐ x ☐

w. ☐ x ☐

x. ☐ x ☐

# Understanding Division

## One for You, One for Me

Get students started on understanding how division works with this small group activity. Give each child 20 Unifix cubes. First, have students count their cubes. Then ask each child to pretend that he's sharing the cubes with a friend. The student divides the cubes until each person has the same number and then counts each pile. Write the corresponding division problem on the board and have students copy it. Next, each child gathers the cubes and repeats the activity, dividing the cubes between four friends, five friends, and ten friends. Each time he writes the corresponding division sentence. The basics of division will really click!

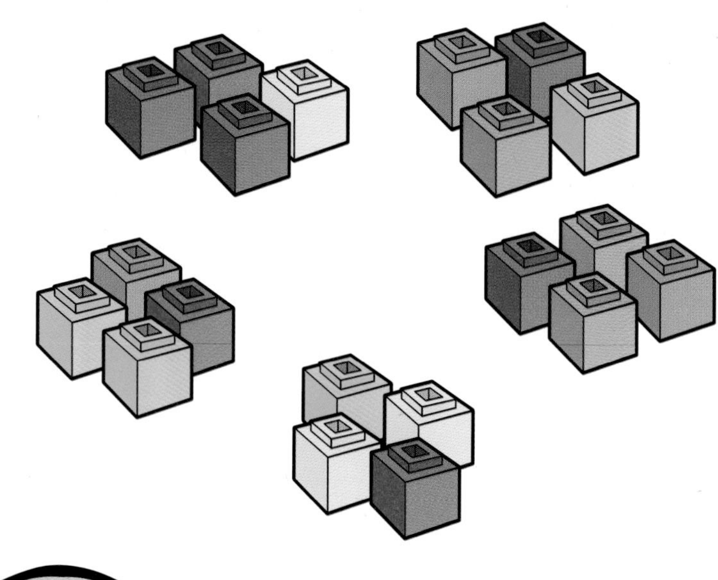

## Jack's Magic Beans

Fee! Fie! Foe! Fum! Students will see that division is fun! Review the story of *Jack and the Beanstalk* with your class. Then give each pair five green pipe cleaners (beanstalks) and 25 dried beans. Explain that one of the things that is magical about Jack's beanstalks is that they each produce an equal number of beans. Then state a certain number of stalks and beans (for example, two stalks and 28 beans) and challenge students to figure out how many beans are on each stalk. Then guide students to write the resulting division sentence. Fee! Fie! Foe! Fum! Let's do some more division!

# Fair Share

Any student who has brothers and sisters knows that when it comes to dividing favorite foods, things aren't always equal! Introduce the concept of remainders with this tasty activity. Prepare a baggie for each student with a large handful of Goldfish crackers, cereal pieces, M&M's minis candies, or other small-size snacks inside. State a problem such as "Your grandma says that you have to divide your snacks equally between yourself and your four cousins. You can eat the leftovers. How many does each cousin get? How many do you get?" Have students put the snacks on a paper towel and then divide them. Once students have the answer, have them write down the division sentence and then munch on the remainder. Continue dividing and eating until your students have had a bellyful of practice. Now that's division that students can sink their teeth into!

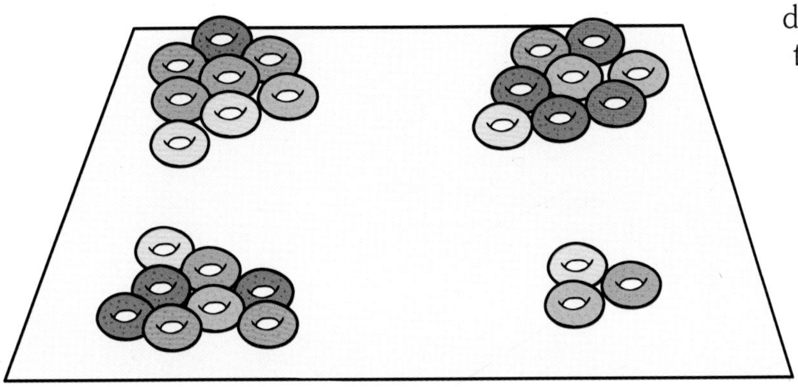

## The Best Job Around

At this learning center, students pretend they work at a local radio station, plus they practice division. Post a sign that says, "Your new job is with radio station WDIV. The station is giving away free movie tickets to a certain number of callers. Your job is to figure out how many tickets each caller gets." At the center place two dice and 66 carnival tickets or colorful slips of paper. Explain that a student rolls the dice once to find out how many callers have won free tickets. She writes down that number. Then she rolls the dice two more times to figure out how many tickets she has to give away. The first number she rolls represents the tens digit and the second number represents the ones digit. She writes down that number. She counts out that number of tickets and divides them. Then she writes the resulting division sentence and any remainder. Congratulations to our lucky winners!

Find more student practice on pages 86–87.

Name ______________________________ Date ____________________

# Stashing Away

Read.
Solve.

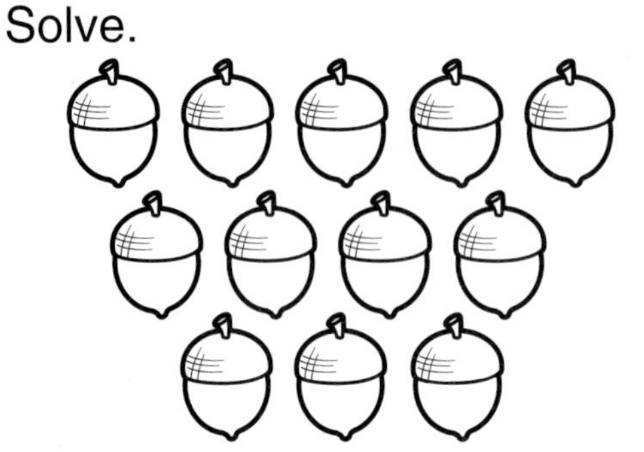

Count.
Circle groups of 3.
How many groups? _____

_____ ÷ 3 = _____ groups

Count.
Circle groups of 5.
How many groups? _____

_____ ÷ 5 = _____ groups

Count.
Circle groups of 1.
How many groups? _____

_____ ÷ 1 = _____ groups

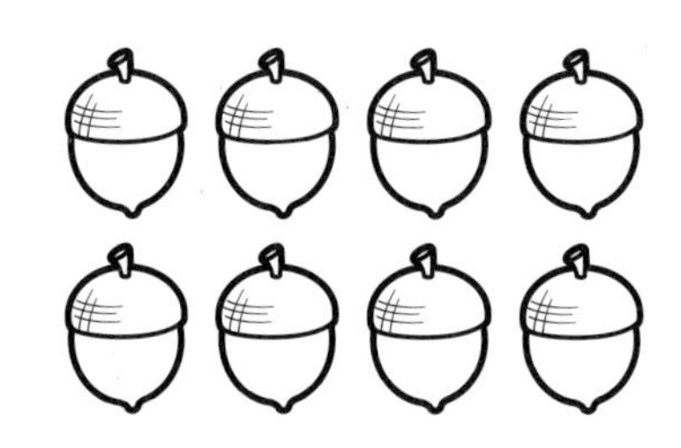

Count.
Circle groups of 2.
How many groups? _____

_____ ÷ 2 = _____ groups

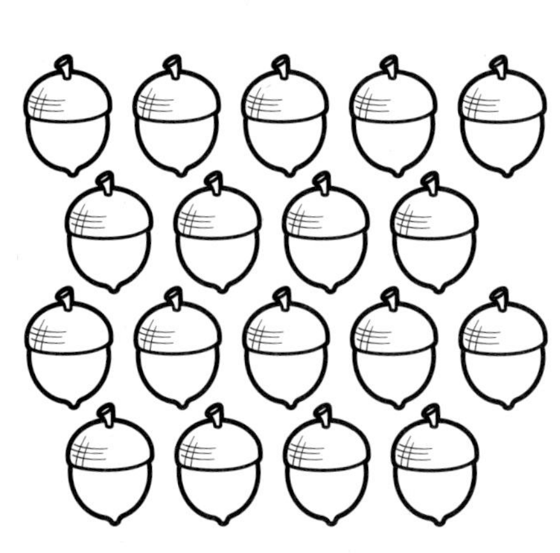

Count.
Circle groups of 2.
How many groups? _____

_____ ÷ 2 = _____ groups

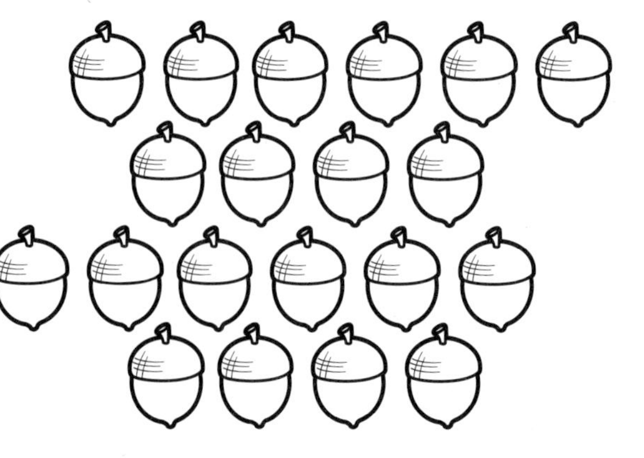

Count.
Circle groups of 4.
How many groups? _____

_____ ÷ 4 = _____ groups

Name ____________________ Date ____________

# Spotted!

Read.
Solve.

A. How many spots in all? ___
How many bugs? ___
How many spots on each bug? ___
___ ÷ ___ = ___ spots on each bug

B. How many spots in all? ___
How many bugs? ___
How many spots on each bug? ___
___ ÷ ___ = ___ spots on each bug

C. How many spots in all? ___
How many bugs? ___
How many spots on each bug? ___
___ ÷ ___ = ___ spots on each bug

D. How many spots in all? ___
How many bugs? ___
How many spots on each bug? ___
___ ÷ ___ = ___ spots on each bug

E. How many spots in all? ___
How many bugs? ___
How many spots on each bug? ___
___ ÷ ___ = ___ spots on each bug

F. How many spots in all? ___
How many bugs? ___
How many spots on each bug? ___
___ ÷ ___ = ___ spots on each bug

# Basic Division Facts (0–9)

| September | | | | | | |
|---|---|---|---|---|---|---|
| Sunday | Monday | Tuesday | Wednesday | Thursday | Friday | Saturrday |
| | | | | 1 | 2 | 3 |
| 4 | 5 | 6 | 7 | 8 | 9 | 10 |
| 11 | 12 | 13 | 14 | 15 | 16 | 17 |
| 18 | 18 | 20 | 21 | 22 | 23 | 24 |
| 25 | 26 | 27 | 28 | 29 | 30 | |

## Great Date

Today's date provides a quick opportunity to practice division facts! Write the date on the board. Then challenge students to come up with all of the division facts that use the date as the dividend. To guide students' thinking, ask them whether the date is divisible by 1, 2, 3, 4, and so forth up to 9. Write resulting division sentences on the board. This activity works well anytime you have a few minutes to spare!

$21 \div 3 = 7$

$21 \div 7 = 3$

## Fact Attack

Want division facts to stick with your students? Try this learning center! Choose an assortment of math facts. Write each dividend and divisor at the top of a craft stick (as shown), and the quotient at the bottom. Then write another fact on the back of the stick. Once you've created a set of sticks, push them into a Styrofoam block so that the answers are hidden. When a student visits the learning center, she chooses a stick, studies the problem, and then states the answer. To check her answer, the students pulls the stick out of the block. Division facts will really stick!

## Ring of Facts

Start a round of division fact practice with this self-checking learning center! On an unlined index card, write a division fact that students are learning, but leave off the answer. On the next card, write the answer to the first problem and then, on the back of the card, write a new problem. Repeat with as many cards as desired. Write the answer for the final fact on the back of the first card. Mix up the cards. As a student visits the center, she determines the answer to a problem and then finds the card with that answer. She flips the card over to find the next problem. Her goal is to sequence the cards so that she uses them all. Let's have another round of practice!

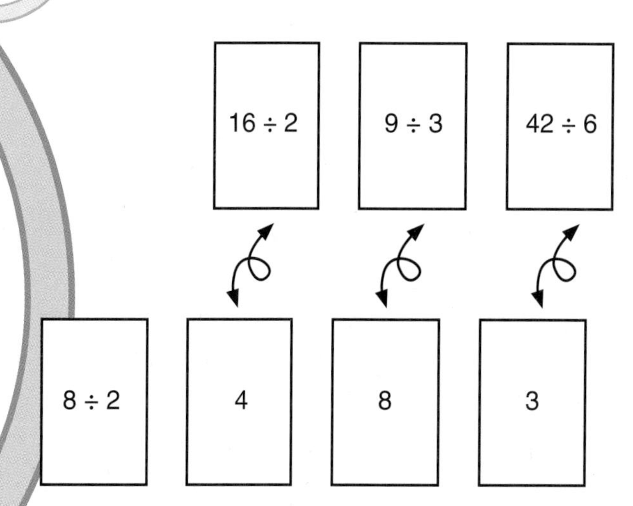


## Grab Bag

Grab your students' attention with this colorful practice activity. Cut out ten circles from each of eight different colors of card stock. Choose one color per set of division facts. Write a problem on the front of each circle and the answer on the back. Then place the colorful circles in a clear gift bag. Use them for a variety of activities:

- To review a single set of facts, a student pulls out one color of circle and answers the problems.
- For mixed practice, a student randomly pulls out any color of circle and answers the problems.
- Challenge a student to close his eyes, pull out a circle, and answer the problem. If his answer is correct, he keeps the circle. The goal is to get one circle of each color.
- Challenge a student to lay several circles on her desk with the answers showing and then determine possible problems that match each answer.
- Challenge a student to continue pulling out circles until he has answered ten problems in a row correctly.

2 ÷ 2 4 ÷ 2 6 ÷ 2 8 ÷ 2

10 ÷ 2 12 ÷ 2 14 ÷ 2

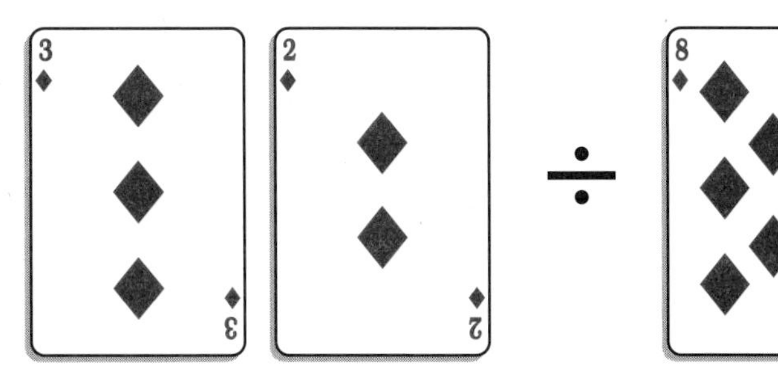


# Diamond Hunt

Help students unearth a treasure trove of division facts with this small-group practice game! Give each student one of each playing card from ace to 9, any suit. Give yourself two similar sets of cards, all diamonds. Explain to your students that you will use your cards to post a division sentence, such as $32 \div 8$, by using the three, two, and eight cards. Explain that each student should use her set of cards to form the answer to the division sentence, reminding students that the ace stands for one. After allowing time for students to find the answer, have students reveal their cards. For each correct answer, declare, "You've found diamonds!" and give the student a point. Keep track of the points. After ten rounds, declare students division experts according to the following scale:

- students with 9 or 10 points = Rock-Solid Diamond Division Masters
- students with 7 or 8 points = Diamond Division Apprentices
- students with 5 or 6 points = Diamond Division Assistant Apprentices
- students with 0 to 4 points = Diamond Division Newbies

60 ÷ 10 =

30 ÷ 10 =

40 ÷ 10 =

# Rip-Roarin' Through Tens

Tear into dividing by tens with this hands-on activity! Give each student nine 1" x 4" paper strips and direct him to write a different tens division fact on each one (but not the answers). Choose a fact to begin with, such as $40 \div 10$. Direct each student to hold up that strip. As a group say, "Forty divided by ten equals," and then gently tear the paper strip between the four and the zero. Then state the answer, "Four!" and hold up that number. Repeat with the other facts. Let 'er rip!

Find more student practice on pages 91–98.

Name_______________________ Date_______________

# Rocketing Raccoons

Color if correct.

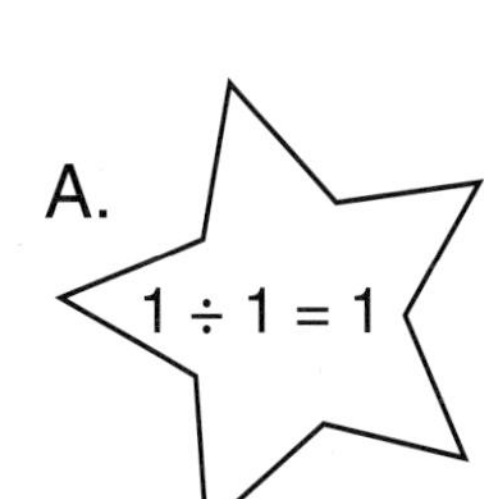

A. $1 \div 1 = 1$

B. $12 \div 2 = 10$

C. $0 \div 7 = 0$

D. $18 \div 2 = 9$

E. $8 \div 1 = 8$

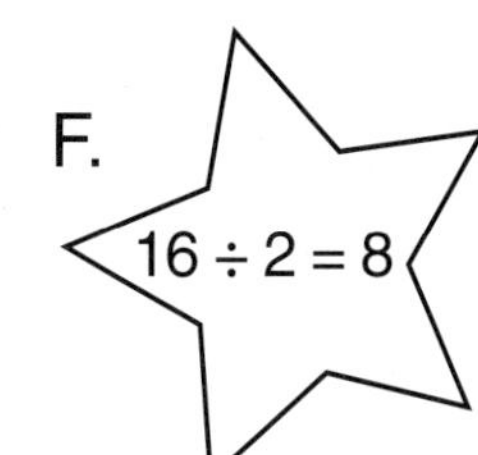

F. $16 \div 2 = 8$

G. $0 \div 3 = 3$

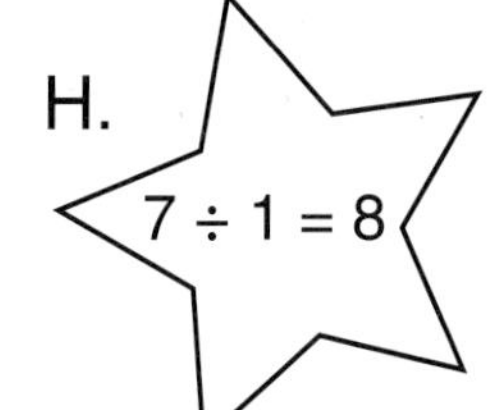

H. $7 \div 1 = 8$

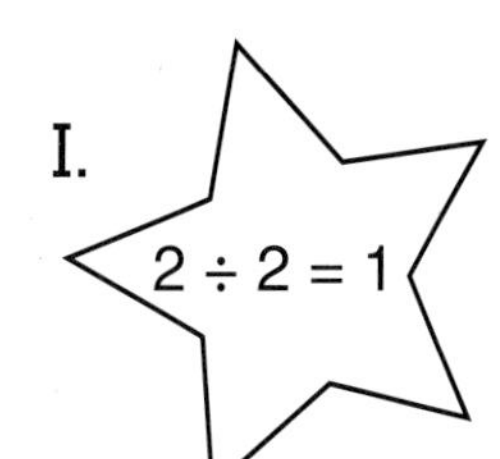

I. $2 \div 2 = 1$

J. $8 \div 2 = 4$

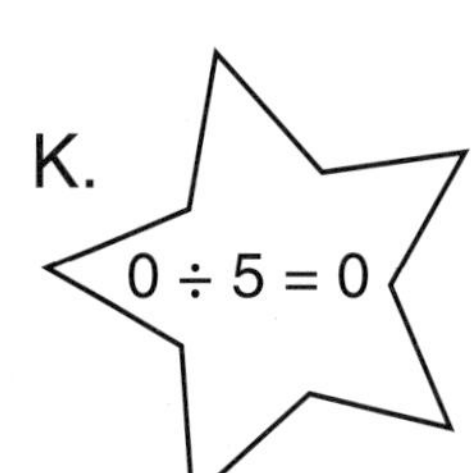

K. $0 \div 5 = 0$

L. $6 \div 1 = 6$

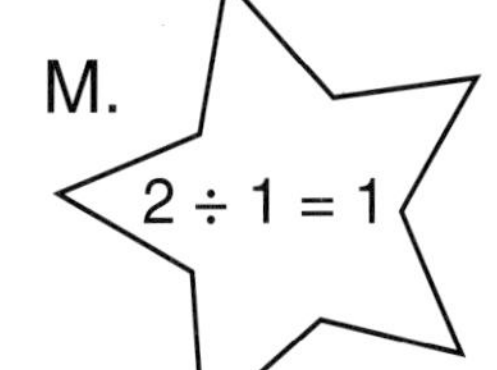

M. $2 \div 1 = 1$

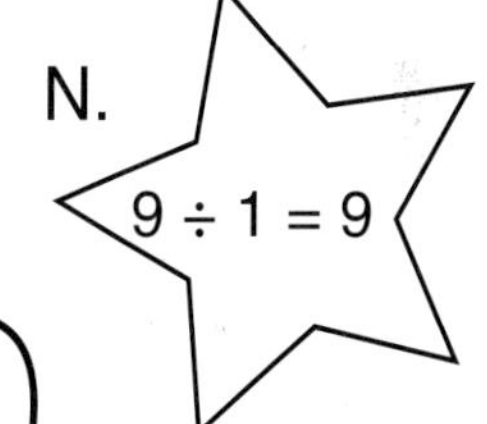

N. $9 \div 1 = 9$

O. $5 \div 1 = 4$

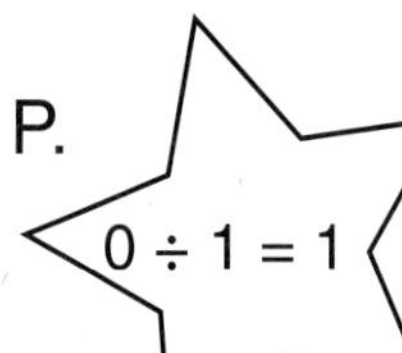

P. $0 \div 1 = 1$

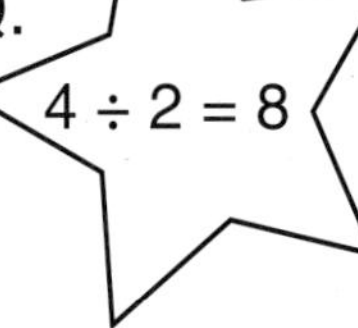

Q. $4 \div 2 = 8$

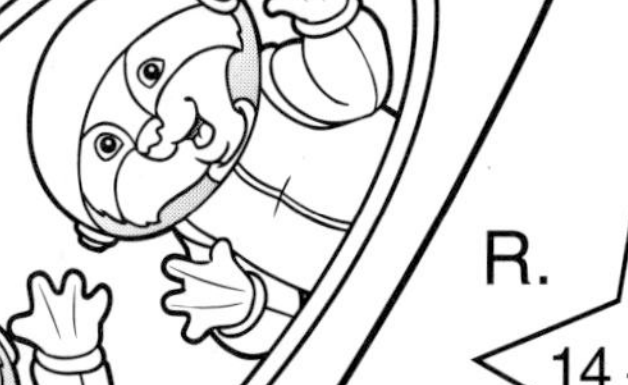

R. $14 \div 2 = 8$

S. $0 \div 4 = 4$

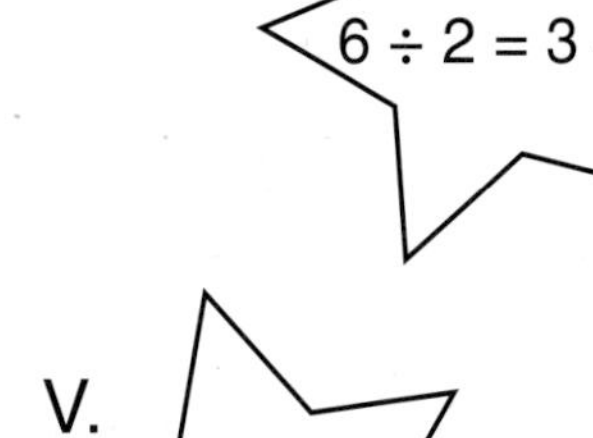

T. $6 \div 2 = 3$

U. $9 \div 1 = 10$

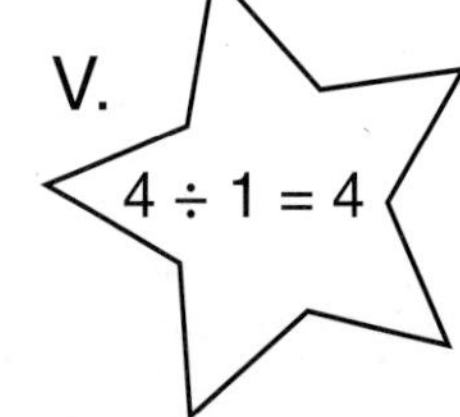

V. $4 \div 1 = 4$

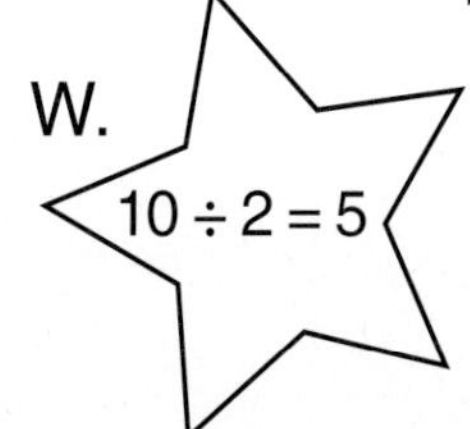

W. $10 \div 2 = 5$

Name____________________________ Date__________________

# Apple Harvest

Write the missing number in each number sentence.
For each number, color an apple in the matching basket.

a. $3 \div \square = 3$

b. $\square \div 6 = 0$

c. $9 \div \square = 9$

d. $0 \div 7 = \square$

e. $12 \div \square = 6$

f. $\square \div 9 = 0$

g. $2 \div \square = 2$

h. $6 \div \square = 3$

i. $0 \div 1 = \square$

j. $\square \div 2 = 1$

k. $5 \div \square = 5$

l. $0 \div 5 = \square$

m. $10 \div \square = 5$

n. $0 \div 2 = \square$

o. $16 \div \square = 8$

p. $4 \div 2 = \square$

q. $6 \div \square = 6$

r. $8 \div \square = 4$

s. $\square \div 3 = 0$

t. $18 \div \square = 9$

u. $7 \div \square = 7$

v. $\square \div 1 = 1$

w. $8 \div \square = 8$

x. $14 \div \square = 7$

y. $0 \div 8 = \square$

z. $\square \div 4 = 0$

0

1

2

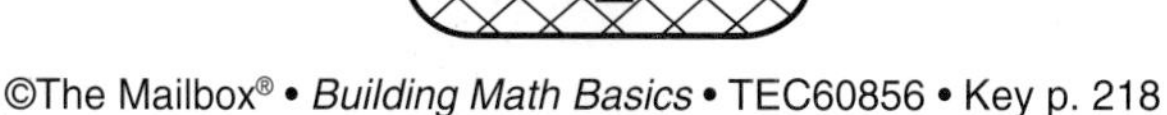

Name ____________________ Date ____________

# Turtles on the Go!

Write a division problem in each row.
Use the wheel chart to help you.
Divide to find the number of racers in each race.
The first one has been done for you.

unicycle = 1 wheel
scooter = 2 wheels
tricycle = 3 wheels
wagon = 4 wheels

| Race | Total Number of Wheels | Problem | Number of Racers |
|---|---|---|---|
| A. unicycles | 4 | 4 ÷ 1 = | 4 |
| B. tricycles | 18 | | |
| C. wagons | 16 | | |
| D. scooters | 18 | | |
| E. wagons | 32 | | |
| F. tricycles | 15 | | |
| G. unicycles | 7 | | |
| H. scooters | 12 | | |
| I. tricycles | 27 | | |
| J. wagons | 20 | | |
| K. tricycles | 12 | | |
| L. wagons | 28 | | |
| M. tricycles | 21 | | |

Name________________________________ Date____________________

# Factory on the Farm

Circle an equal number of eggs in each group.
Write and solve a number sentence for each egg crate.

A. Circle 2 equal groups.

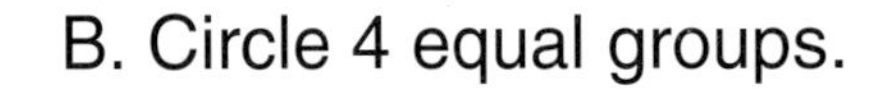

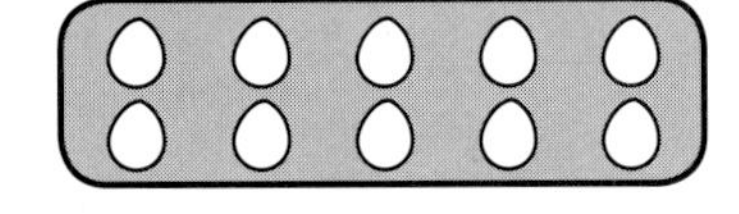

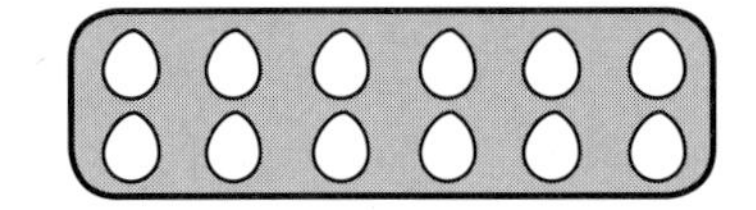

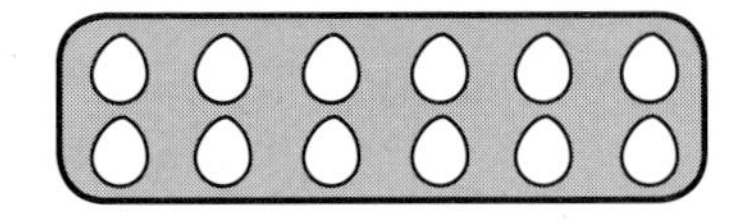

____ ÷ ____ = ____

D. Circle 3 equal groups.

E. Circle 4 equal groups.

F. Circle 3 equal groups.

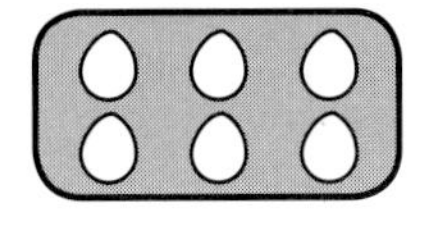

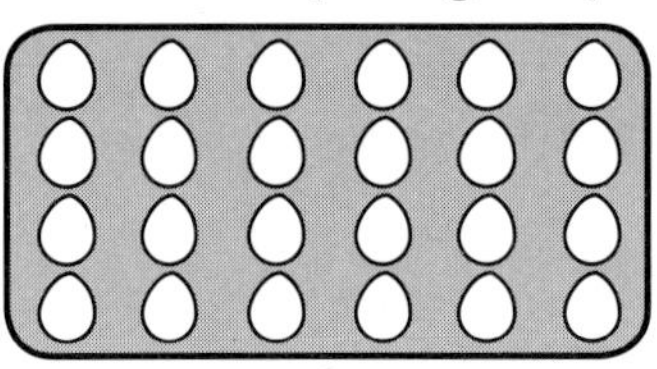

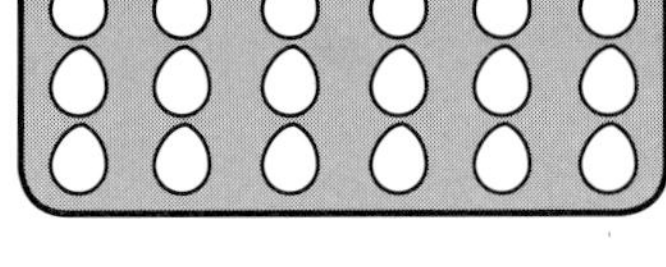

____ ÷ ____ = ____

____ ÷ ____ = ____

____ ÷ ____ = ____

G. Circle 4 equal groups.

H. Circle 2 equal groups.

I. Circle 4 equal groups.

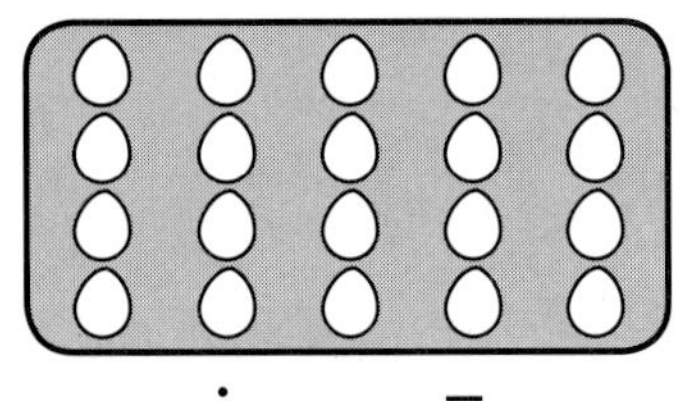

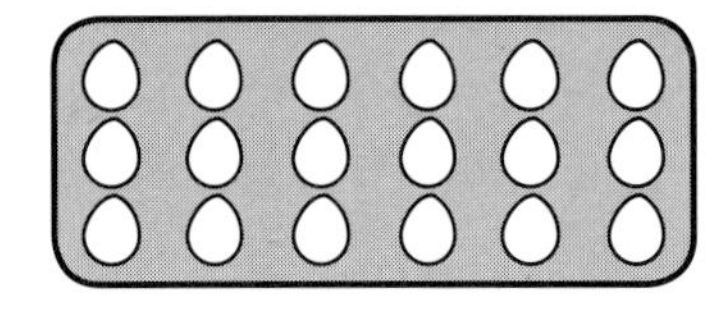

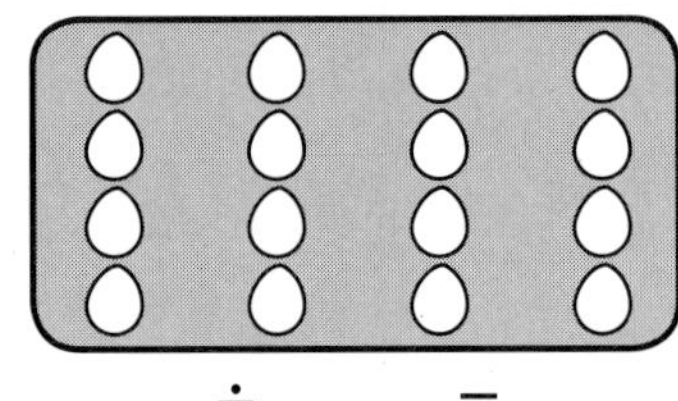

____ ÷ ____ = ____

____ ÷ ____ = ____

____ ÷ ____ = ____

Name____________________ Date__________

# Snack Attack!

Circle the number that can be divided equally into the number on the ant.
Write the problem on the lines.
Solve the problem.

A. 

4 3 6

___ ÷ ___ = ___

B. 

3 7 4

___ ÷ ___ = ___

C. 18

4 5 6

___ ÷ ___ = ___

D. 48

6 7 9

___ ÷ ___ = ___

E. 10

3 4 5

___ ÷ ___ = ___

F. 42

4 5 6

___ ÷ ___ = ___

G. 21

3 4 5

___ ÷ ___ = ___

H. 20

3 4 6

___ ÷ ___ = ___

I. 6

4 5 6

___ ÷ ___ = ___

J. 32

4 5 6

___ ÷ ___ = ___

K. 28

3 4 5

___ ÷ ___ = ___

L. 6

1 7 8

___ ÷ ___ = ___

M. 25

3 4 5

___ ÷ ___ = ___

N. 45

4 5 6

___ ÷ ___ = ___

O. 14

2 3 4

___ ÷ ___ = ___

Name______________________________ Date______________________

# Undersea Circus

Write the missing number in each shell.
Use the code.
Divide.

A. 12 ÷ ◯ = ______

B. 35 ÷ ◯ = ______

C. 16 ÷ ◯ = ______

D. 10 ÷ ◯ = ______

E. 54 ÷ ◯ = ______

F. 24 ÷ ◯ = ______

G. 24 ÷ ◯ = ______

H. 30 ÷ ◯ = ______

I. 36 ÷ ◯ = ______

J. 6 ÷ ◯ = ______

K. 20 ÷ ◯ = ______

L. 12 ÷ ◯ = ______

M. 40 ÷ ◯ = ______

N. 15 ÷ ◯ = ______

O. 4 ÷ ◯ = ______

P. 36 ÷ ◯ = ______

Q. 42 ÷ ◯ = ______

R. 20 ÷ ◯ = ______

S. 30 ÷ ◯ = ______

T. 45 ÷ ◯ = ______

U. 25 ÷ ◯ = ______

V. 28 ÷ ◯ = ______

W. 8 ÷ ◯ = ______

X. 18 ÷ ◯ = ______

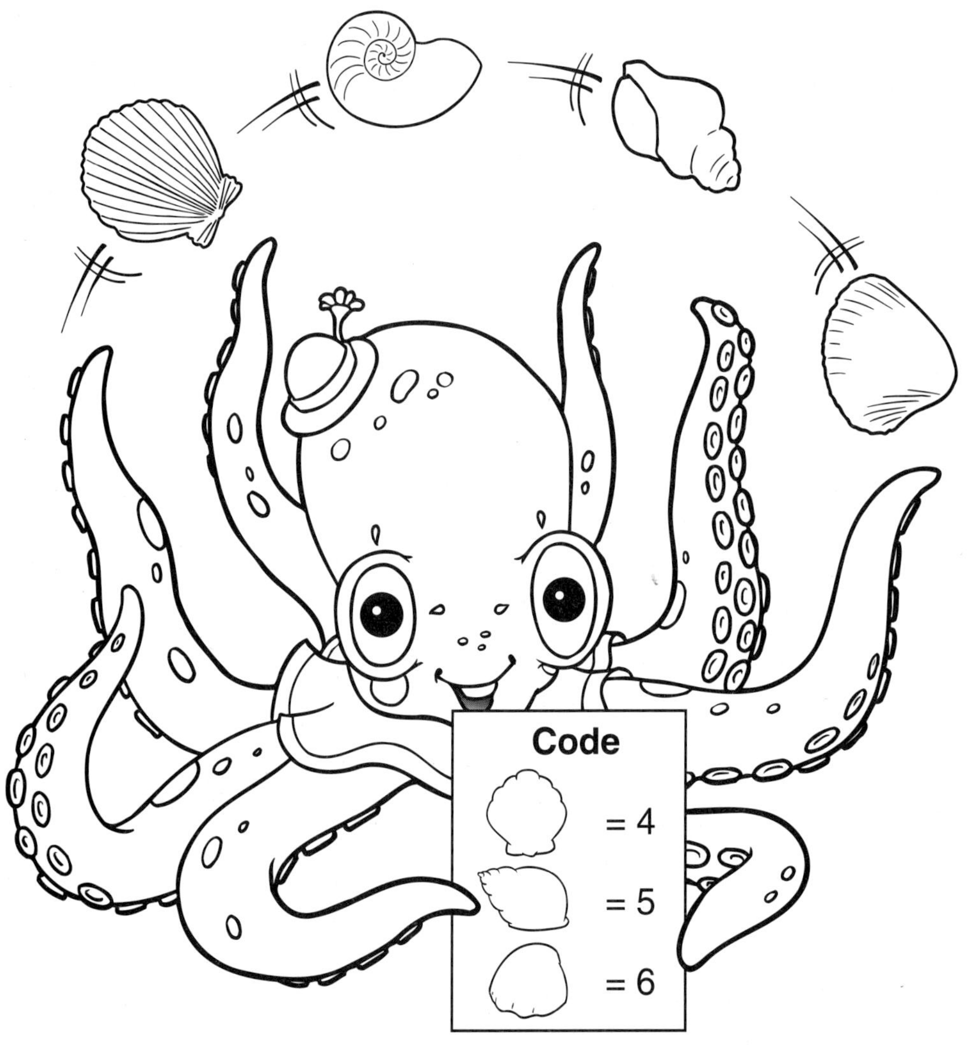

Name________________________________ Date____________________

# Round 'em Up!

Complete the number sentences.
Cross out each number in the lasso as it is used.

| | | | |
|---|---|---|---|
| A. $\square \div 8 = 5$ | B. $18 \div \square = 9$ | C. $\square \div 7 = 4$ | D. $\square \div 8 = 8$ |
| E. $9 \div \square = 9$ | F. $\square \div 9 = 4$ | G. $\square \div 7 = 3$ | H. $0 \div 8 = \square$ |
| I. $\square \div 8 = 3$ | J. $\square \div 7 = 7$ | K. $48 \div 8 = \square$ | L. $\square \div 9 = 5$ |
| M. $\square \div 7 = 2$ | N. $72 \div 9 = \square$ | O. $\square \div 8 = 4$ | P. $56 \div \square = 8$ |
| Q. $9 \div 3 = \square$ | R. $\square \div 9 = 9$ | S. $20 \div \square = 5$ | T. $\square \div 7 = 5$ |
| U. $\square \div 9 = 3$ | V. $54 \div \square = 6$ | W. $\square \div 4 = 4$ | X. $30 \div 6 = \square$ |

Name________________________ Date______________

# Pretty As a Picture

Divide.
Color according to the code.

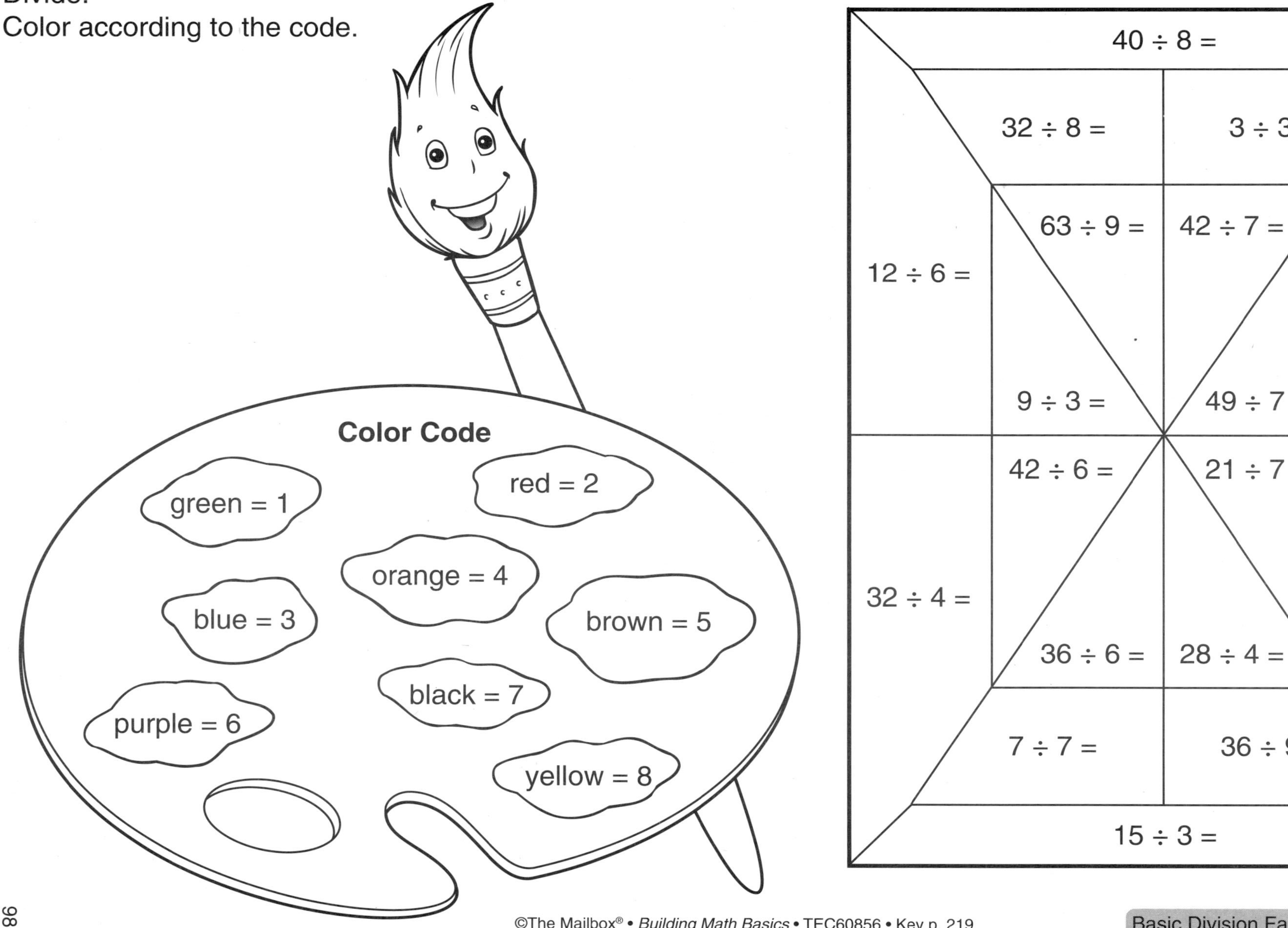

6 x 5 = 30

30 ÷ 5 = 6

# Multiplication and Division Fact Families

5 x 6 = 30

30 ÷ 6 = 5

## Quite a Quilt!

Sew up students' knowledge of fact families with this colorful activity! Give each child four three-inch paper squares in various colors. Also give the child a slip of paper with one multiplication or division math sentence (but not the answer) written on it. Challenge her to determine the answer, write the entire math sentence on a paper square, and then write three more multiplication or division math sentences using those three numbers (one on each of the remaining squares). Finally, have the child glue her squares to an eight-inch square of wrapping paper or construction paper. Post the completed squares side by side on a bulletin board to create a colorful class quilt of fact families.

x

=

6 x 9 = 54

54 ÷ 6 = 9

54 ÷ 9 = 6

9 x 6 = 54

## "Tree-mendous" Leaves

When it comes to fact families, are your students ready to branch out? Try this! Write a list of the facts from several fact families so that you have one fact for each child. Cut the list apart and give one fact to each child. Have each student cut one paper leaf from green construction paper and then write his fact on it. Then challenge the student to find the three other children who are in his fact family. Each group cuts a tree trunk from brown construction paper and a leafy top from green construction paper; then the group members glue their fact leaves to the leafy top. Finally, they program their tree trunk with the numbers in the fact family and glue their tree together. You won't "be-leaf" how much fun students will have with their fact family trees!

# Stacks of Practice

These fun fact books help students see how fact families fit together! To create a book, a student stacks two sheets of white paper. She holds the sheets vertically and slides the top sheet upward about one inch. Then she folds the papers forward to make four graduated layers and staples at the fold. Next, she selects two numbers that are each less than ten and determines their multiplication and division facts. She writes one fact on the bottom of each booklet tab. Then, inside the booklet, she uses one or more rubber stamps in varying sizes to illustrate each fact. Multiplication and division stack up perfectly!

$6 \times 4 = 24$

$4 \times 6 = 24$

$24 \div 6 = 4$

$24 \div 4 = 6$

Find more student practice on pages 101–102.

## Let's Go for a Stroll!

| My Number | Your Number | Our Facts |
|---|---|---|
| 7 | 9 | $7 \times 9 = 63$<br>$9 \times 7 = 63$<br>$63 \div 7 = 9$<br>$63 \div 9 = 7$ |
| 2 | 3 | $2 \times 3 = 6$<br>$3 \times 2 = 6$<br>$6 \div 2 = 3$<br>$6 \div 3 = 2$ |

Here's an activity sure to promote a little fact family fun! Create a recording sheet with three columns: one labeled "My Number," one labeled "Your Number," and the last labeled "Our Facts." Give a copy to each child. Also give each child an index card with a number from 1 to 9 written on it. Challenge students to find a partner and write down one another's numbers on the recording sheet. Next, they figure out the fact family for the two numbers and write the facts in the final column. Then students stroll on to find a new partner and a new set of facts!

Name______________________ Date______________

# Guarding the Towers

Use the numbers on each castle to write the fact family.

**A** 6 42 7

____ x ____ = ____
____ x ____ = ____
____ ÷ ____ = ____
____ ÷ ____ = ____

**B** 56 8 7

____ x ____ = ____
____ x ____ = ____
____ ÷ ____ = ____
____ ÷ ____ = ____

**C** 6 48 8

____ x ____ = ____
____ x ____ = ____
____ ÷ ____ = ____
____ ÷ ____ = ____

**D** 8 9 72

____ x ____ = ____
____ x ____ = ____
____ ÷ ____ = ____
____ ÷ ____ = ____

**E** 6 54 9

____ x ____ = ____
____ x ____ = ____
____ ÷ ____ = ____
____ ÷ ____ = ____

**F** 63 9 7

____ x ____ = ____
____ x ____ = ____
____ ÷ ____ = ____
____ ÷ ____ = ____

Name ______________________ Date ______________

# "Chews-ing" to Chew

Use the numbers on each set of gumballs to write a fact family on each machine.

A.

_____ x _____ = _____

_____ x _____ = _____

_____ ÷ _____ = _____

_____ ÷ _____ = _____

2 4 8

B.

_____ x _____ = _____

_____ x _____ = _____

_____ ÷ _____ = _____

_____ ÷ _____ = _____

3 1 3

C.

_____ x _____ = _____

_____ x _____ = _____

_____ ÷ _____ = _____

_____ ÷ _____ = _____

12 4 3

D.

_____ x _____ = _____

_____ x _____ = _____

_____ ÷ _____ = _____

_____ ÷ _____ = _____

4 20 5

# Multiplying Multidigit Numbers by One-Digit Numbers

| Start | | | | | | | | | | | | | Finish |
|---|---|---|---|---|---|---|---|---|---|---|---|---|---|
| Start | | | | | | | | | | | | | Finish |

## Fast-Paced Factor Scramble

Get students revved up for some speedy multiplication practice with this partner game! Ahead of time, draw a simple trail game with two rows of 14 spaces. Label the first space "Start" and the last "Finish." Provide pairs of students with three dice, scratch paper, game markers, and a timer. The first player rolls the dice and records the numbers on his paper. The second player repeats this process. Then one player sets the timer for two minutes. Each student races to arrange his numbers into different multiplication problems with a two-digit number multiplied by a one-digit number. He then works each problem. After two minutes, players check one another's answers to determine which player has the problem with the biggest product. That player moves ahead three spaces; the other player moves ahead two. In case of a tie, each student moves ahead three spaces. Students will be on the fast track to speedy multiplication mastery!

## Bull's-Eye!

Here's a class game designed to get students' multiplication skills right on target! Draw two bull's-eye targets on the chalkboard, each with a circle in the center and two rings around it. For the first target, divide the outside ring into seven sections and write a two-digit number in each. Then divide the inside ring into seven sections and write a one-digit number in each. Write a multiplication sign in the center circle. Repeat for the second target, using the same numbers in a different order. To play, each student, in turn, chooses a one-digit and a two-digit number, writes them on the board to form a multiplication problem, and then crosses them off on the target. Then the student solves the problem. After checking the answer, the student writes the product beside the target to create the team's running tally. Continue until each student has had a chance to play, plugging new numbers into each target as needed. The team with the bigger tally wins!

36, 19, 54, 27, 45, 12, 61
2, 8, 4, 1, 3, 5, 7
×

$$\begin{array}{r} {}^{7} \\ 19 \\ \times\ 8 \\ \hline 152 \end{array}$$

# Sunny-Side Up!

Students' multiplication skills will shine with this sunny activity!In advance, create several oaktag patterns in the shape of a pair of sunglasses measuring approximately 2" x 6". Gather the following for each student: scrap paper, six 2" x $8^1/_2$" yellow construction paper strips, one six-inch yellow construction paper circle, one 2" x 6" piece of red construction paper, glue, scissors, a black marker, and a piece of white chalk. Randomly program each paper circle by writing three one-digit numbers along the outside edge.

Distribute the materials. Explain that each student will use the three numbers to create and solve six different problems that have a two-digit number times a one-digit number. The student writes and solves these on the scratch paper. Once he has checked the answers, he copies each problem onto one end of a paper strip. Then he glues the strips to the paper circle with the problems fanned out to make a sun shape. Finally, he traces a sunglasses pattern on the red paper and writes his name on it in chalk. He cuts out the sunglasses and glues them to the yellow circle. Display the sun shapes on a bulletin board with the title "Our Multiplication Skills Really Shine!"

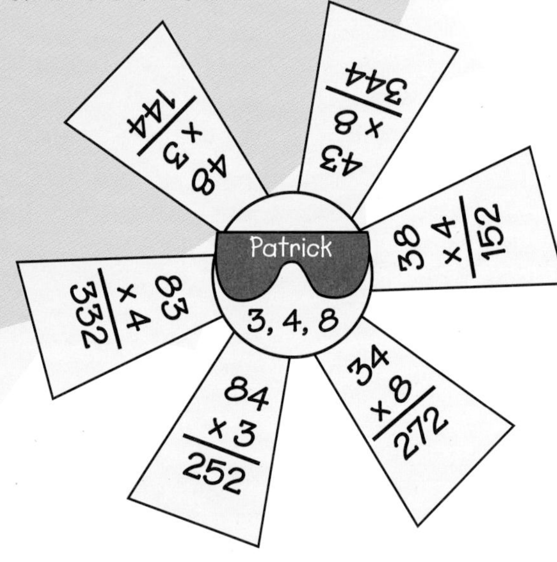


42
x 8
402
336
326

# Red Light, Green Light

Signal students to practice their multiplication skills with this traffic light–themed learning center. Draw a basic traffic light shape on a half sheet of white paper. Add a multiplication sign to the left of the middle light and a line above the bottom light to mimic a multiplication problem. Photocopy ten lights. Fill in a two-digit or three-digit number in each top light and a one-digit number in each center light. If desired, color or outline the lights in red, yellow, and green. Then cut out ten green construction paper circles to match the bottom light. Write the answer to each problem on a circle. As students visit the center, challenge them to work each problem on scrap paper and then check their work by finding the matching answer on a green light. Students will stop, learn, and go!

# The Cupboard Is Bare!

The goal of this whole-class game is for each team to raid the opposing team's cabinet by solving multiplication problems! In advance, program the backs of 30 paper plates with the numbers 0 to 9 for each team, using a different color for each of two teams. Then have each student illustrate the front of one or two plates to show his favorite foods.

To play the game, each team, in turn, selects four plates from its cupboard. Team members display the plates in the form of a life-size multiplication problem by having three students stand side by side holding the plates at waist level to create a three-digit number and having one student kneel below holding another plate to create a one-digit number. Set a timer for 30 seconds. The opposing team writes down the problem and solves it. Choose one team member to give the answer. If the student solves the problem correctly before the buzzer goes off, the team raids the other team's cabinet by taking the four plates. If the answer is incorrect, the first team keeps the plates. Play continues until one team is out of plates or as time allows. Anyone hungry?

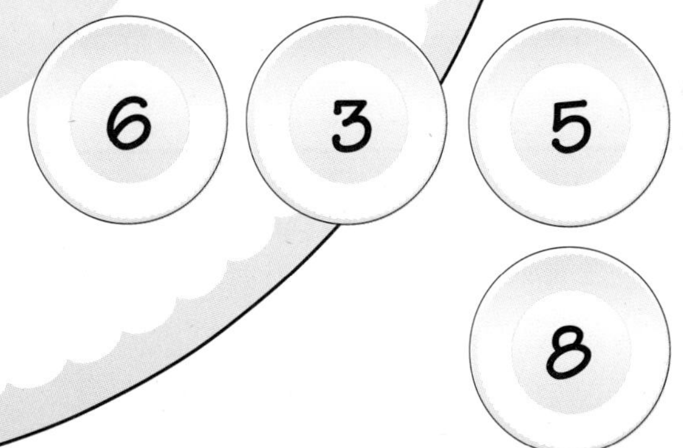


# The Great Multiplication Race

Students race to the finish line and practice lots of multiplication along the way with this small-group game! In advance, draw a trail game with 31 spaces. Write each number from 100 to 130 in order, one per space, and add a finish line. Provide each team of two players with a game piece, one die, a calculator, and scratch paper. Each team puts its game piece on the first space. Explain that one team member writes down the number on the space her piece is on. Then she rolls the die and multiplies the number on the space by the number on the die, using scratch paper to determine the answer. A teammate uses a calculator to check the answer. If it's correct, the team moves its game piece ahead the number of spaces on the die. If the answer is incorrect, the team does not move forward. Then the other team takes a turn. The first team to multiply its way across the finish line wins!

Find more student practice on pages 106–108.

Name ______________________ Date ______________

# A Wild and "Hare-y" Ride

Write the missing number in each box.

Name ____________________ Date ____________

# Groundskeeping at the Golf Course

Solve each problem.

A. $124 \times 6$

B. $306 \times 3$

C. $427 \times 2$

D. $254 \times 4$

E. $185 \times 7$

F. $339 \times 5$

G. $272 \times 8$

H. $169 \times 4$

I. $437 \times 3$

J. $312 \times 2$

K. $132 \times 4$

L. $520 \times 4$

M. $618 \times 5$

N. $119 \times 8$

O. $277 \times 3$

**Why did the golf course hire Gopher to mow the lawn?**

To solve the riddle, match the letters above to the numbered lines below.

They'd rather see him mow a lawn than ___ ___ ___ ___ ___
(3,090 744 528 1,295 744)

___ ___ ___ ___ ___ ___ ___ ___ ___!
(676 831 2,080 1,295 1,311 952 831 952 1,295)

Name ______________________________ Date ______________

# Ticket Time

Solve each problem.
Color the ticket with the matching answer.

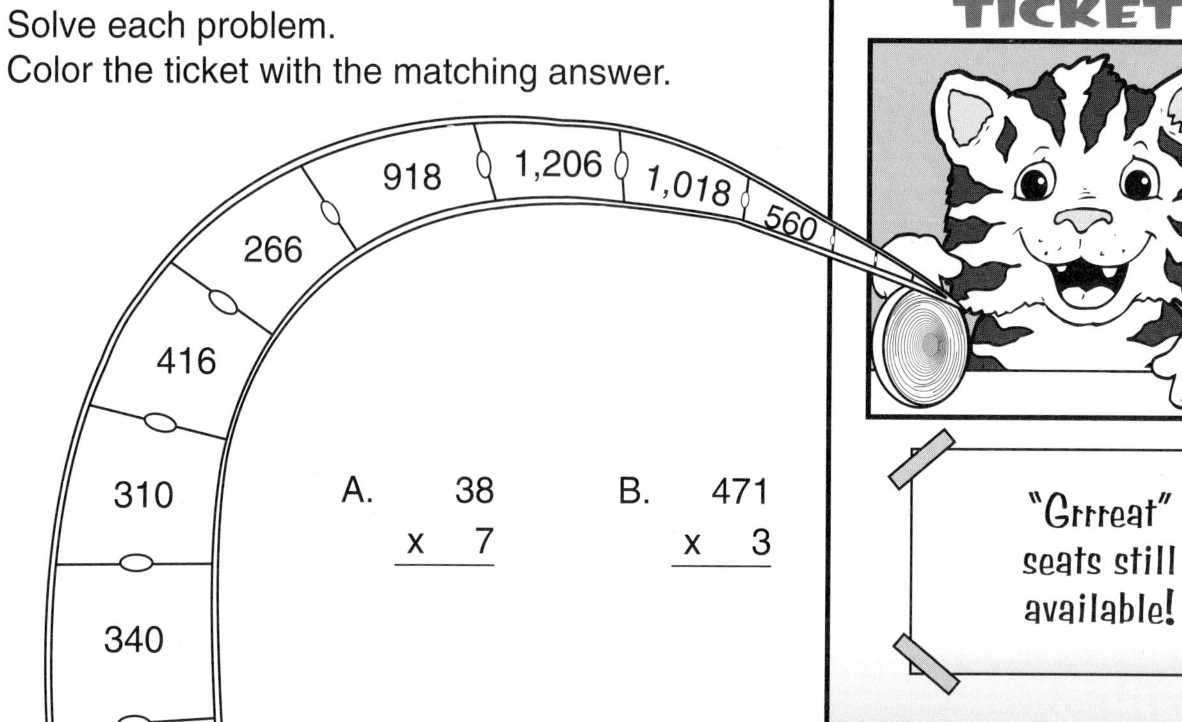


| | | | |
|---|---|---|---|
| A. $38 \times 7$ | B. $471 \times 3$ | | |
| C. $509 \times 2$ | D. $62 \times 5$ | E. $58 \times 6$ | F. $274 \times 4$ |
| G. $70 \times 8$ | H. $341 \times 4$ | I. $603 \times 2$ | J. $45 \times 9$ |
| K. $712 \times 3$ | L. $52 \times 8$ | M. $153 \times 6$ | N. $68 \times 5$ |

# Story Problems

## Picturing Problems

Increase the appeal of word problems with this picture-perfect idea! Laminate a photograph of each child; then attach a strip of magnetic tape to the back of each snapshot. Place each photo in a corner of a magnetic dry-erase board or chalkboard. To begin, write on the board a word problem such as the one shown and ask a student volunteer to arrange the photos to illustrate the problem. Have each student solve the problem on a sheet of paper. Then ask another student volunteer to write the correct number sentence on the board. Erase the board and remove the photos. Then repeat the activity with different problems and photos as time allows.

9 children are at the park. Cindy, Lee, Pam, and Frank join them. How many students are at the park in all?

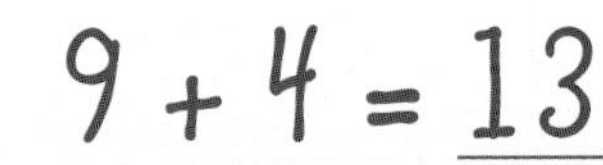

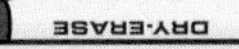

## Protected Problems

A simple sheet protector clearly makes this center idea one you'll use over and over! Collect several story problem skill sheets, such as those on pages 111–114, and make two copies of each sheet. Place one copy of each sheet into its own plastic sheet protector. Use the remaining copy of each sheet to create an answer key. Store the keys in a folder and place them at a center with the protected sheets, a wipe-off marker, and a damp cloth. Students calculate the answers to the problems and then wipe away their work after checking their answers. To keep student interest high, replace the skill sheets at regular intervals. Students will get the story problem practice they need, and you'll save paper!

## Transparent Stories

Students' enthusiasm for story problems will be transparent with this small-group activity. Give each group a set of 8 number cards, a transparency sheet, and a wipe-off marker. Instruct each group to stack its cards facedown; then have two group members each draw a card. Next, have each group use the two numbers to create an addition or a subtraction story problem. Direct one child from each group to write the word problem on the transparency. Have students place the cards in a discard pile and repeat the activity with two new cards. Instruct the groups to continue in this manner until they have used all of their cards. Then ask each group, in turn, to place its sheet on the overhead projector and challenge the class to solve the problems.

12 children are on the bus. 9 get off at the first stop. How many children are left?

9

12

15 pigs live at the Oink Oink Farm. 7 pigs got hungry and munched on some corn. How many pigs did not eat corn?

## Problem Presentations

These daily presentations are just right for keeping students' word problem skills sharp! Every morning, write an addition or a subtraction fact on a card and select a group of students to be presenters. (The number of students is determined by the largest number in the math fact.) Work with the group for a few minutes to help it develop a creative manner in which to present its number sentence to the class. Next, invite the group to present its number sentence and have the remaining students solve the problem. Then ask a group member from the presenting group to share the answer. Plan to repeat the activity with a different fact (and other students) the following day. Lights, camera, action!

**Find more student practice on pages 111–114.**

Name ____________________ Date ____________

# Start Your Engines

Read.
Choose the best operation to solve each problem.
Write the sign on the car's flag.
Solve each problem on another sheet of paper.

1. There are 42 cars in the race. 15 cars crash. How many cars are left?

2. Chuck won 59 races last year. He has won 24 races this year. How many races has he won in all?

3. The first half of the race is 65 miles. The rest of the race is 53 miles. How many miles is the race in all?

4. There are 38 cars on the track. 16 cars stop in the pit. How many cars are left on the track?

5. 21 cars stop for gas. 12 of those cars get new tires. How many cars do not get tires?

6. Chuck uses 25 gallons of gas in the first part of the race. He uses 14 gallons in the rest of the race. How many gallons does he use in all?

Name ______________________ Date __________

# Mr. Crow's Produce Stand

Read.
Choose the best operation to solve each problem.
Write the sign in the box.
Solve each problem on another sheet of paper.

1. Mr. Crow sells 216 carrots. He sells 129 peppers. How many more carrots does he sell? ☐

2. On Thursday, Mr. Crow sells 164 pounds of fruit. On Friday, he sells 98 pounds. How many pounds of fruit does he sell in all? ☐

3. Mr. Crow picks 309 ears of white corn. He picks 521 ears of yellow corn. How many more ears of yellow corn does he pick? ☐

4. Last week Mr. Crow had 240 shoppers. He had 258 shoppers this week. How many shoppers did he have in all? ☐

5. Mr. Crow sells 345 pansies. He sells 431 tulips. How many plants does he sell in all? ☐

6. Mr. Crow sells 150 melons. He sells 87 peaches. How many more melons does he sell? ☐

7. Mr. Crow has 437 squashes. He sells 214 of them. How many squashes are left? ☐

8. Mr. Crow sells 133 red onions. He sells 142 yellow onions. How many onions does he sell in all? ☐

Name ____________________ Date ____________________

# Dora's Doughnuts

Read.
Multiply.
Show your work on another sheet of paper.

1. Dora makes 8 doughnuts per batch. She makes 5 batches. How many doughnuts does she make in all?

2. Dora needs 3 eggs for each batch of doughnuts. She makes 9 batches. How many eggs does she need in all?

3. Six customers come to the store. Each one buys 2 doughnuts. How many doughnuts are sold in all?

4. Dora sells 9 kinds of doughnuts. She sells 6 of each kind. How many doughnuts does she sell in all?

5. Dora works 5 days this week. She works 7 hours each day. How many hours does she work in all?

6. Dora has 4 workers. Each worker works 6 hours. How many hours do they work in all?

7. Dora puts 8 doughnuts on a tray. She has 8 trays. How many doughnuts can she put out?

8. Five shoppers buy doughnut holes. Each shopper buys 9 doughnut holes. How many doughnut holes are sold in all?

Name ______________________ Date ______________

# Laying Bricks

Read.
Divide.
Show your work.

1. Harry Hippo lays 7 bricks in 63 seconds. How long does it take him to lay one brick?

    _____ seconds

2. Four boxes of bricks need to be placed. There are 32 bricks in all. How many bricks are in each box?

    _____ bricks

3. Each brick has 3 holes. There are 18 holes in all. How many bricks are there?

    _____ bricks

4. Harry lays 9 rows of bricks. He uses 54 bricks. How many bricks are in each row?

    _____ bricks

5. Harry uses 5 bags of mortar for each job. He has used 35 bags. How many jobs has he completed?

    _____ jobs

6. Harry keeps 3 tools in each box. He has 21 tools. How many toolboxes does he have?

    _____ toolboxes

7. Harry works 6 hours each day. One project takes him 36 hours. How many days does it take him?

    _____ days

8. Harry eats 2 snacks every hour he works. One day he ate 12 snacks. How many hours did he work?

    _____ hours

# Fractions

## Container Fractions

Grab some divided containers to help youngsters master fractions. In advance, gather containers divided into equal sections, such as clear or foam deli boxes (halves), drink holders (fourths), and muffin pans (sixths). Also have on hand a supply of large pom-poms. Have students count the number of sections in a chosen container. Explain that this number represents the denominator, or the bottom number in a fraction. Then add a number of pom-poms, each one in a different section of the container. Have students count the pom-poms. Explain that this number represents the numerator, or the top number of the fraction. Finally, place the items at a center and invite students to work in pairs to make and identify fractions.

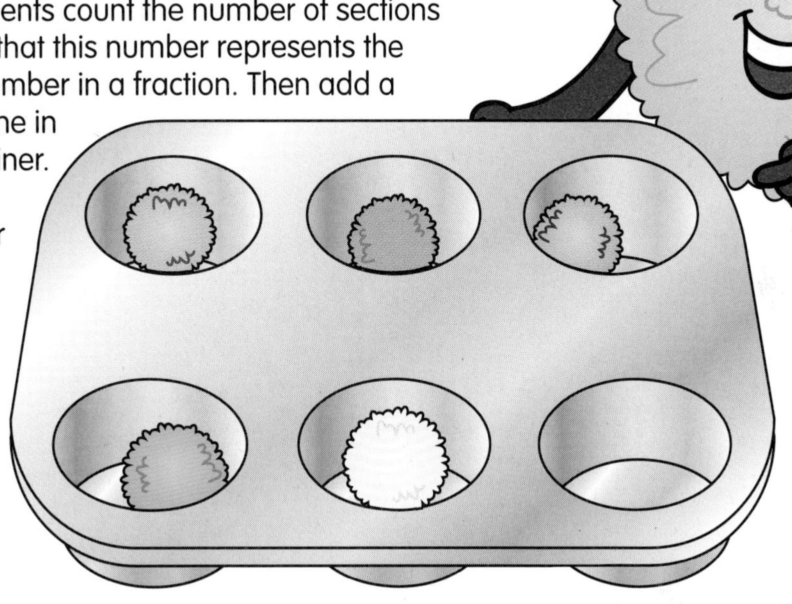


## Fraction Stamp

Count on students to give this center activity their stamp of approval. To prepare, make a set of fraction cards. To make a card, write "group of" and a whole number on one side of an index card. On the back, write a fraction that describes a part of the number. Create a desired number of cards. Place the cards, assorted stampers or bingo daubers, paper, and pencils at a center. Each child folds her paper in half twice, unfolds it, and traces the fold lines. Next, she chooses a card, reads the whole number, and stamps that number of times in one section of her paper. Then she turns her card over, reads the fraction, copies it onto her paper, and circles the appropriate fraction of stamp prints. She chooses different cards and completes the process three additional times. Stamped and approved!

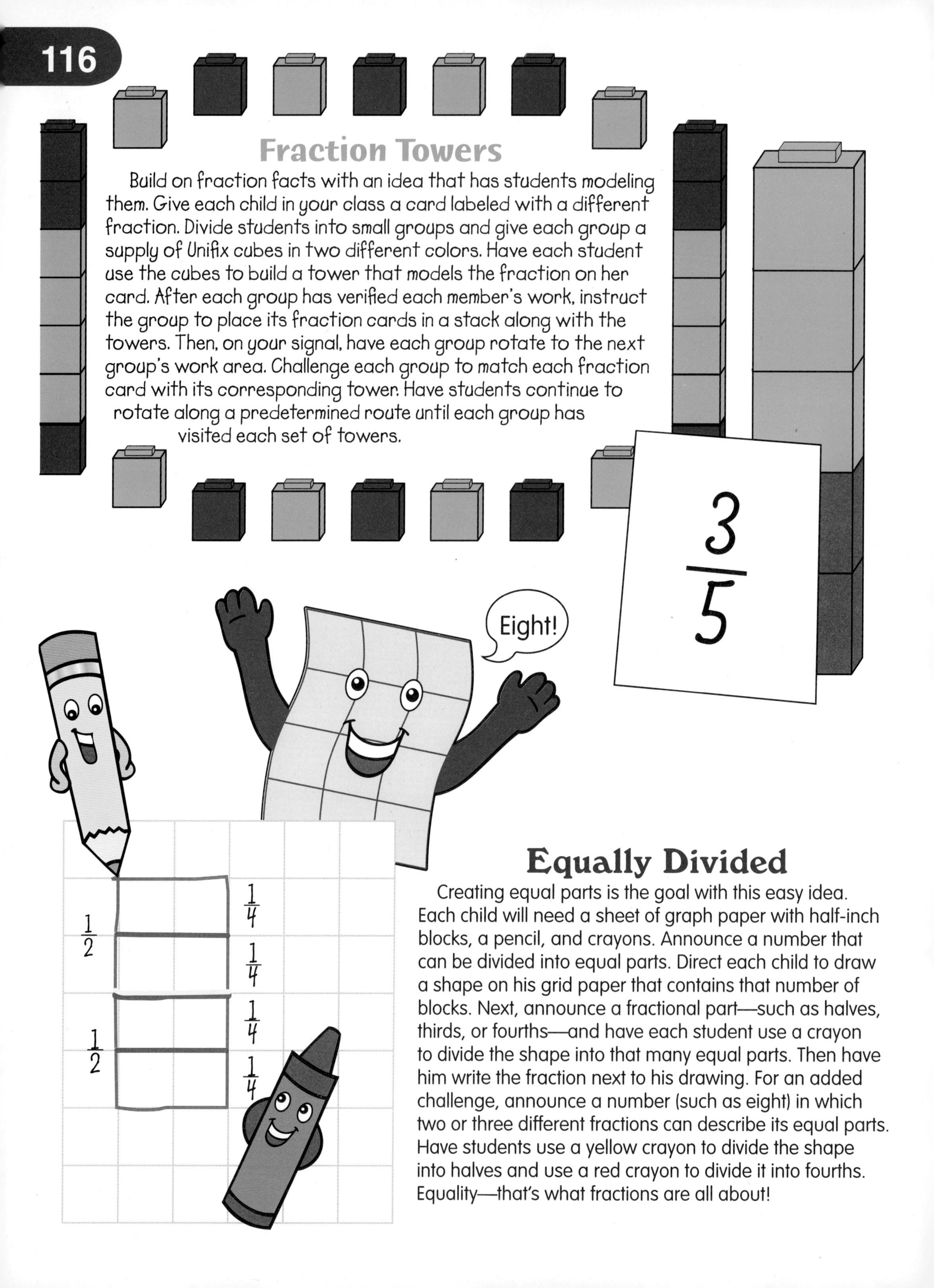


## Fraction Towers

Build on fraction facts with an idea that has students modeling them. Give each child in your class a card labeled with a different fraction. Divide students into small groups and give each group a supply of Unifix cubes in two different colors. Have each student use the cubes to build a tower that models the fraction on her card. After each group has verified each member's work, instruct the group to place its fraction cards in a stack along with the towers. Then, on your signal, have each group rotate to the next group's work area. Challenge each group to match each fraction card with its corresponding tower. Have students continue to rotate along a predetermined route until each group has visited each set of towers.

## Equally Divided

Creating equal parts is the goal with this easy idea. Each child will need a sheet of graph paper with half-inch blocks, a pencil, and crayons. Announce a number that can be divided into equal parts. Direct each child to draw a shape on his grid paper that contains that number of blocks. Next, announce a fractional part—such as halves, thirds, or fourths—and have each student use a crayon to divide the shape into that many equal parts. Then have him write the fraction next to his drawing. For an added challenge, announce a number (such as eight) in which two or three different fractions can describe its equal parts. Have students use a yellow crayon to divide the shape into halves and use a red crayon to divide it into fourths. Equality—that's what fractions are all about!

# Fraction Lotto

Students listen for and identify fractions with this game that is a whole "lotto" fun! Write 25 different fractions—such as $\frac{1}{2}$, $\frac{3}{5}$, and $\frac{2}{8}$—on the board. (The fractions should reflect your students' abilities.) Provide each child with a blank copy of a gameboard (page 118). Instruct him to randomly program each of the 16 squares with a different fraction from the board. As students are programming their cards, write each of the 25 fractions on a separate small piece of paper and place the pieces in a container. Also give each child 16 game markers. To play the game, draw a piece of paper from the container and announce the fraction. If a student has the fraction on his board, he covers it. The first student to cover all the numbers on his board announces, "Fractions!" To win the game, he must read aloud the fractions on his board for verification.

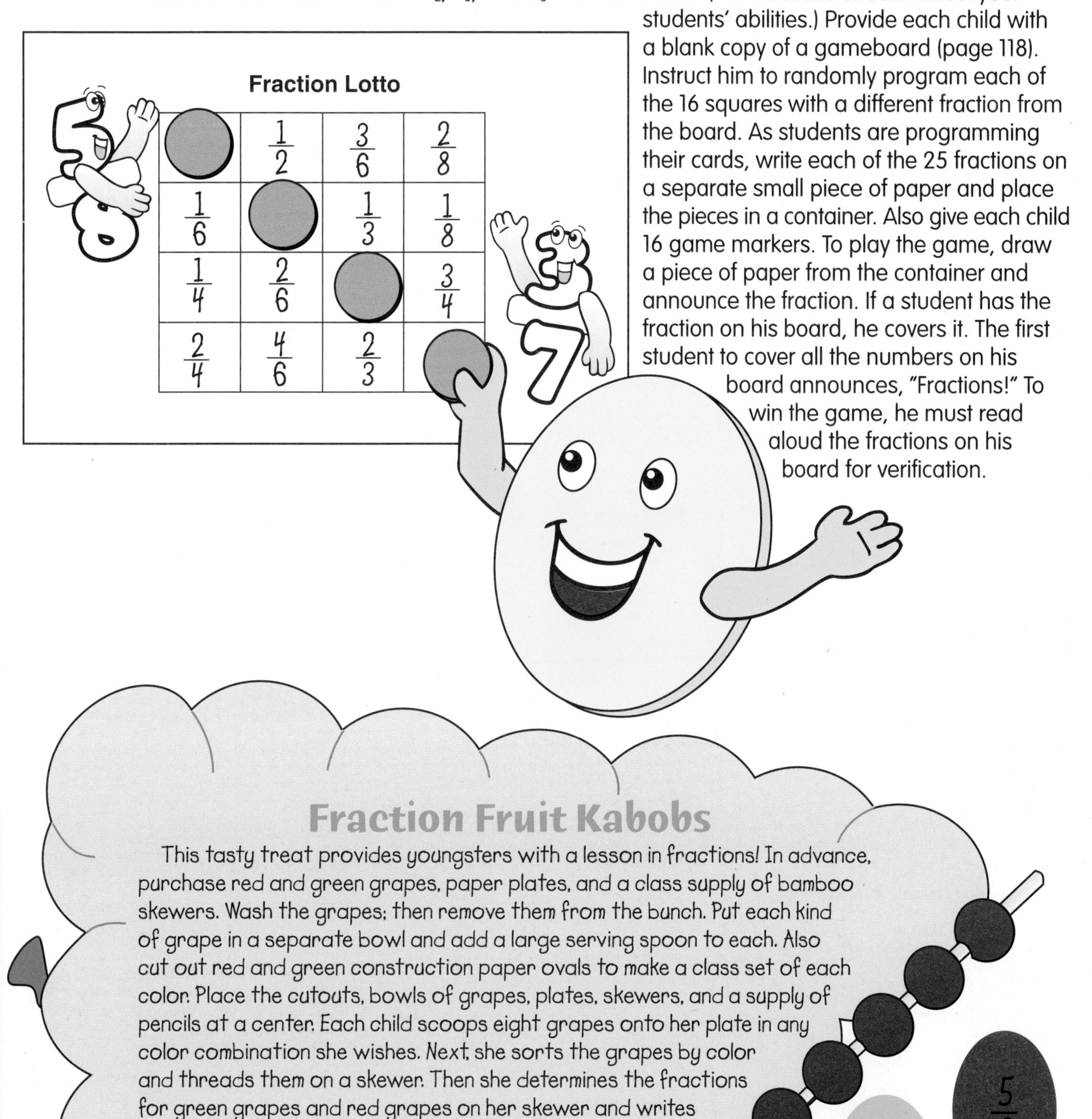


## Fraction Fruit Kabobs

This tasty treat provides youngsters with a lesson in fractions! In advance, purchase red and green grapes, paper plates, and a class supply of bamboo skewers. Wash the grapes; then remove them from the bunch. Put each kind of grape in a separate bowl and add a large serving spoon to each. Also cut out red and green construction paper ovals to make a class set of each color. Place the cutouts, bowls of grapes, plates, skewers, and a supply of pencils at a center. Each child scoops eight grapes onto her plate in any color combination she wishes. Next, she sorts the grapes by color and threads them on a skewer. Then she determines the fractions for green grapes and red grapes on her skewer and writes each fraction on its corresponding grape cutout. After a student has had a classmate verify her work, she eats the grapes in fractional portions. Yum!

$\frac{3}{8}$ $\frac{5}{8}$

Find more student practice on pages 119–123.

## Lotto Gameboards

Use with "Fraction Lotto" on page 117.

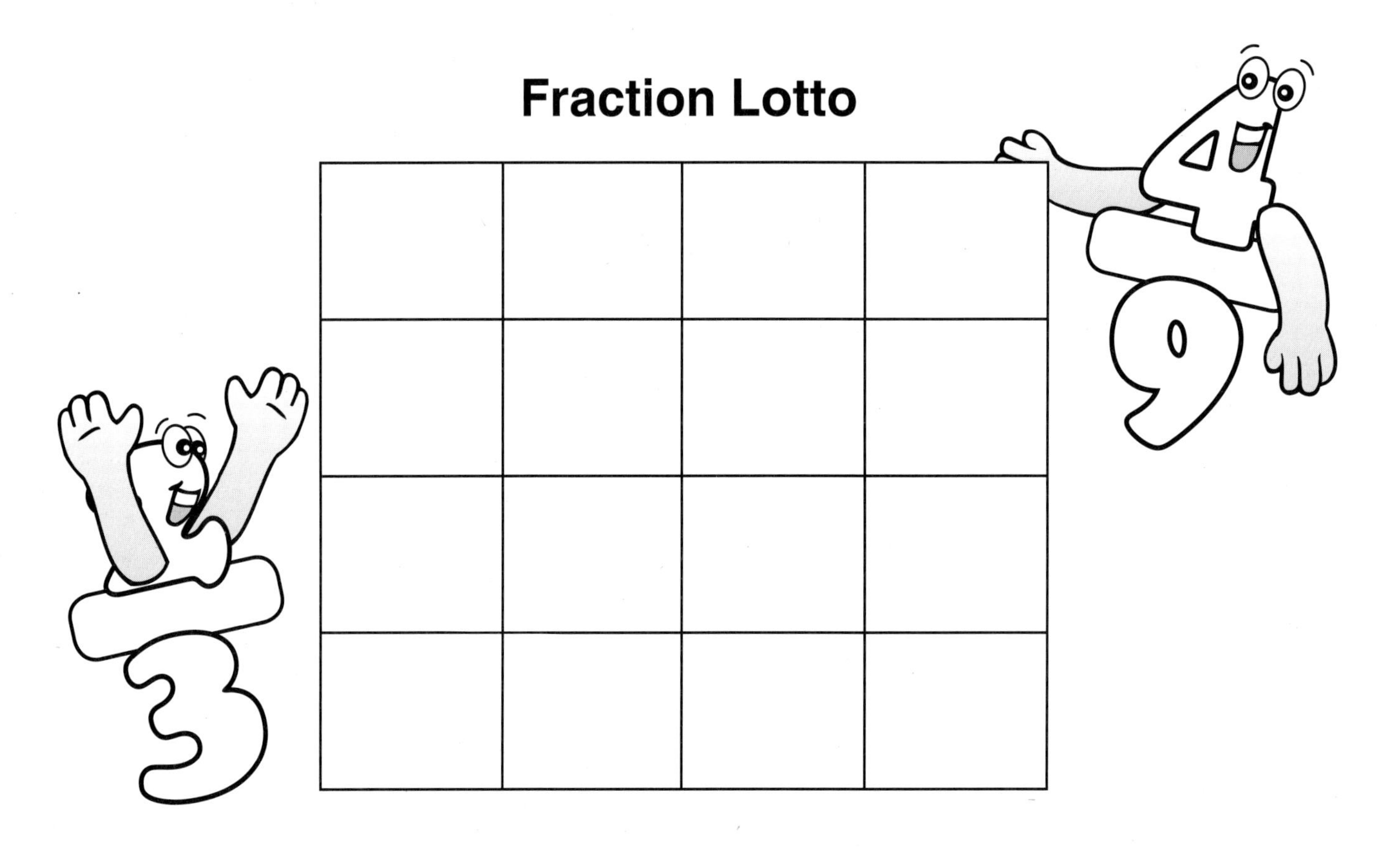

Name ______________________ Date ______________

# Top Rock

Help Gus Goat get to the top of the mountain.
To show the path, color the figures that are divided into equal parts.

Top Rock

Word Box
halves
thirds
fourths
eighths

For each colored rock, write the word that tells how the figure is divided. Use the word box.

Name ____________________ Date ____________________

# Pepperoni Pizzas

Write a matching fraction in each blank.

1.

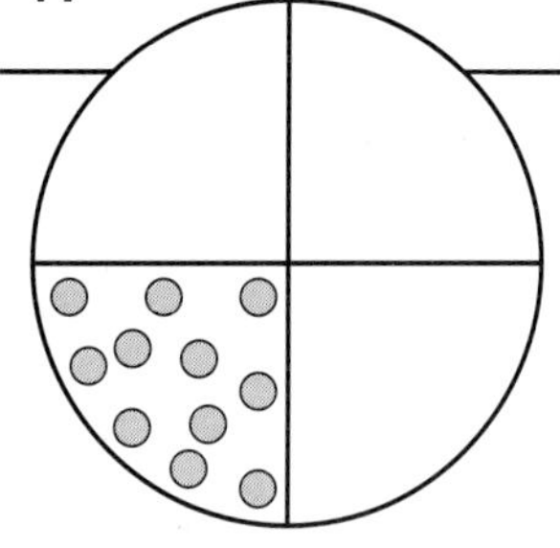

___ has pepperoni

___ has no pepperoni

2.

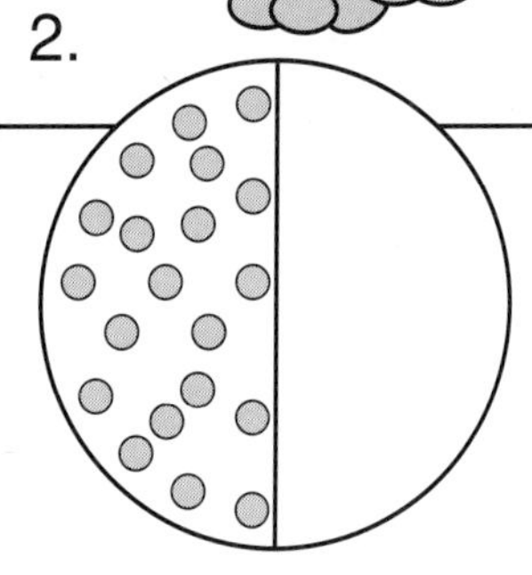

___ has pepperoni

___ has no pepperoni

3.

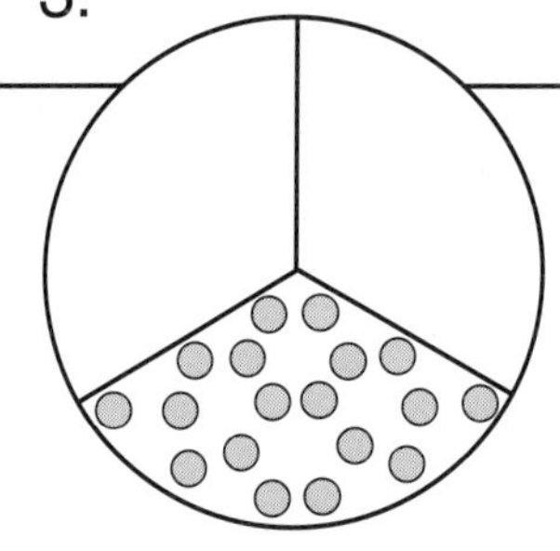

___ has pepperoni

___ has no pepperoni

4.

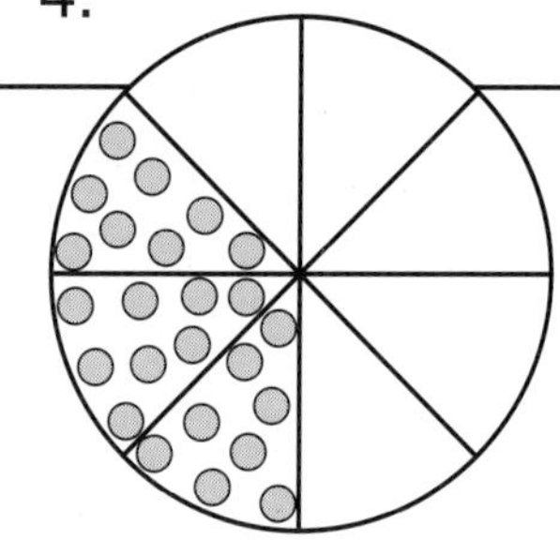

___ has pepperoni

___ has no pepperoni

5.

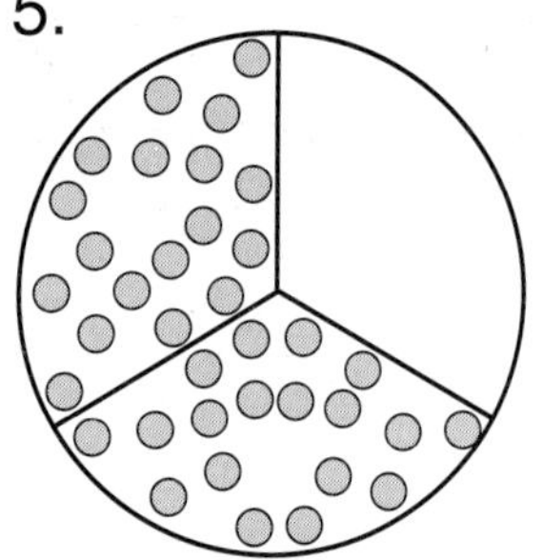

___ has pepperoni

___ has no pepperoni

6.

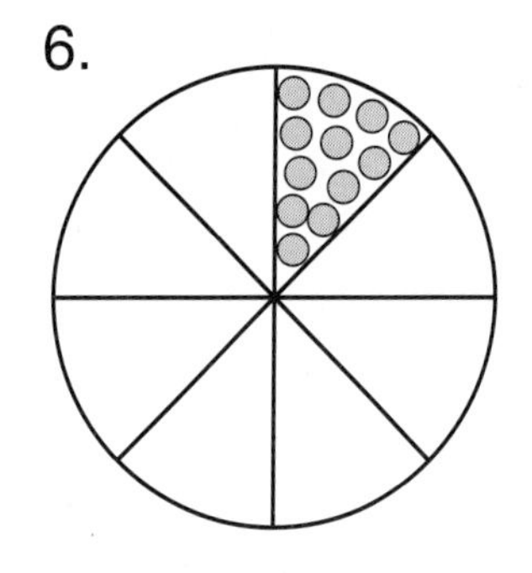

___ has pepperoni

___ has no pepperoni

7.

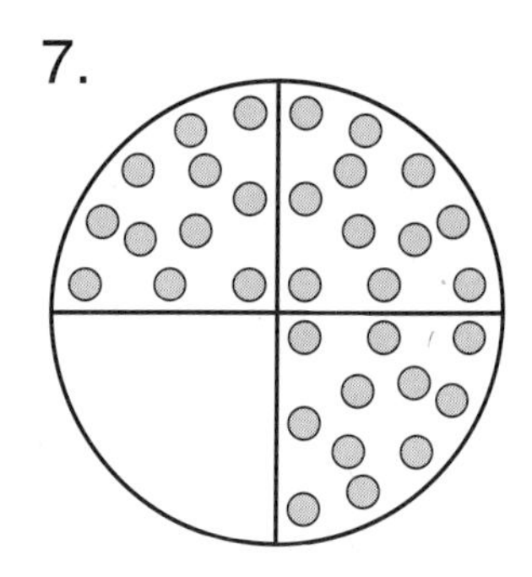

___ has pepperoni

___ has no pepperoni

8.

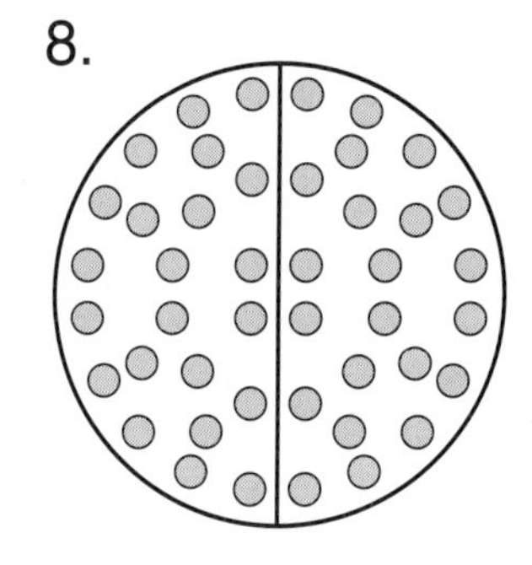

___ has pepperoni

___ has no pepperoni

9.

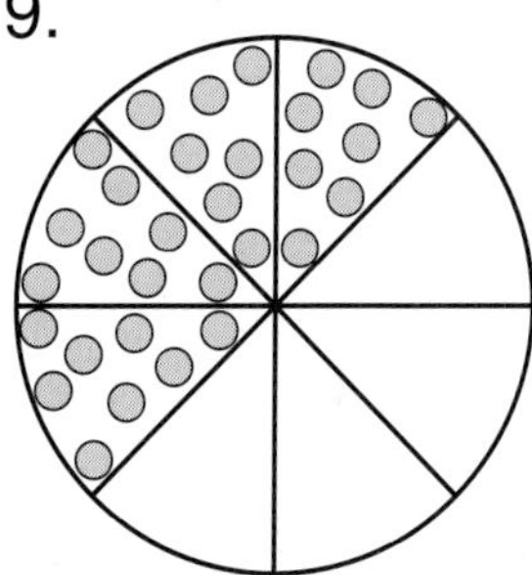

___ has pepperoni

___ has no pepperoni

10.

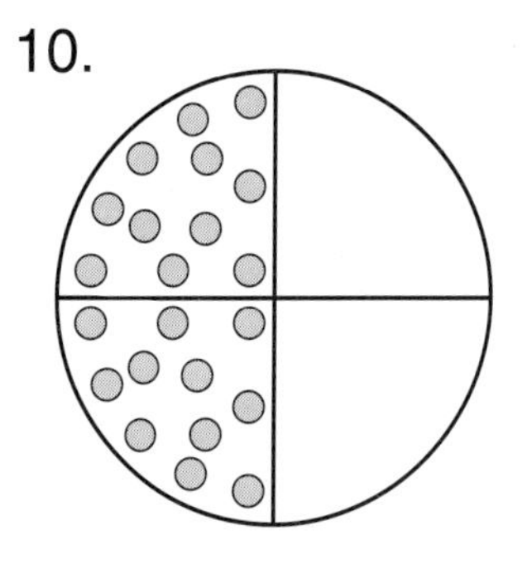

___ has pepperoni

___ has no pepperoni

11.

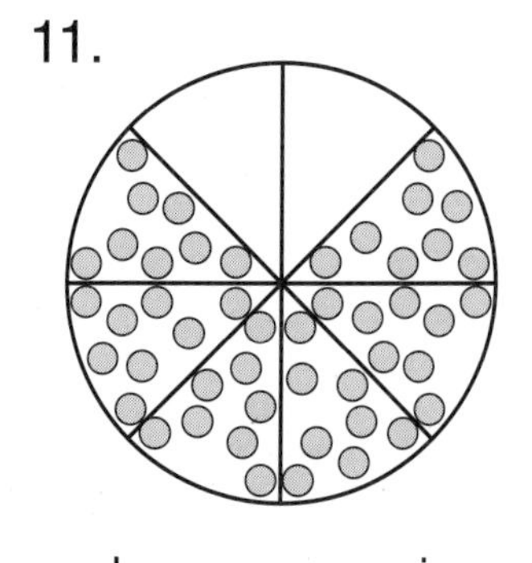

___ has pepperoni

___ has no pepperoni

12.

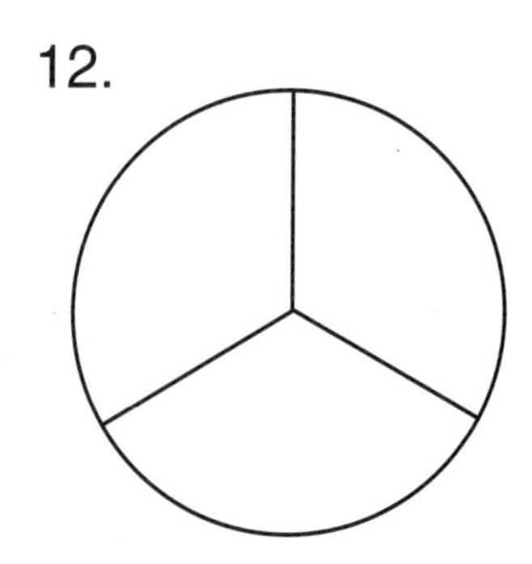

___ has pepperoni

___ has no pepperoni

Name ______________________ Date ______________

# Fish Shop Finds

What fraction of fish are in each tank?
Write a matching fraction in each blank.
Use the fish code.

Fish on Sale Today!

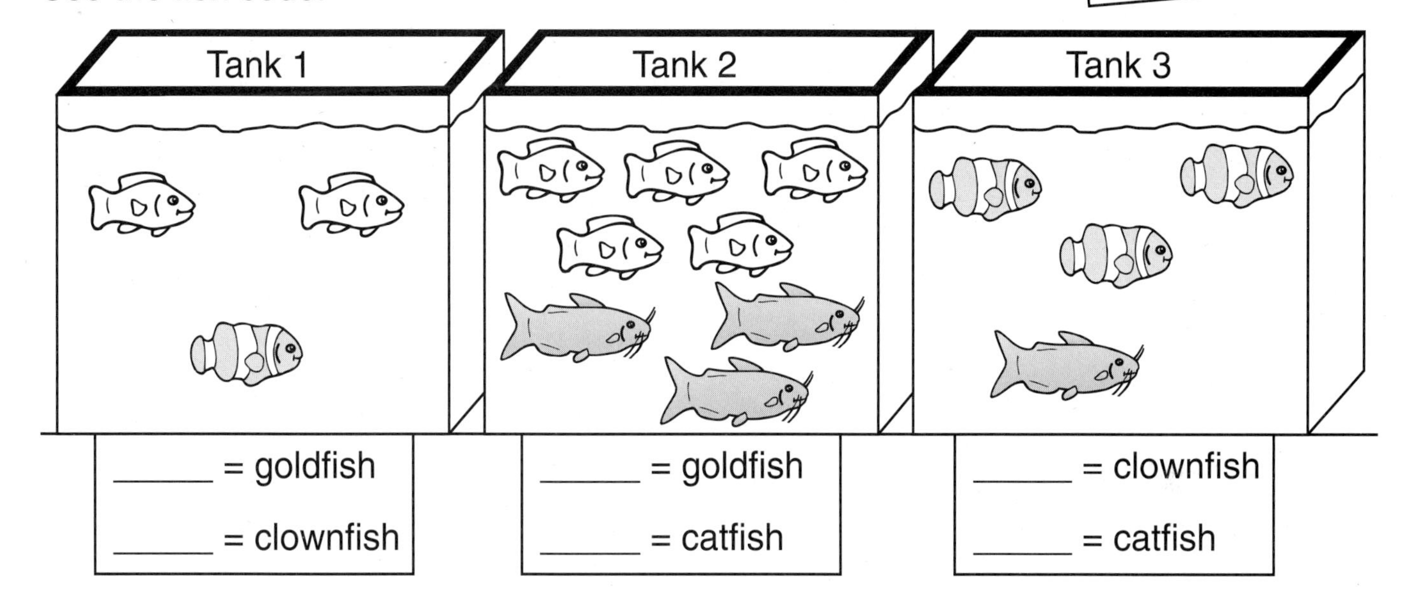


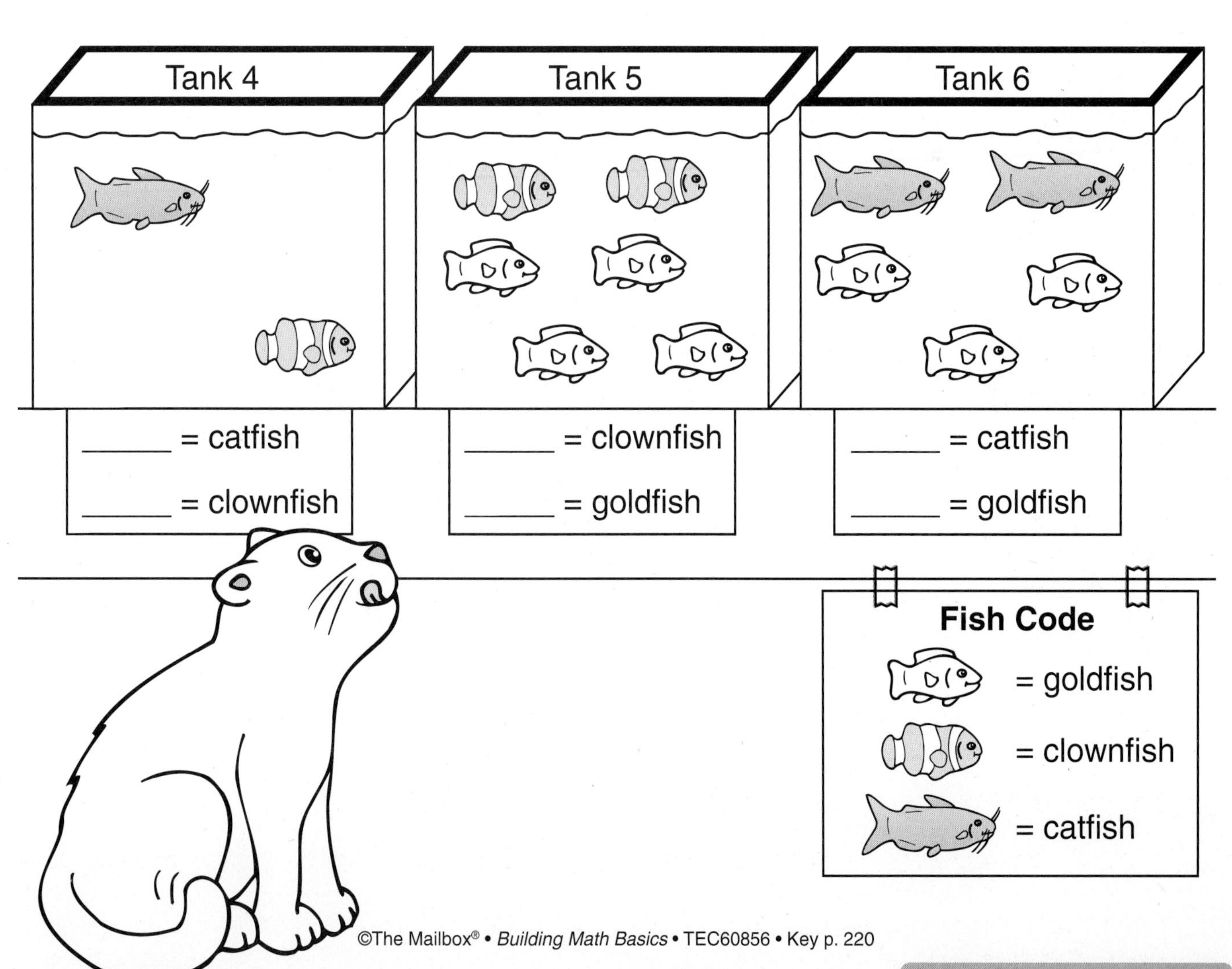

Name ____________________ Date ____________________

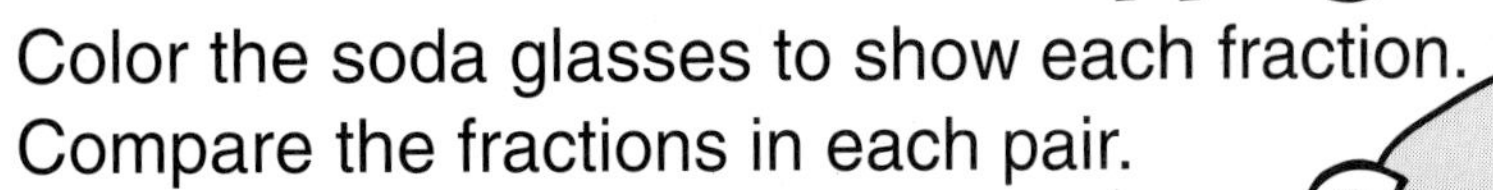

# Sipping Sodas

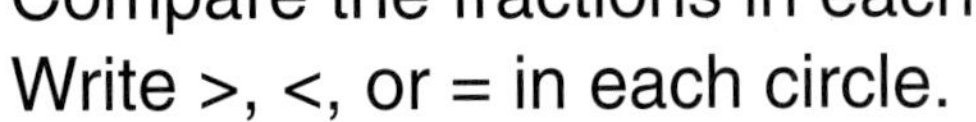

Color the soda glasses to show each fraction.
Compare the fractions in each pair.
Write >, <, or = in each circle.

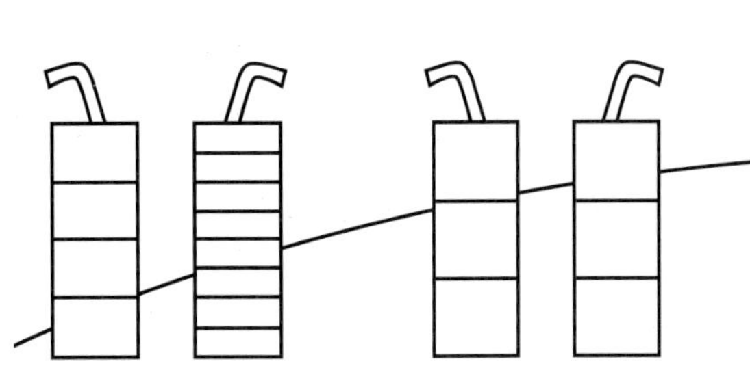

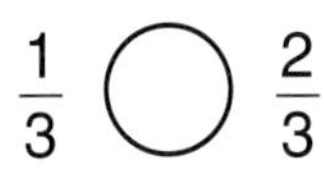

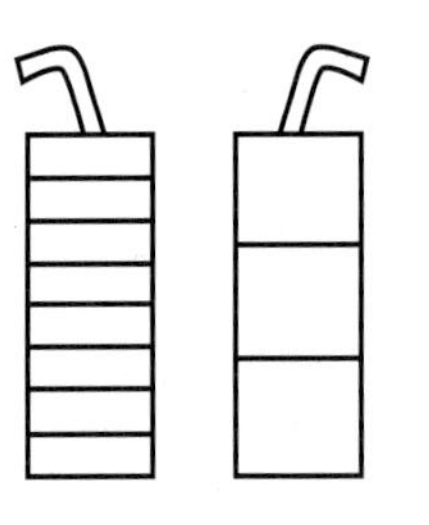
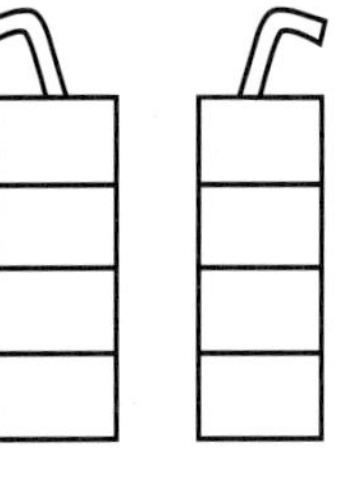
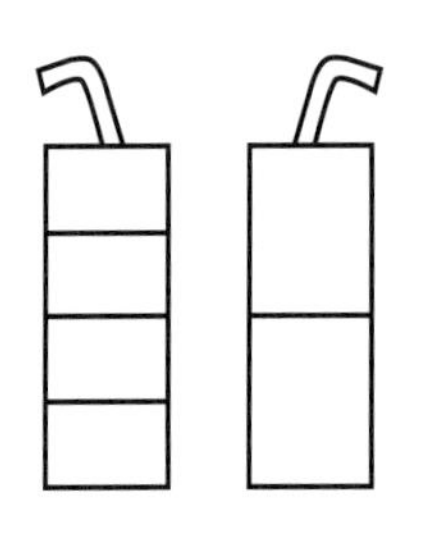
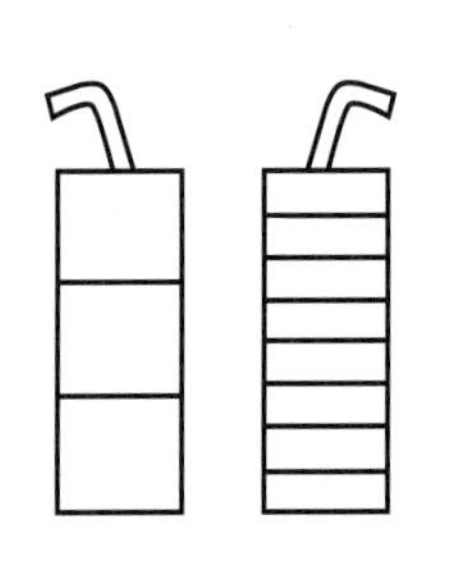
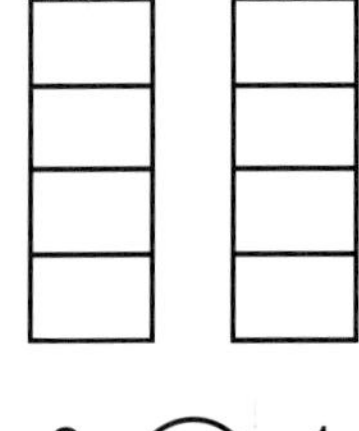

$\frac{1}{4}$ ◯ $\frac{2}{8}$ | $\frac{1}{3}$ ◯ $\frac{2}{3}$ | $\frac{1}{2}$ ◯ $\frac{2}{4}$ | $\frac{5}{8}$ ◯ $\frac{3}{4}$ | $\frac{4}{4}$ ◯ $\frac{1}{3}$

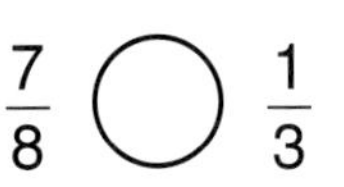
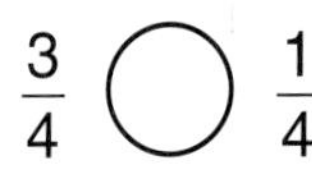

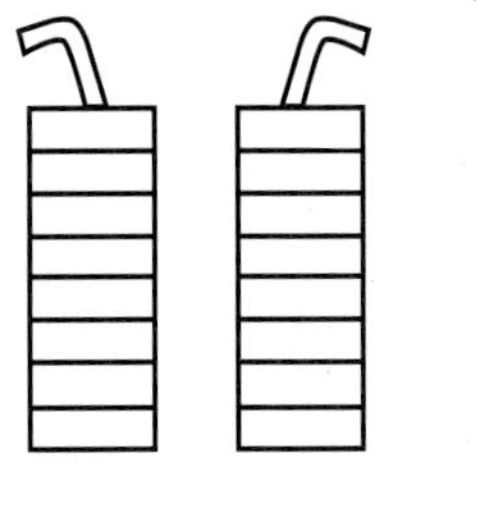
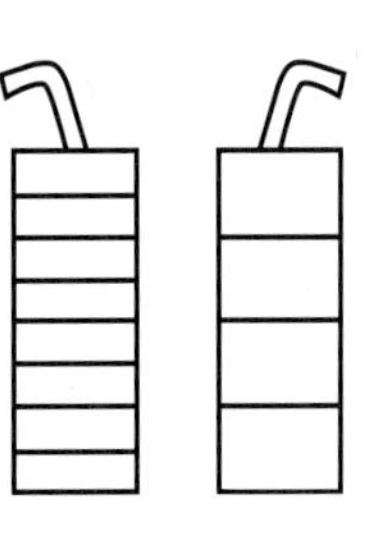
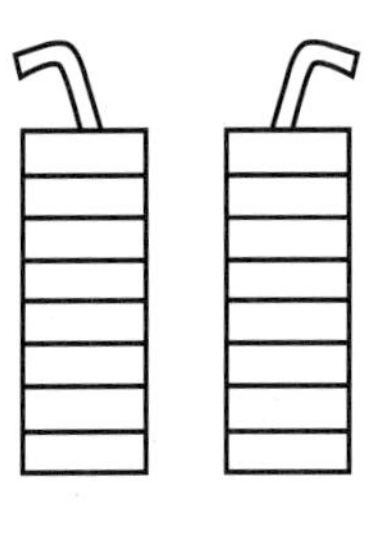
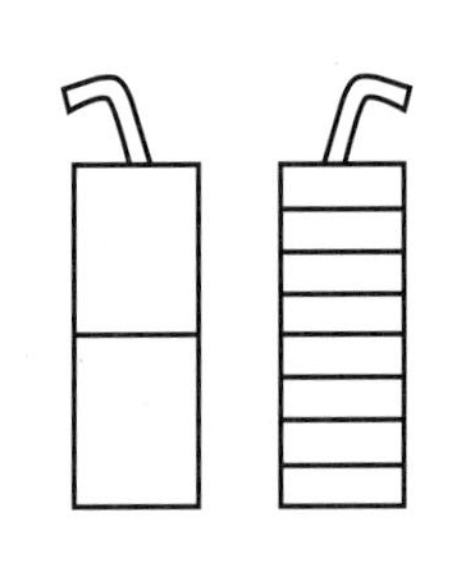
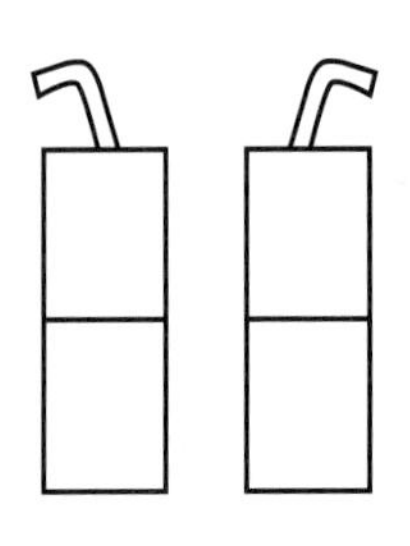

$\frac{7}{8}$ ◯ $\frac{1}{3}$ | $\frac{1}{4}$ ◯ $\frac{2}{4}$ | $\frac{4}{4}$ ◯ $\frac{2}{2}$ | $\frac{3}{3}$ ◯ $\frac{3}{8}$ | $\frac{3}{4}$ ◯ $\frac{1}{4}$

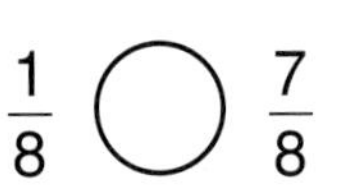
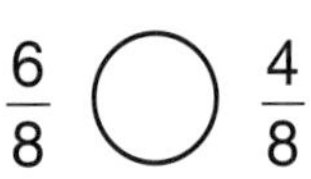

$\frac{1}{8}$ ◯ $\frac{7}{8}$ | $\frac{6}{8}$ ◯ $\frac{3}{4}$ | $\frac{6}{8}$ ◯ $\frac{4}{8}$ | $\frac{1}{2}$ ◯ $\frac{4}{8}$ | $\frac{2}{2}$ ◯ $\frac{1}{2}$

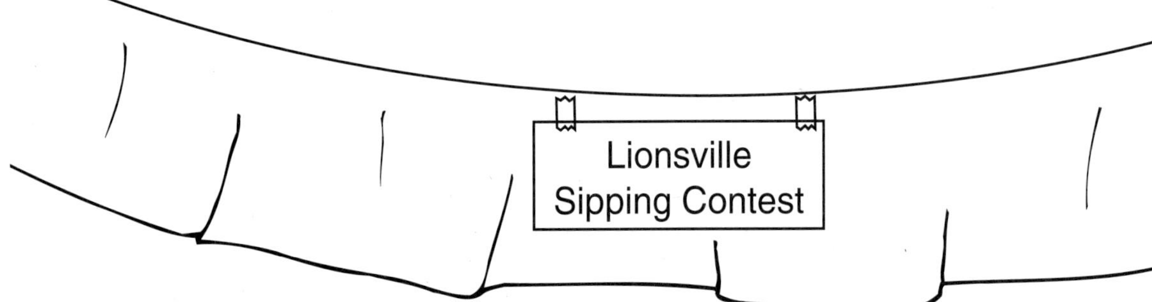

Name ______________________ Date ______________

# Spare Tires

Color the front tire on each car to match the fraction.
Write a fraction to show the colored bolts on the back tire.
If the fractions are the same, color the car.

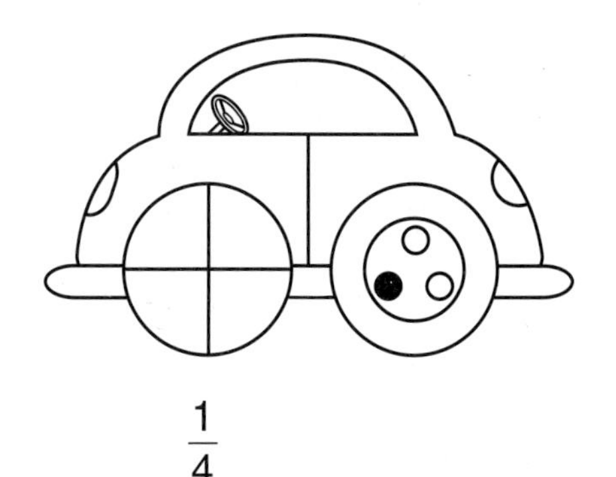

$\frac{1}{4}$ ______ ______

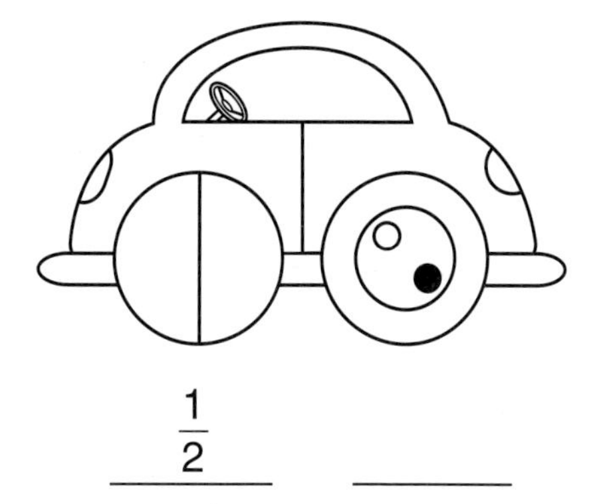

$\frac{1}{2}$ ______ ______

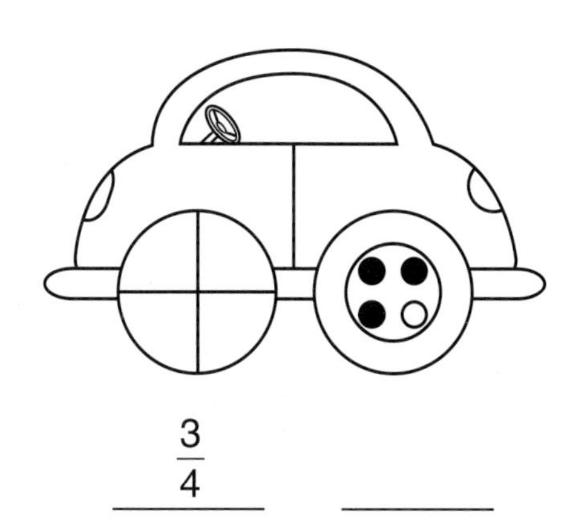

$\frac{3}{4}$ ______ ______

$\frac{1}{6}$ ______ ______

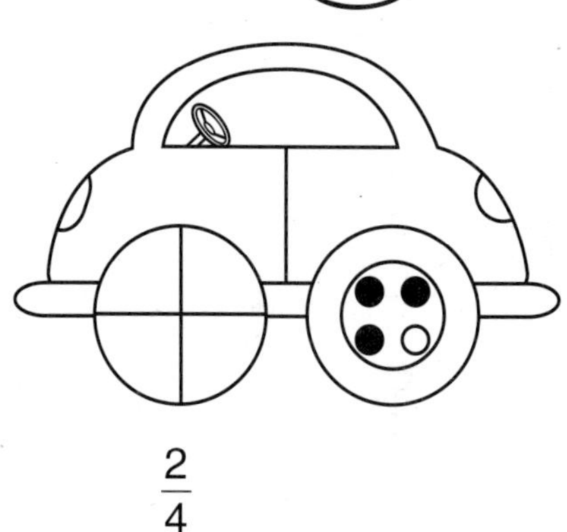

$\frac{2}{4}$ ______ ______

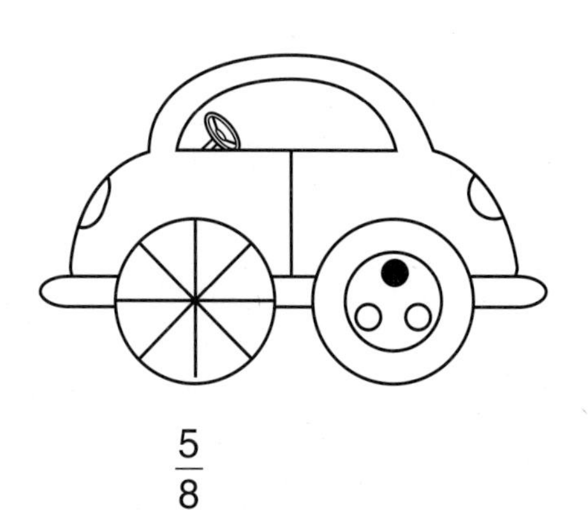

$\frac{5}{8}$ ______ ______

$\frac{4}{6}$ ______ ______

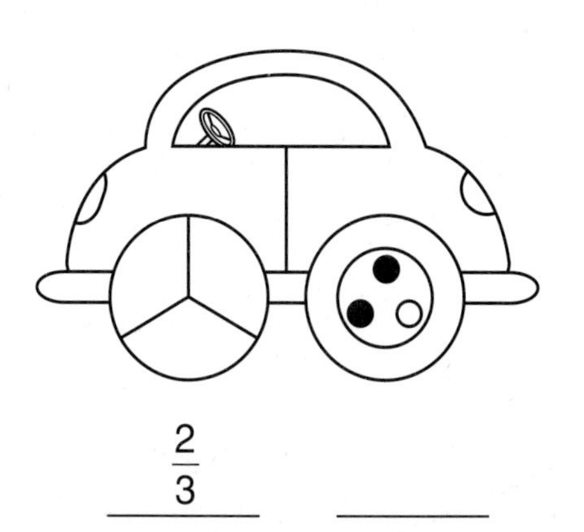

$\frac{2}{3}$ ______ ______

## Match the Masterpieces

This idea shapes up students' understanding of plane figures. Prepare index cards for student use by listing on each card a different combination of 12 shapes (see the example). Stock a center with the prepared cards and several pattern blocks and circle templates that correspond to those on the card. Also supply colorful construction paper, drawing paper, scissors, glue, and pencils.

To use the center, a student selects an index card and then finds a block or template for each of the shapes listed on the card. She traces the appropriate number of shapes onto construction paper and cuts out the shapes. Next, she glues them in any desired design on a sheet of drawing paper. She writes a description of her picture on her paper and signs her name. She also signs her name on the back of the index card before putting it back with the others.

When every student has completed the center, post the pictures and place the index cards nearby. Choose a card and challenge students to find the pictures with the matching shape combinations. Then flip the card to verify their matches.

## Pasta Polygons

Get students using their noodles with this idea for polygon practice! Give each student a supply of uncooked fettuccine noodles. Direct him to break some in half and others into fourths. Remind students that a polygon is a closed plane figure with straight sides. Then ask a volunteer to use four noodle pieces to construct a quadrilateral, or four-sided polygon, on an overhead projector. If the student forms a specific quadrilateral, such as a square, rectangle, trapezoid, parallelogram, or rhombus, be sure to discuss its special characteristics. Next, direct each student to use his noodles to construct a different four-sided polygon on his desk. Scan students' desks for accuracy. Invite a student with a correct answer to model it on the projector. Assist students with naming the quadrilateral, if applicable. Continue in this manner until students have created a variety of quadrilaterals. Then explore triangles, pentagons, hexagons, and octagons in a similar manner. Wow, polygons are "pasta-rrific"!

# Polygon Puzzles

Try this puzzle challenge to add interest to polygon practice! In advance, create a labeled reproducible tangram puzzle as shown and then make a copy for each child. To begin, give each child a copy and explain that a tangram is a puzzle made up of seven separate polygons (two large triangles, one medium triangle, two small triangles, one square, and one parallelogram). Have each child cut out the puzzle pieces. Then have students make new polygons using pieces of the tangram puzzle. To do this, have each child form a square using the pieces *M* and *E*. After allowing students time to work, draw the solution on the board. Then post a chart like the one shown and ask students to work independently through each additional puzzle. If desired, have each student compare her work with a neighbor's to check her solutions. Can you make a square with *B, A,* and *T*? I can!

Make each of these shapes using the tangram pieces listed:

1. Square with *A* and *T*
2. Triangle with *B, E, A,* and *T*
3. Rectangle with *A, N,* and *T*
4. Parallelogram with *M* and *E*
5. Square with *B, A,* and *T*
6. Triangle with *T, A,* and *N*
7. Trapezoid with *A, N,* and *T*
8. Pentagon with *A, N, T, B,* and *E*

*(See the possible solutions below.)*

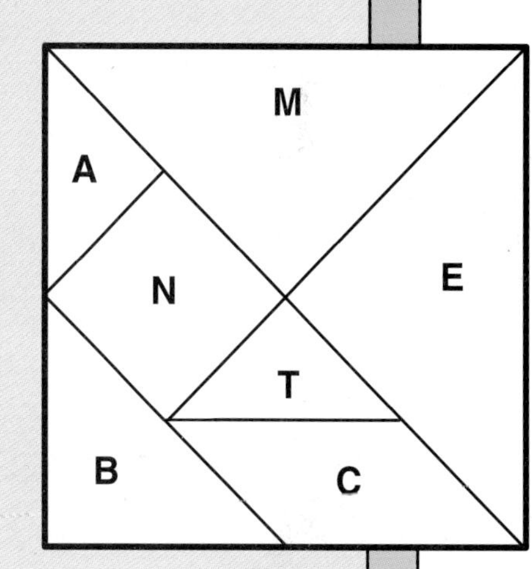


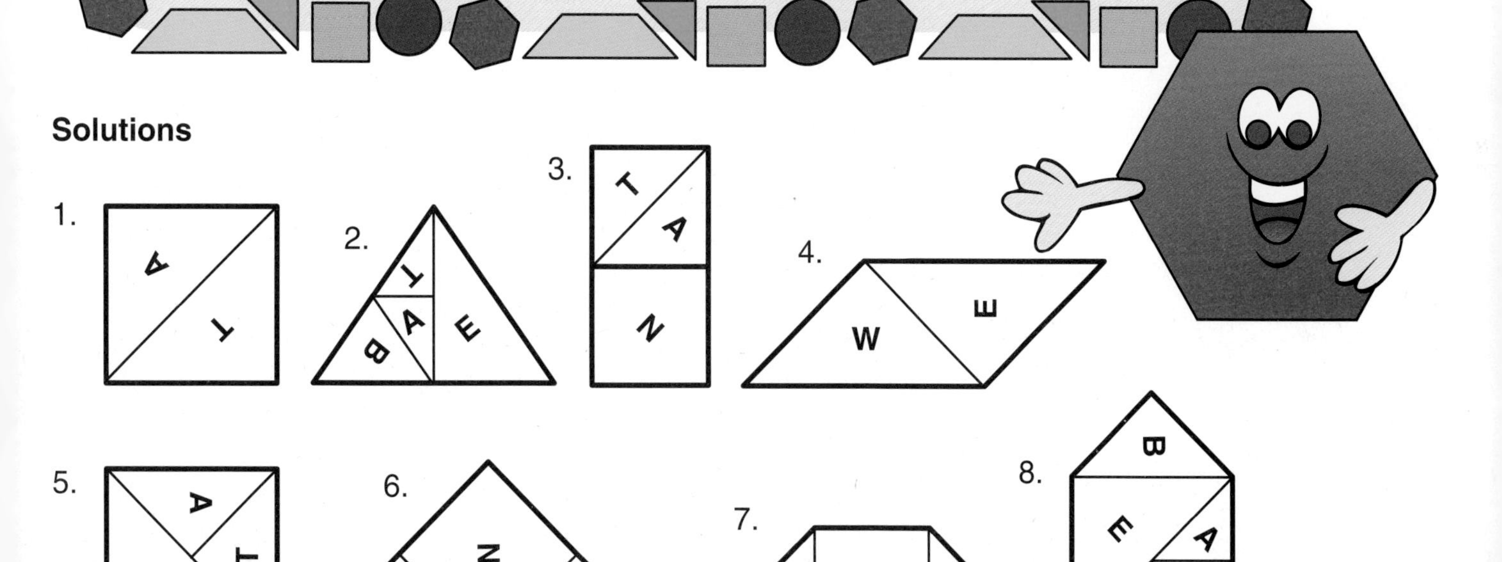


Find more student practice on page 126.

Name ________________________ Date ________________

# Mr. Owl's Artwork

Outline each shape by the code.

# Solid Figures

## Shape It!

Get students in touch with solid figures! Post a list of solid figures and display a model of each one. After reviewing each shape with the class, divide students into as many groups as you have solid figures. Challenge each group to form each shape with a supply of modeling clay. When the models are complete, have the group display the creations on a table, label each figure with a different letter, and create an answer key like the one shown. Later, send each group to a display other than its own and challenge the group members to name each solid figure and then check their answers with the key. At your signal, have students rotate to a different display. Continue in this manner until the groups have visited the remaining displays. This kinesthetic learning process is sure to give students a good feel for solid shapes!

sphere
cone
cylinder
cube
square pyramid
rectangular prism

Answer Key
A. rectangular prism
B. square pyramid
C. cube
D. sphere
E. cylinder
F. cone

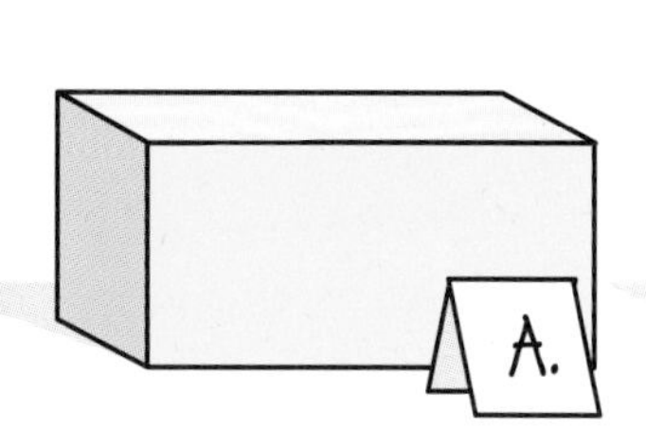


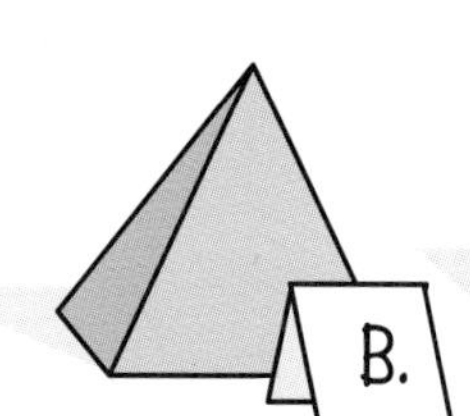


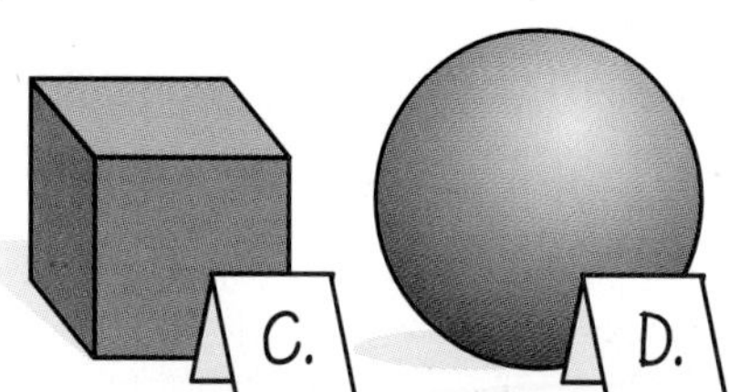

## Solid Stations

Encourage students to observe the attributes of solid figures with this small-group activity. In advance, set up for each geometric solid a separate station that consists of several objects of the corresponding shape, a sheet of chart paper, and a marker. To begin, divide students into as many groups as there are stations and send each group to a station. Each group identifies the geometric solid represented at its station. Then the group members list the common attributes of the objects—such as the number of faces, edges, and corners—along with whether the shape rolls or slides and whether it can be stacked. Then have each group use the items on the list to help them write on the chart paper a definition for their shape. When each group has finished, gather students and invite each group to share its findings with the class. Then display the charts as handy references.

# What Is It?

Here's a partner center that's easy to make and fun for kids! Copy each clue shown onto a separate blank card. Make the cards self-checking by labeling the back of each one with the corresponding solid figure. Place the cards clue side up at a center along with a supply of blank cards. In turn, have each student read a clue for his center partner. After his partner answers, have the student flip the card to check the answer. Direct students to continue in the same manner until they have solved all the clues. Then challenge each partner to choose a solid figure, write a new clue for it on a blank card, and write the answer on the back. Encourage each child to share his clue with his partner. After verifying the new clues, add them to the center for the next pair of students to solve.

**Clues**

It looks like the object that holds a scoop of ice cream. *(cone)*

It rolls and looks like a soccer ball. *(sphere)*

It looks like a shoebox. It has six faces. *(rectangular prism)*

It looks like a Pringles potato chip can. It has two faces. *(cylinder)*

It can be stacked. It looks like an alphabet block. *(cube)*

It has five faces. It cannot be stacked. *(square pyramid)*

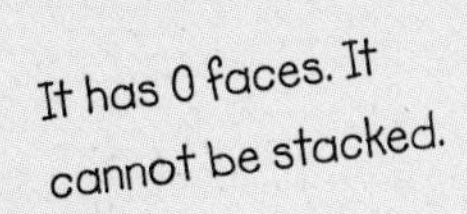


sphere

## Solid Shape Detectives

Put your students hot on the trail of solid shapes! To prepare, create a blank recording sheet (similar to the one shown) for various solid figures that are evident in your classroom, school, or playground. To begin, give each student (or pair of students) a copy of the list and challenge her to find items in the classroom to write on her list. Then lead students through the hallways, library, and cafeteria to look for additional items. Next, take students outside to the playground to look for more items. Follow up the search by inviting youngsters to share their findings and encouraging them to give reasons for their answers. For a visual display, label a large sheet of paper for each solid shape. Then, on each poster, list the items that your youngsters found.

| cone | sphere | rectangular prism | cylinder | cube | square pyramid |
|---|---|---|---|---|---|
| traffic cone<br>megaphone | ball<br>globe<br>orange | book<br>brick<br>videotape | paint can<br>drum<br>kaleidoscope | dice<br>block<br>sugar cube | toy tent |

Find more student practice on pages 129–131.

Name ______________________ Date ____________

# The Juggling Monkey

Cut out the solid figures.
Glue to match.

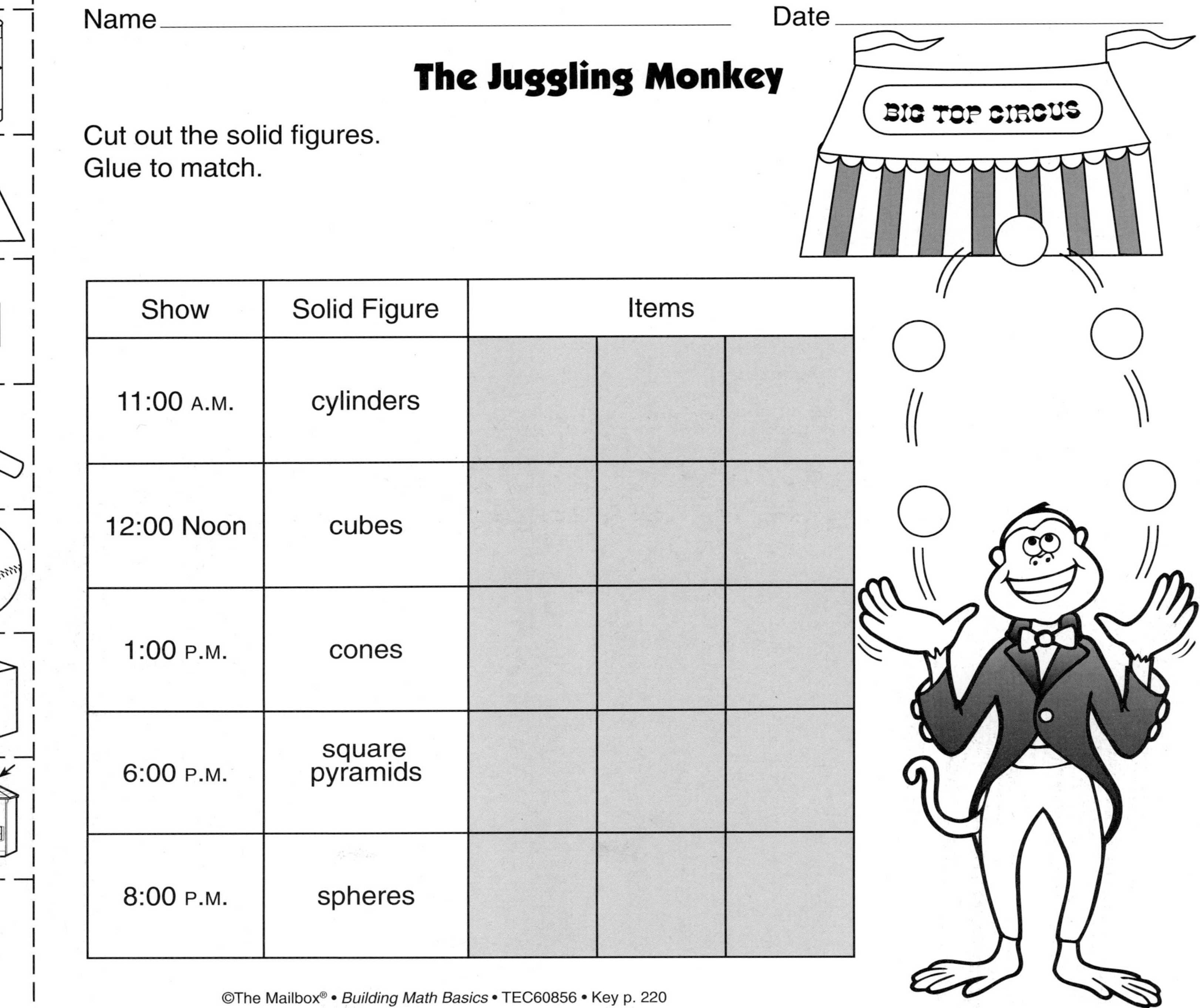


| Show | Solid Figure | Items | | |
|---|---|---|---|---|
| 11:00 A.M. | cylinders | | | |
| 12:00 Noon | cubes | | | |
| 1:00 P.M. | cones | | | |
| 6:00 P.M. | square pyramids | | | |
| 8:00 P.M. | spheres | | | |

Name ______________________________ Date ____________________

# A Closer Look

Study each solid figure.
Complete the notes on each sheet of paper.

1. 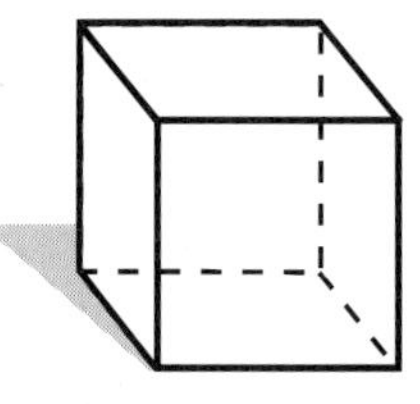

_____ faces
_____ edges
_____ corners

2. 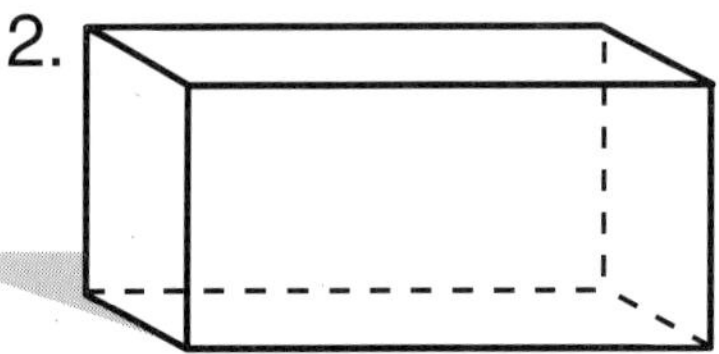

_____ faces
_____ edges
_____ corners

3. 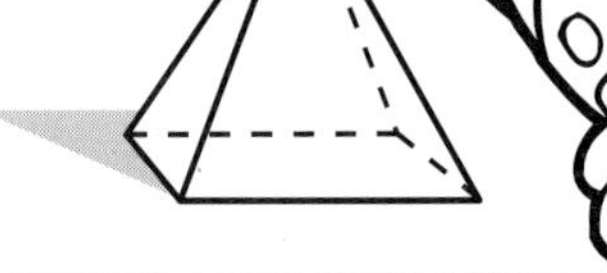

_____ faces
_____ edges
_____ corners

4. 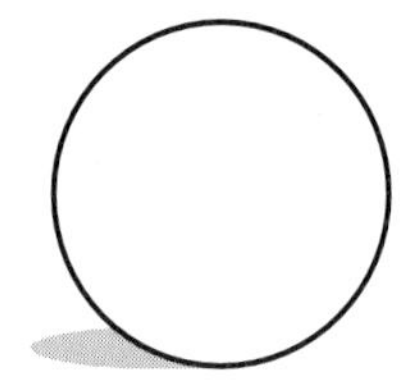

_____ faces
_____ edges
_____ corners

5. 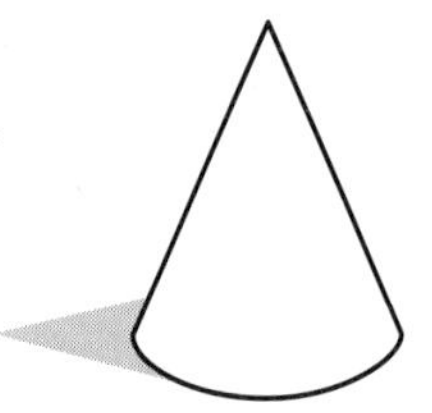

_____ faces
_____ edges
_____ corners

6. 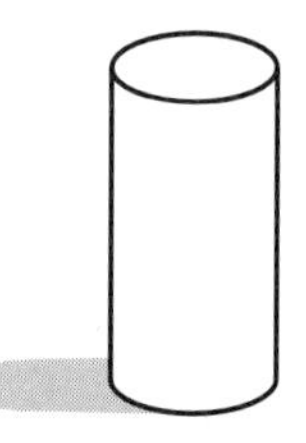

_____ faces
_____ edges
_____ corners

Write the name of each solid figure in the matching blank below.
Use the notes and the word bank.

1. ____________________  4. ____________________
2. ____________________  5. ____________________
3. ____________________  6. ____________________

**Word Bank**
sphere
cube
square pyramid
cylinder
cone
rectangular prism

Name________________________ Date____________

# Shaping Up

Study the solid figures.
Color each plane figure that matches a face of the solid figure.

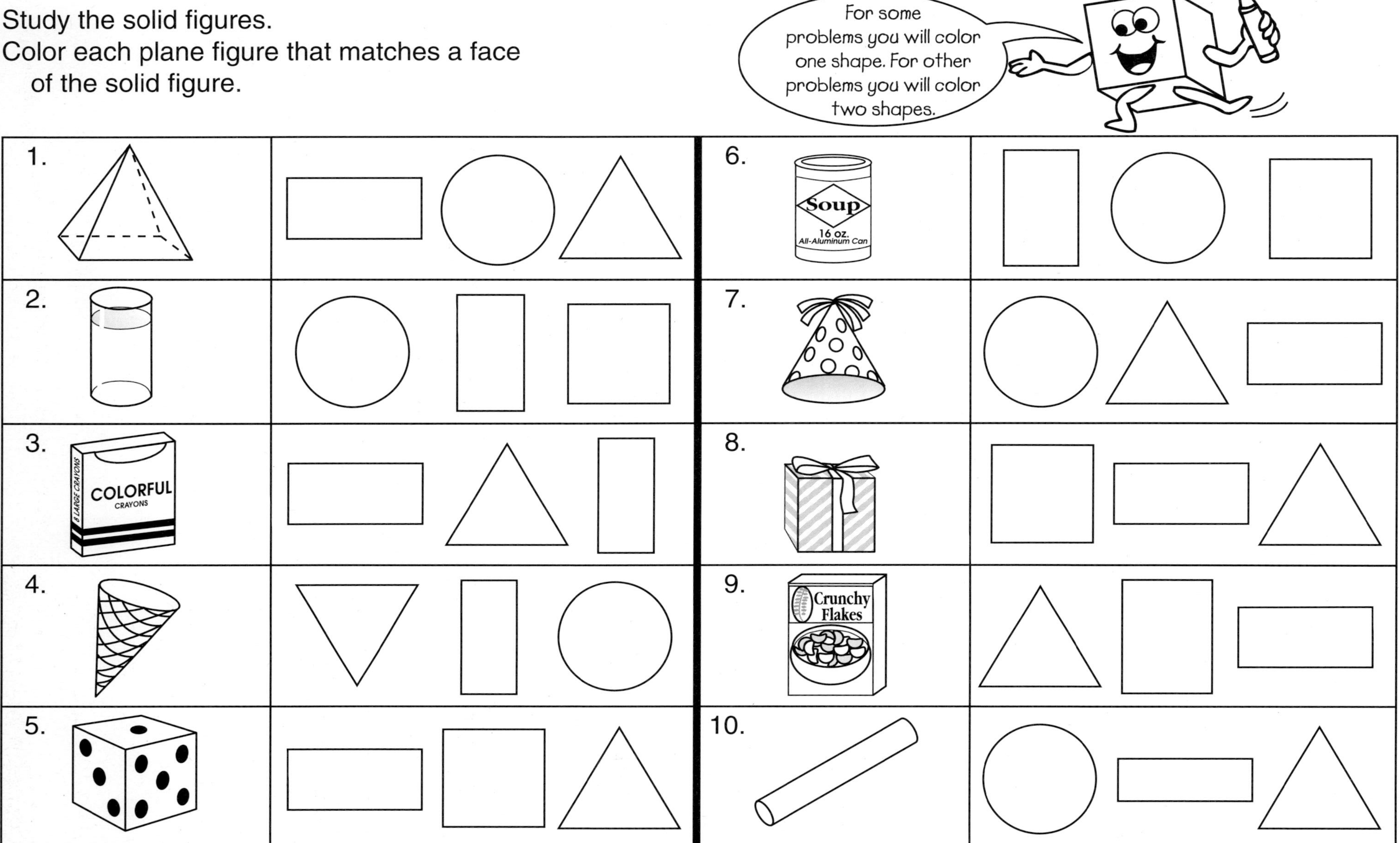

# Congruence

## Perfect Pairs?

This partner activity has students sizing up congruent figures. Have two students sit facing each other at a table. Direct each child to stand up a folder in front of him to use as a screen. Then give each student a Geoboard and a rubber band. To play, Partner 1 makes a secret shape on his Geoboard. Then he gives clues, such as the figure's number of sides or number of right angles, as his partner tries to match the figure. Partner 1 continues giving clues (up to five), making each one more specific as his partner adjusts his figure with the additional information. Then both students reveal their Geoboards to see whether their figures match. If needed, Partner 2 adjusts his shape accordingly to make it congruent. Then Partner 2 takes a turn creating a secret shape in a similar manner.

## Congruent Castle

Reinforce congruent-shape recognition with this royal center game. Color, cut out, and laminate a copy of the gameboard on page 133. Also cut out the shapes at the bottom of page 133 and glue each one to a different section on a construction paper square to make a spinner as shown. Place the spinner, the gameboard, and two game markers at a center.

To begin, Player 1 spins the spinner and moves her marker to the next corresponding congruent shape on the gameboard. Player 2 checks her partner's choice. If the shapes are congruent, Player 1 leaves her marker on the space. If the shapes are not congruent, Player 1 returns her marker to its previous position and her turn is over. Then Player 2 takes a turn in a similar manner. The game continues until one of the royal subjects reaches the castle tower and is declared the winner!

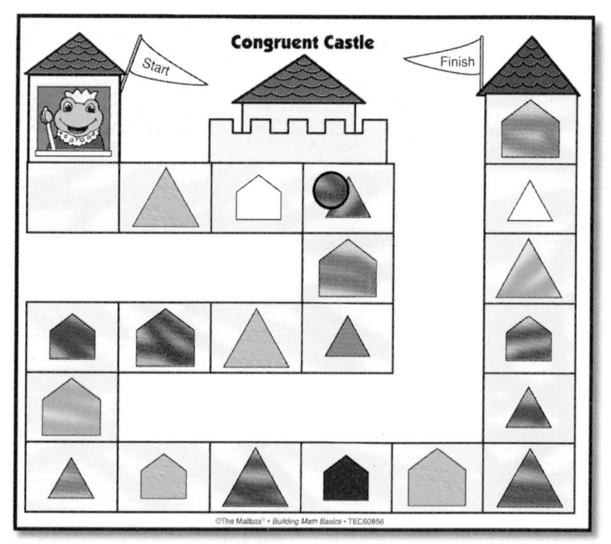


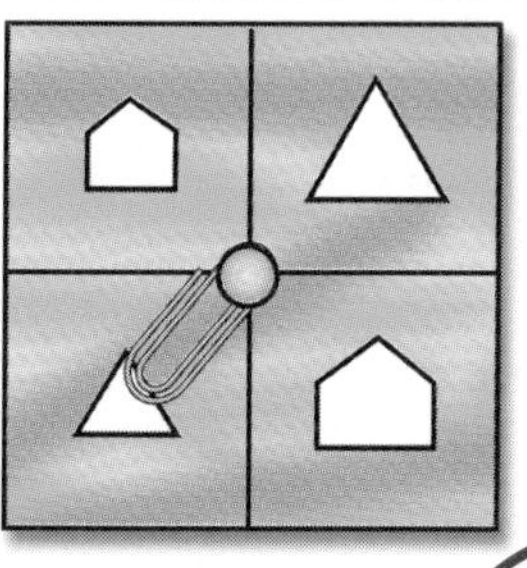

Find more student practice on page 134.

## Gameboard and Shape Patterns

Use with "Congruent Castle" on page 132.

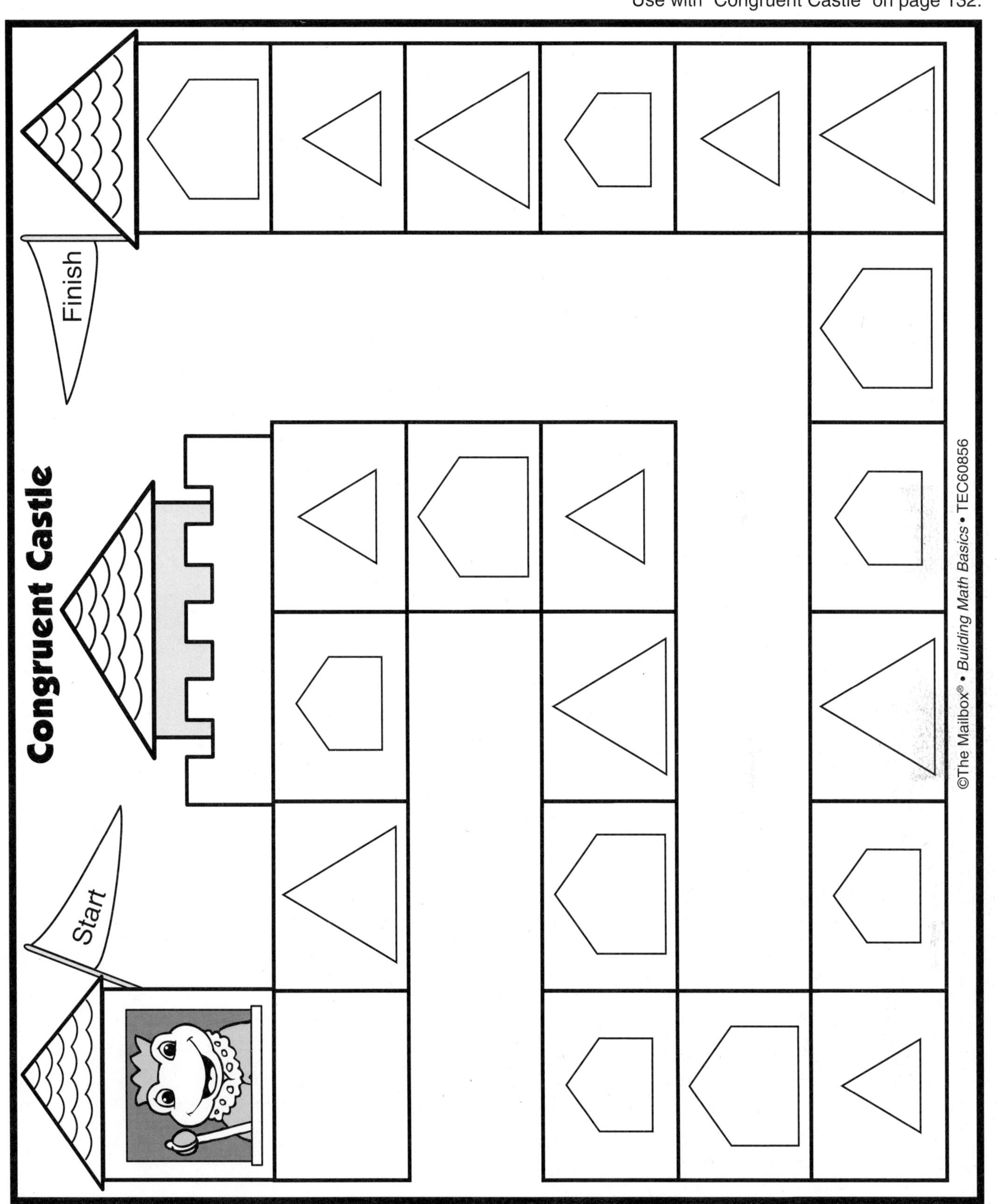

Name ______________________ Date ______________

# Pots of Shape Soup

Color the two congruent shapes on each kettle of soup.

1\.
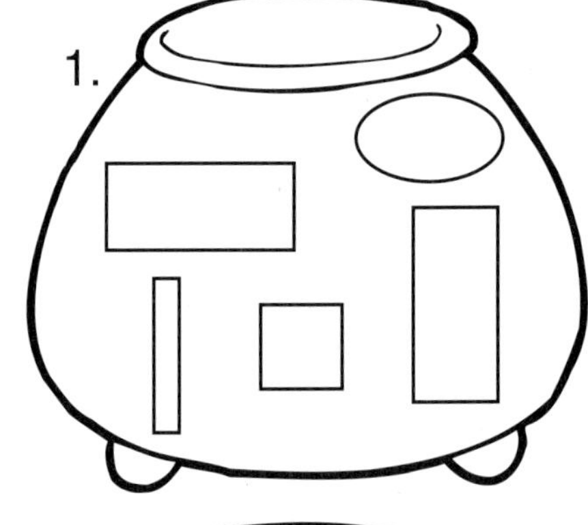

2\.
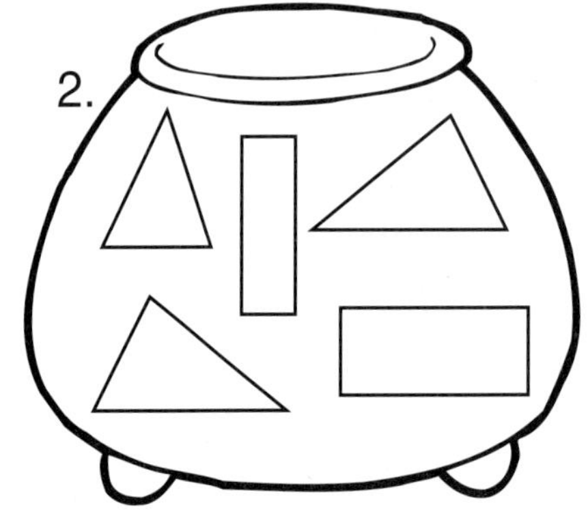

3\.

4\.
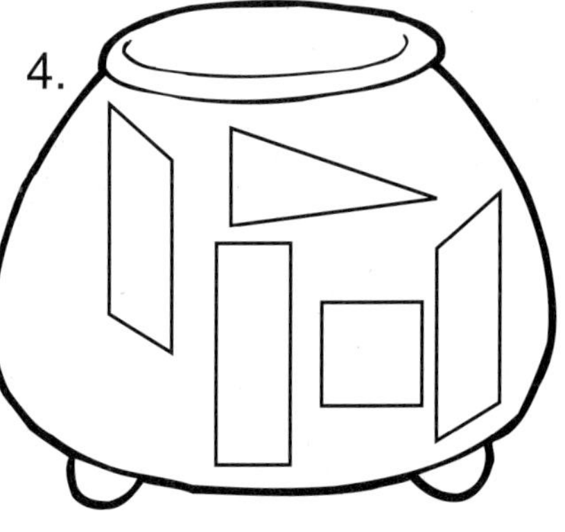

5\.
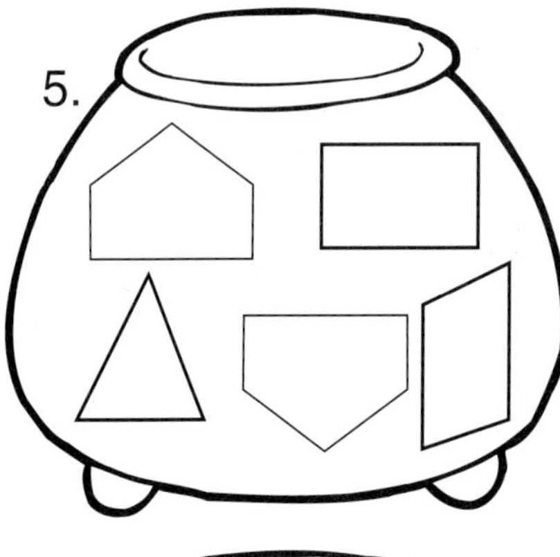

6\.
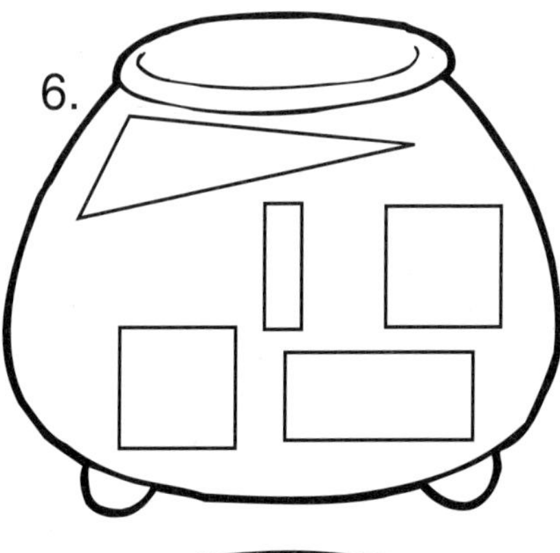

7\.
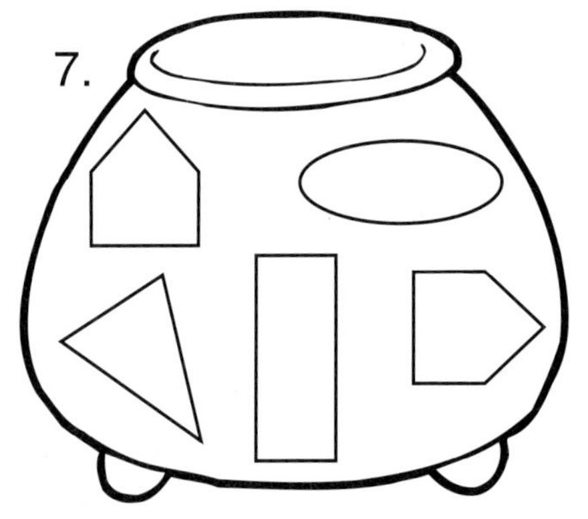

8\.
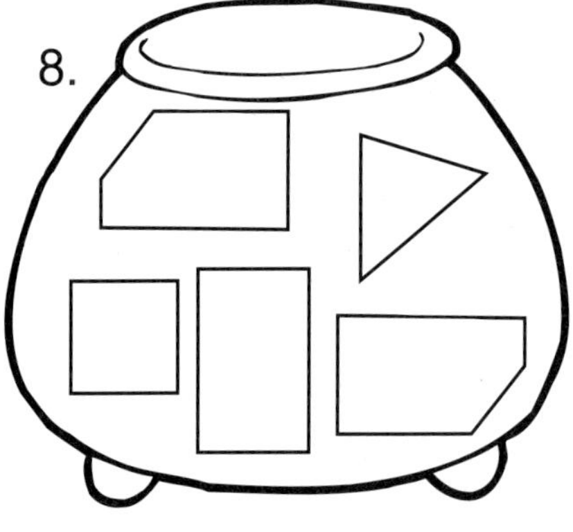

9\.
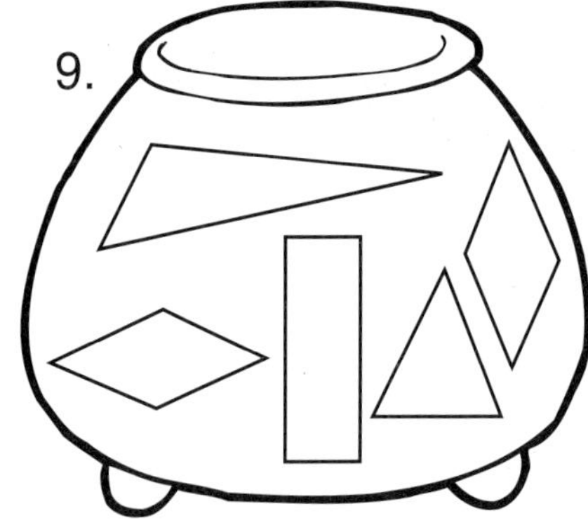

10\.
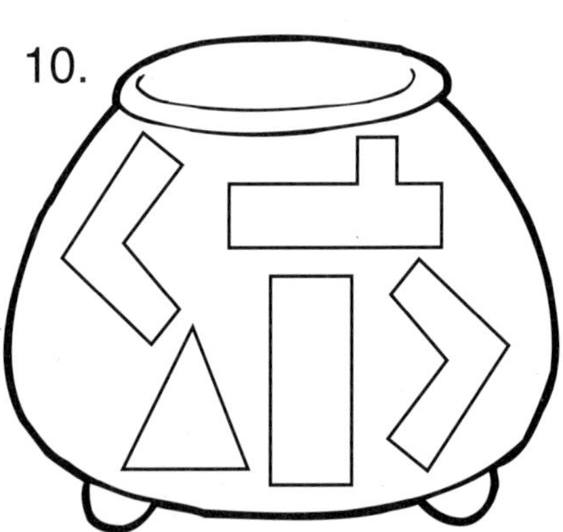

11\.
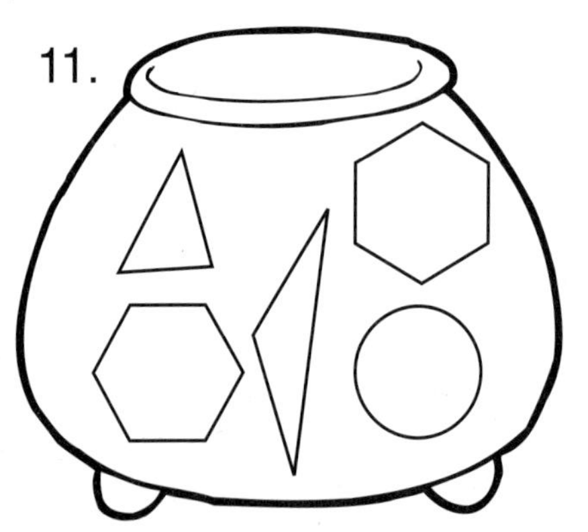

12\.
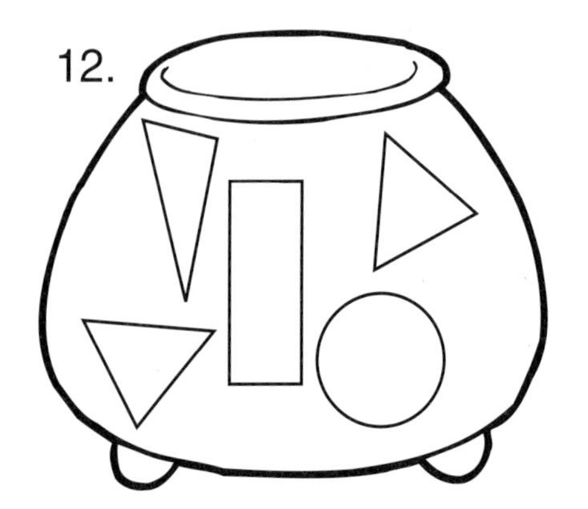

# Symmetry

Symmetry

## Round and Round

Introduce symmetry by using a familiar manipulative for a new purpose! Give each child a fraction circle that represents fourths and one that represents a whole. Have each student place the fourths pieces atop the whole circle. Then direct him to find a line of symmetry in his circle by taking off two adjacent fourths pieces. Have him examine the pieces to determine whether the halves showing are a reflection of one another. Repeat with other fraction circles. What a fun round of symmetry practice!

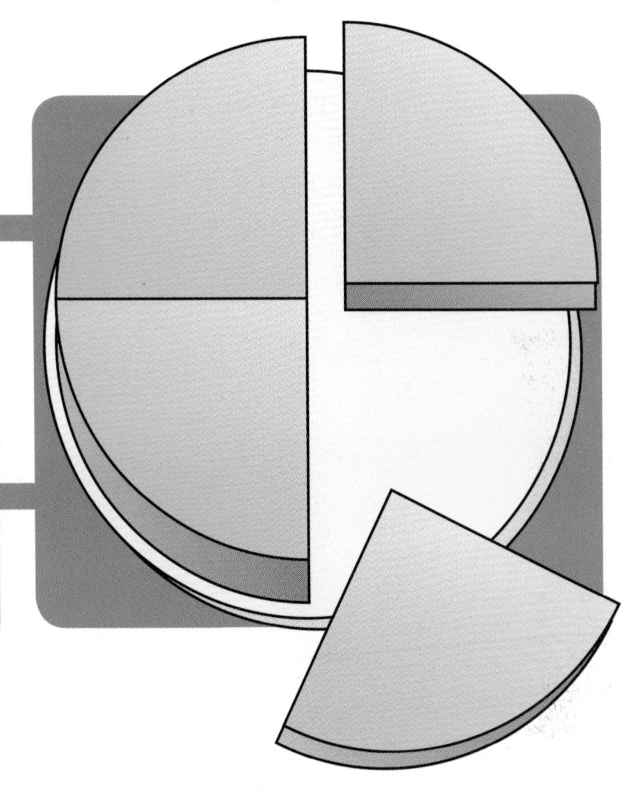

## Flutter and Fly

Here's a quick craft project that's sure to get students all aflutter about symmetry! Give each child two contrasting colors of construction paper, a 2" x 8" strip of brown or black construction paper, scissors, glue, and access to a hole puncher. Each student chooses one sheet of construction paper, folds it in half, and cuts out a butterfly shape. Next, while the paper is still folded, she uses the hole puncher to punch through both layers of paper to add a design to the butterfly. Then she cuts out a second butterfly shape that is slightly larger than the first. She unfolds the first shape and glues it onto the larger one. Finally, she uses the paper strip to add a butterfly body, a head, and antennae. Use yarn to hang the butterflies from a bulletin board titled "We're All Aflutter About Symmetry!"

## Snack Attack

Give symmetry a tasty twist with this "snack-tastic" activity! Give each child a plastic knife, a small piece of waxed paper, and a small snack cake such as a Twinkie cake. Show students that the cake has a vertical line of symmetry, and demonstrate by cutting your own cake on this line. After showing it to students, gently press the cake back together. Then show students that the cake also has a horizontal line of symmetry, and cut on that line. Discuss ways to cut the cake that would not result in symmetrical pieces. Then lead students through assorted cuts on their cakes. Once the lesson is learned, invite students to gobble up their symmetrical snack. If desired, repeat with mini doughnuts, cupcakes, fruit slices, or other symmetrical treats. Yum!

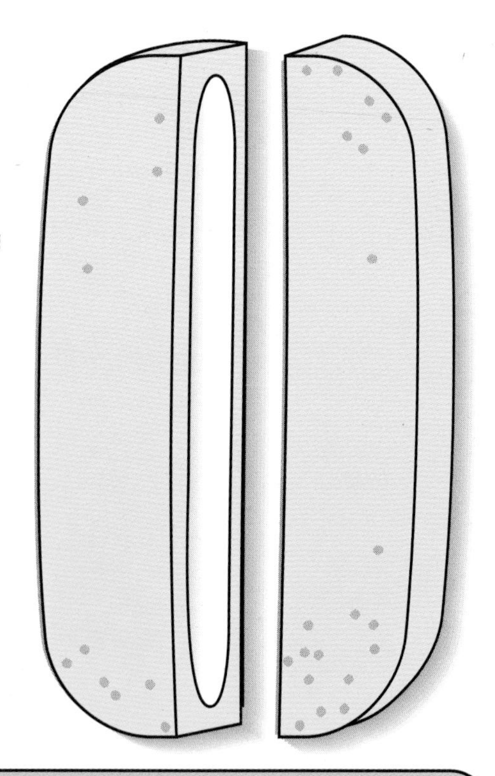

## Pizza, Please!

Serve up a slice of symmetry with this made-to-order activity! Discuss with students that a circle has an infinite number of lines of symmetry. Demonstrate by folding and unfolding a paper circle in half several times in different ways. Then explain that one way to establish a single line of symmetry is to add characteristics to the circle. Distribute tan, red, and yellow construction paper; several other colors of paper scraps; scissors; and glue. Challenge students to create paper pizzas that have just one line of symmetry. Instruct each child to cut a large circle from the tan paper (crust), a slightly smaller circle from the red paper (sauce), and a slightly smaller circle from the yellow paper (cheese).

Next, to make toppings, each student folds a construction paper scrap in half and cuts through both layers to make two identical pieces. Then she folds her cheese circle in half, creases the paper to create a line of symmetry, and unfolds it. To add the toppings in a symmetrical pattern, the student dabs glue on one side of the top of the cheese circle and then gently refolds the circle so that some of the glue transfers to the other side. Then she unfolds the circle and presses one of the two identical toppings onto each spot of glue. She repeats this process until all of the toppings have been placed. Finally, she layers her cheese, sauce, and crust circles and glues them together. Invite students to compare their pizzas and show that each has only one line of symmetry. Special order coming right up!

Find more student practice on page 137.

Name ______________________ Date ______________

# This Little Piggy Went Swimming

Cut out the object cards at the bottom of the page.
Glue the objects that are symmetrical in the pool.
Glue the objects that are not symmetrical outside of the pool.

# Transformations

## Ready to Flip?

Grab these kid-created card games to teach students about flips, slides, and turns! Give each student six different pattern block pieces, six index cards, and access to a supply of die-cut pattern block cutouts. Using an overhead projector, place one shape on the projector and demonstrate a flip. Then have each student choose a pattern block piece and demonstrate a flip. Repeat with slides and turns.

To make a game card, the student horizontally positions an index card. He glues a shape near the left edge of the card and labels the bottom of the card "slide." Then, on the back of the card, he glues a second shape in the slide position, near the right edge of the card. He initials the back of the card. Have him repeat the steps so that he makes two cards each for slides, flips, and turns. Check students' cards.

To play the game, pair students. Each player stacks his cards so that the answers are facedown. Player 1 chooses a card from the other player's stack and uses a pattern block in that shape to perform the transformation. Then he turns the card over to check the result. If the answer is correct, he keeps the card. Play continues until each player has captured the other player's cards. What terrific transformations!

slide

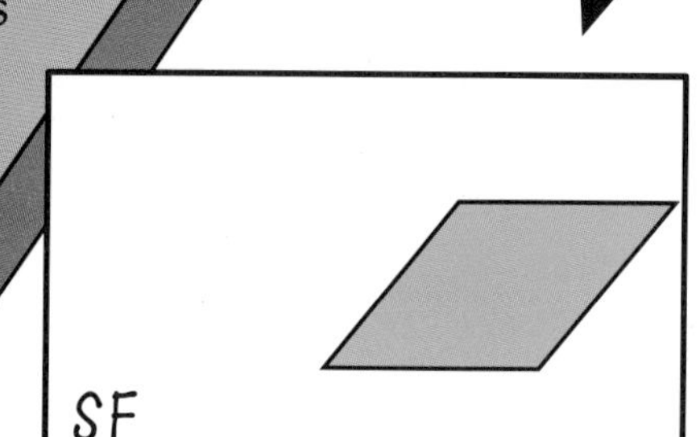


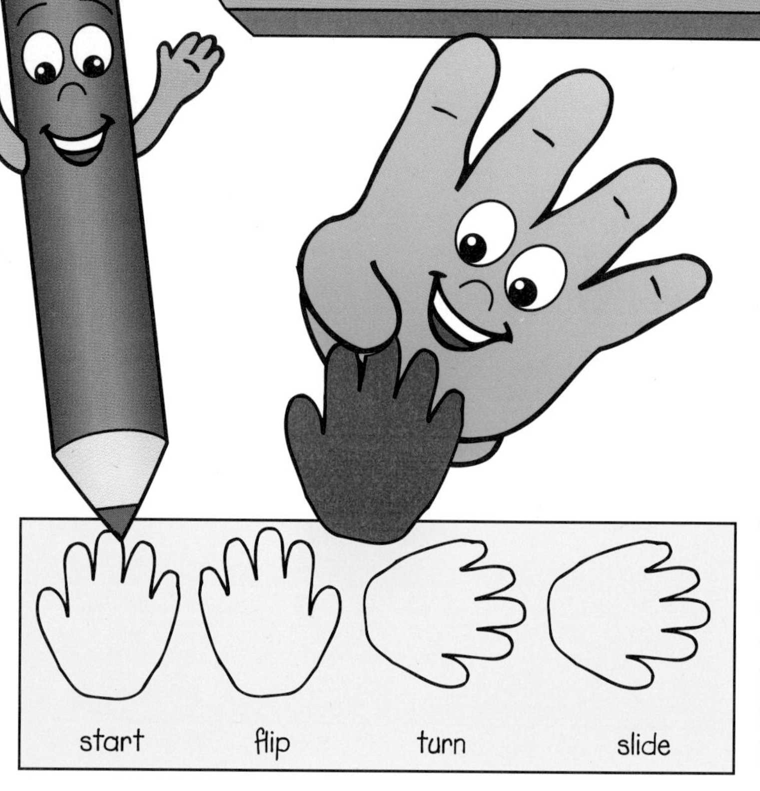


## Give 'em a Hand

Invite each student to try her hand at flips, slides, and turns! Give each child a 6" x 6" tag board square and a 6" x 18" construction paper rectangle. Each child traces her hand on the tagboard square and cuts it out to make a template. Next, she traces the template on the left-hand side of the construction paper and labels it "start." Then direct her to flip, slide, or turn the template and then trace it again and label it. Have her repeat this step several times. Post the tracings on a bulletin board titled "Give Us a Hand! We Know Our Flips, Slides, and Turns!"

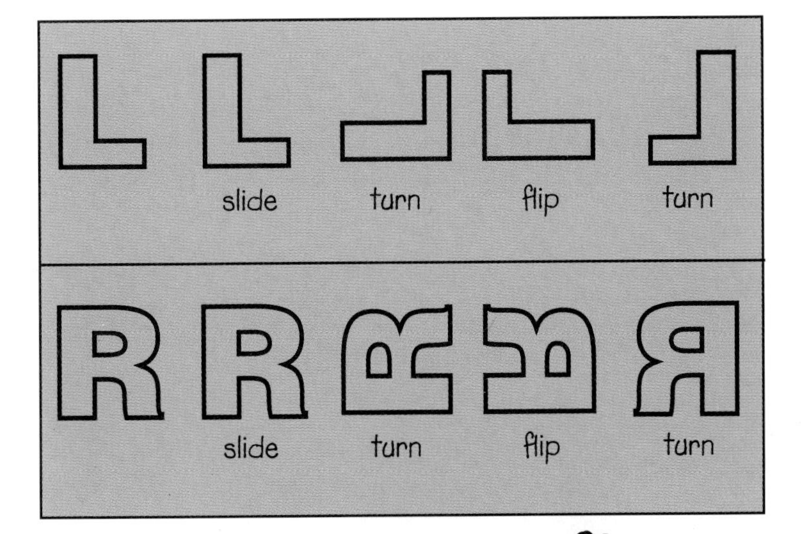


## Personalized Math

With a flip, slide, or turn of their initials, students create designs that are all their own! Provide each student with a 12" x 18" sheet of construction paper, crayons, and either stencils or die-cut letters of her initials. The child folds the construction paper to 6" x 18" and then unfolds it and places it on her desk horizontally. In the upper left corner, she traces the initial of her first name. Then instruct her to flip, slide, and turn the letter, tracing the letter each time until the top half of the paper is filled. Have her repeat this process on the bottom half of the paper with the initial of her last name. She colors her design to further personalize it. What letter-perfect practice!

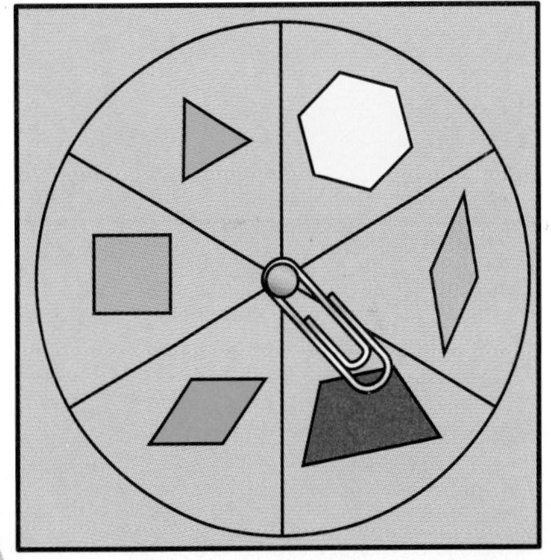

## It's Your Spin!

Take students for a spin—and practice flips, slides, and turns—with this small-group game. Ahead of time, make a game spinner that shows a pattern block shape in each space. Then make another spinner with three spaces: one labeled "flip," one labeled "slide," and one labeled "turn." Finally, make a recording sheet that lists each shape down the side and each transformation at the top and add lines to create a grid. Make a class supply of the page. Place the copies and a set of pattern blocks in the game area.

To play the game, a child spins the pattern block spinner and chooses that shape from the supply. Then he spins the transformations spinner. He uses his block to demonstrate the transformation that he spins. If his partners agree that he performed the transformation correctly, he marks off that square on his grid. Students take turns spinning and demonstrating. The child who has marked off the most spaces on his grid after a specified number of turns is the winner. Let's spin some more!

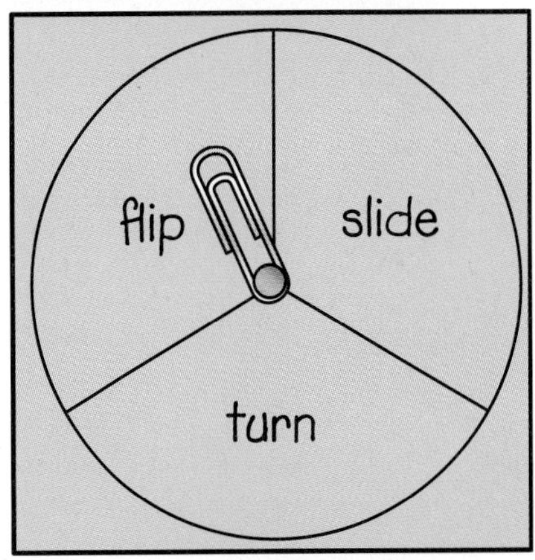


Find more student practice on page 140.

Name ____________________ Date ____________

# Number Line Dancing

Study the examples.
Write "flip," "slide," or "turn" between each pair of pictures.

flip slide turn

A.   ________ 

F.  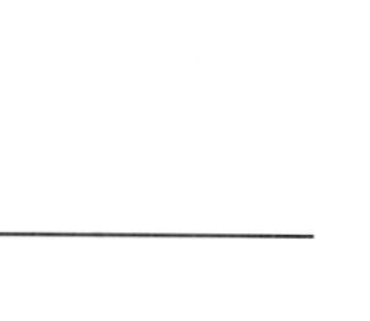 ________ 

B.   ________ 

G.   ________ 

C.   ________ 

H.   ________ 

D.   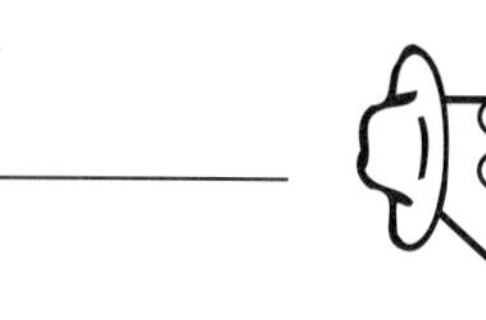 ________ 

I.   ________ 

E.  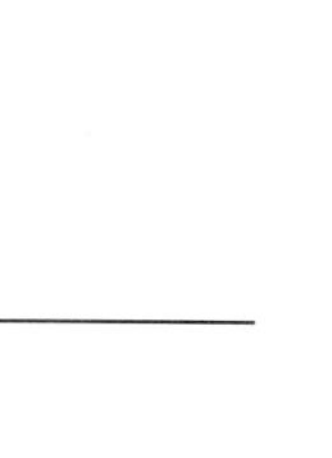 ________ 

J.  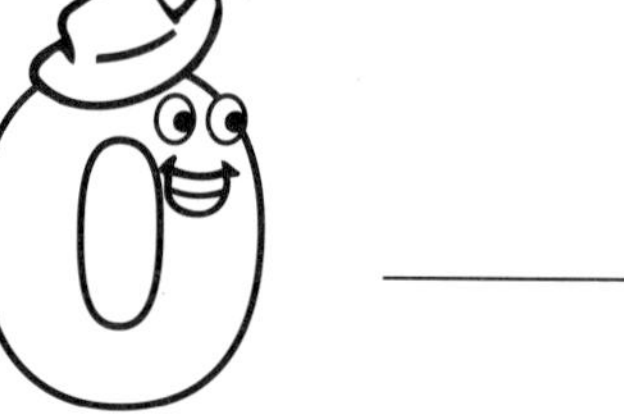 ________ 

# Measuring Length

## Stringin' Along

Tie up nonstandard measurement skills with this simple partner activity. Give each pair of students a foot-long length of string, a piece of paper, and a pencil. Instruct each duo to choose five classroom items and compare the length of the string to each item. For each item, direct the pair to record the item's name and whether it is longer, shorter, or the same length as the string. Later, have each pair share its findings with the class as you record them on a poster like the one shown.

1. table–longer than the string
2. pencil–shorter than the string
3. ruler–same length as the string
4.

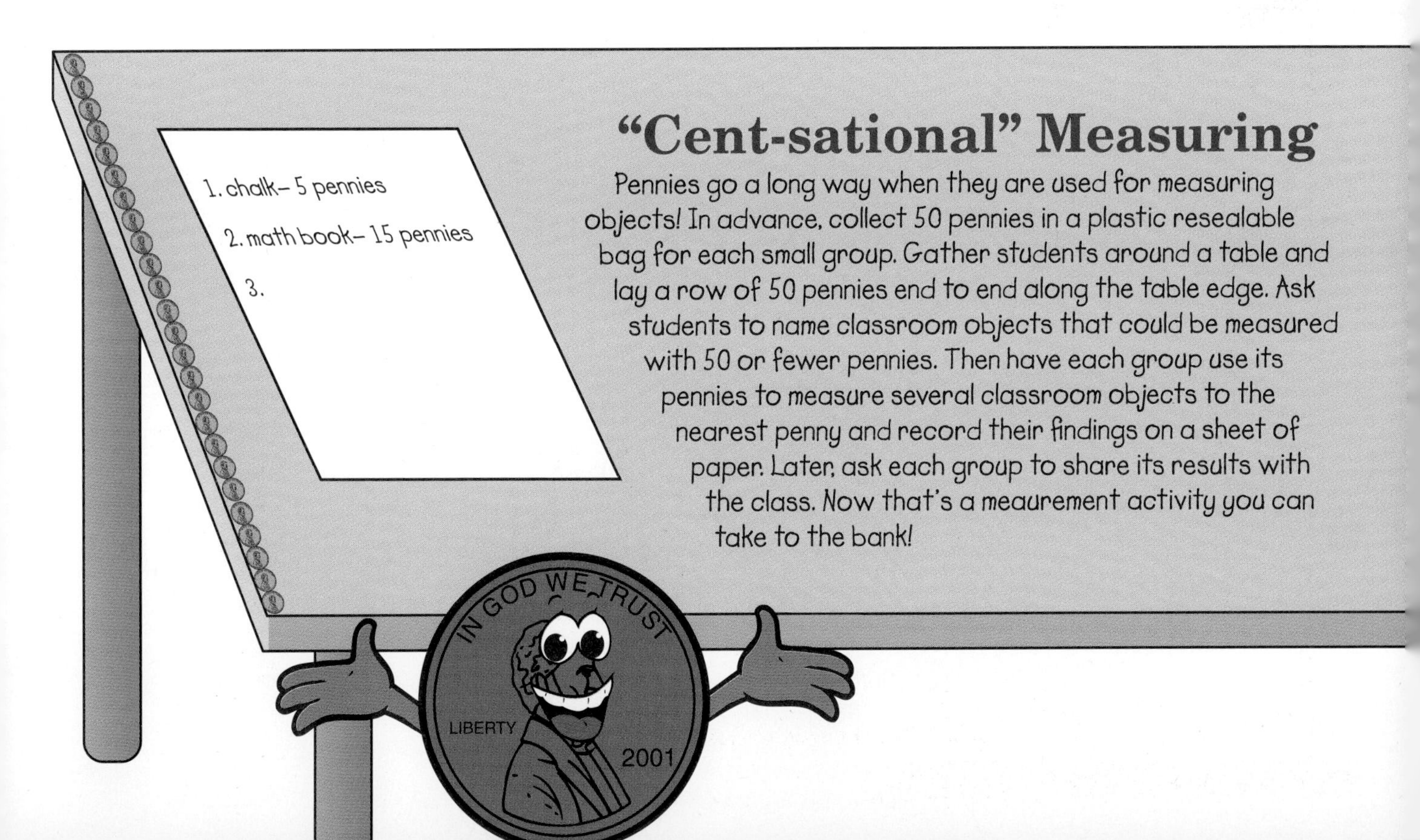


## "Cent-sational" Measuring

Pennies go a long way when they are used for measuring objects! In advance, collect 50 pennies in a plastic resealable bag for each small group. Gather students around a table and lay a row of 50 pennies end to end along the table edge. Ask students to name classroom objects that could be measured with 50 or fewer pennies. Then have each group use its pennies to measure several classroom objects to the nearest penny and record their findings on a sheet of paper. Later, ask each group to share its results with the class. Now that's a meaurement activity you can take to the bank!

# Partner Measurement

Provide plenty of hands-on measurement practice with this partner activity. Create a recording sheet similar to the one shown and make a copy for each student pair. Give each twosome a copy of the recording sheet, measuring tape, and a pencil. Have each child, in turn, estimate the length of his partner's arm and record it in the corresponding area of the chart. Next, have each child measure his partner's arm and then record the results. Direct the pair to repeat the procedure with the remaining parts. When the chart is complete, encourage each pair to compare its estimations and the actual measurements.

**Partner Measurement**

| | Tim (Student) | | Frank (Student) | |
|---|---|---|---|---|
| | Estimate | Actual | Estimate | Actual |
| Arm | 75 in. | 20 in. | 40 in. | |
| Hand | | | | |
| Leg | | | | |
| Foot | | | | |

Tape Measure

6 inches

14 inches

27 inches

# Measuring and Matching

Matching these cards to yarn lengths gives your students great measurement practice! To prepare the center, give each child an index card, a length of yarn (shorter than 36 inches), scissors, and access to a yardstick. Direct the child to measure his yarn and cut it to a desired length, helping him measure to the nearest inch. Next, have the child write the measurement on his index card. Collect the cards and yarn lengths and display them at a center, along with the yardstick. Instruct each child who visits the center to use the yardstick to match each yarn length to an index card. Then have him choose a partner to check his work. Now that's an activity that really measures up!

# Measurement Hunt

Inches, feet, or yards—which is the best unit of measurement? Students will decide the answer to that question with this group activity! In advance, prepare a recording sheet like the one shown. Provide each group with a copy of the recording sheet, a pencil, and three lengths of string: one that is one inch, one that is one foot, and one that is one yard. Then challenge each group to find three classroom items for each length that would best be measured with that unit. Have each group record its items in the corresponding chart columns. Next, gather students and have each child label three cards as shown. Then have each group, in turn, name one object from its list and tell which length it chose. Direct each remaining student to hold up a card to indicate how he would measure the stated object. Compare students' choices with the group's and discuss the reasoning for the choices. Continue in this same manner until each group has shared all of its items.

| Which is best? | | |
|---|---|---|
| Inches | Feet | Yards |
| crayon<br>book | chair<br>pillow | table<br>door |

Feet

Inches

Yards

## Double Duty

Students take a spin at measuring with standard and metric units at this center. In advance, create a spinner and a recording sheet like the ones shown. Place the spinner and copies of the recording sheet at a center along with pencils, rulers (with inches and centimeters), a box containing at least ten objects of different lengths, and an answer key that tells the length of each object in inches and centimeters.

To use the center, each child in a pair chooses an object from the box. Player 1 spins the spinner and announces the measurement unit. Each player measures her object using the resulting unit and writes the object's name and its measurement on a copy of the recording sheet. Then each player returns her item to the box and chooses a different object. Player 2 spins and the duo repeats the activity. The pair continues in this manner until each child has spun five times and measured each object. Finally, have students use the answer key to check their work.

Name Janet

book 10 in.

marker 16 cm.

inches (in.)

centimeters (cm)

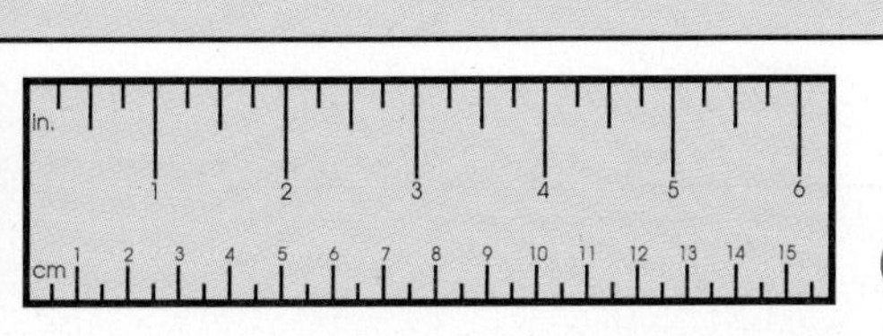

Find more student practice on pages 144–145.

Name ____________________ Date ____________________

# Run, Lion! Run!

How far is each football from the first down?
Use a ruler to measure to the nearest inch or half inch.
Write the answer on the football.

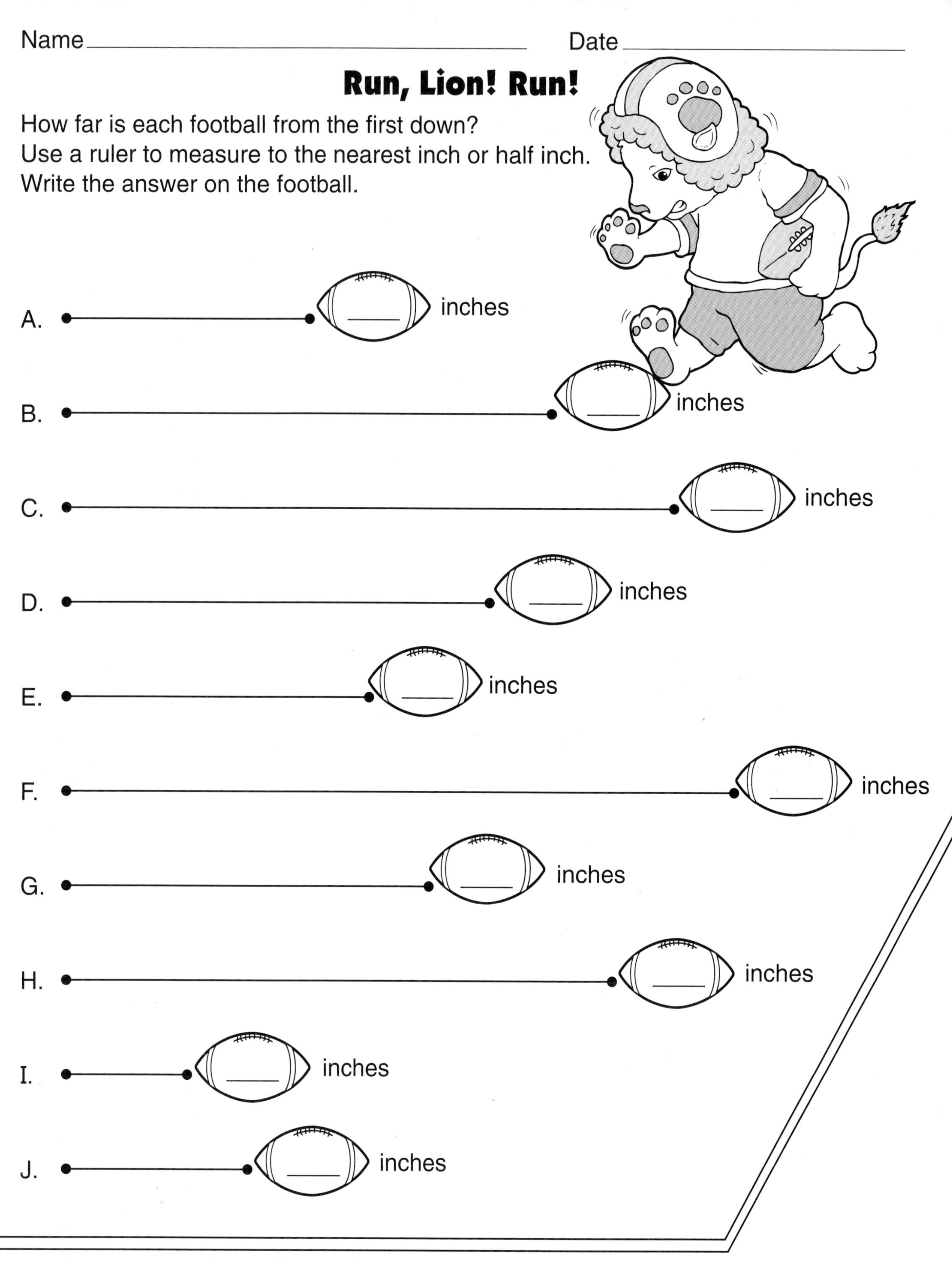

A. ______ inches

B. ______ inches

C. ______ inches

D. ______ inches

E. ______ inches

F. ______ inches

G. ______ inches

H. ______ inches

I. ______ inches

J. ______ inches

Name ____________________ Date ____________________

# For Good Measure

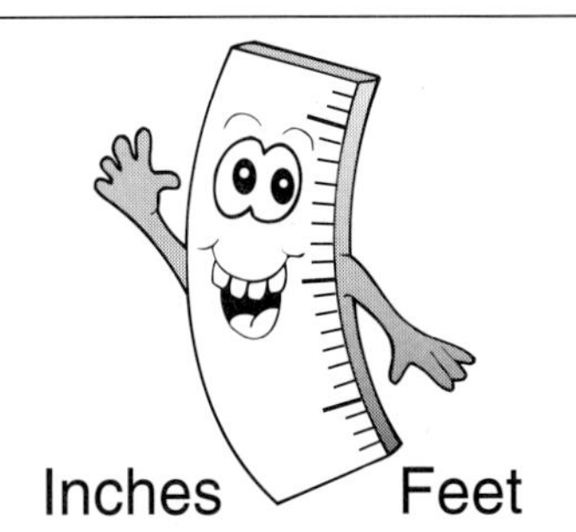

Choose the unit of measure that would work best.
Circle the letter in the matching column.

| | Inches | Feet |
|---|---|---|
| 1. length of a bed | G | A |
| 2. length of a pencil | S | P |
| 3. length of a glue stick | Y | M |
| 4. length of a piece of chalk | A | O |
| 5. height of a tree | N | L |
| 6. length of a book | D | Q |
| 7. length of a bus | P | A |
| 8. height of a door | C | T |
| 9. height of a student | O | R |
| 10. length of a spoon | E | P |
| 11. height of a house | F | A |
| 12. length of a paintbrush | Z | E |

**Where can you find a ruler that is three feet long?**
To solve the riddle, match the circled letters to the numbered lines below.

___ ___    ___    ___ ___ ___ ___    ___ ___ ___ ___
4 8    1    3 11 9 6    2 7 5 10

# Area and Perimeter

## Snaking the Perimeter

Introduce your youngsters to perimeter with this "sssuper" idea! Create a paper clip snake by clipping together 40 small paper clips. Present the snake to the class with great fanfare as you explain that it can help determine the perimeter of an object. Next, position the snake around a classroom object, such as a book. Ask a student volunteer to count the number of paper clips used; then record the measurement on the board. Then pair students and have each twosome make its own paper clip snake and use it to measure different classroom objects. To conclude, have students share their findings, comparing the measurements for any identical objects.

Ms. Perry's Shoe

| Space | Perimeter |
|---|---|
| reading center | 24 shoes |
| sandbox | 32 shoes |
| class garden | 40 shoes |
| hallway | |
| rug | |

Mr. Sampson (janitor)

Mrs. Henderson (principal)

Ms. Perry (librarian)

## Step by Step

Youngsters will put their best foot forward with this perimeter activity. In advance, trace a shoe from each of several school workers (such as the principal, the janitor, and the librarian) onto tagboard. Label each shoe pattern with the person's name and then cut it out. Divide students into groups and give each group a shoe cutout and a recording sheet like the one shown. Have each group use the cutout to measure the perimeter of each location and then record the findings on the sheet. Later, discuss the similarities and differences of each group's results. No doubt your students will be eager to repeat the activity using their own footprints!

# Pleasing Perimeter

This sweet measurement activity is a treat! Program a blank sheet of paper with several different sizes of rectangles or squares as shown. Give each child a copy of the sheet and a small resealable snack bag containing several small snack items such as M&M's candies or Cheerios cereal. Have each child line one shape with the snack as shown. Then instruct her to calculate the perimeter of the shape by adding the total number of snack items. Repeat the activity with each remaining shape. If desired, invite students to munch on their snacks as you check their work. Finding perimeter never tasted so good!

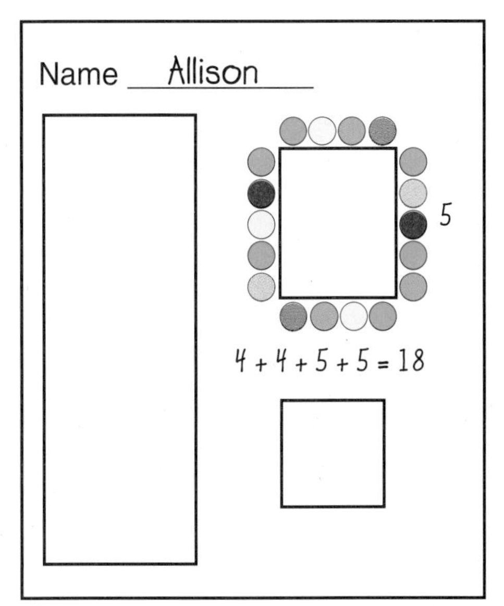

# Sizing Up Books

Book reviews and nonstandard measurement go together with this center activity. Create a recording sheet like the one shown and make a copy for each child. Each week, ask five students to each share a favorite book with the class. Number the books from 1 to 5. Then place the books, a ruler, the recording sheets, and an answer key at a center. For each book, a student estimates and measures its perimeter, recording the measurements on her paper. Then she uses the answer key to check her work before perusing these great book selections. Wow! What a great way to provide perimeter practice and encourage students to read!

Perimeter

| Book | Estimate | Actual |
|---|---|---|
| 1 | 36 | 20 |
| 2 | | |
| 3 | | |
| 4 | | |
| 5 | | |

## Cracker by Cracker

Introduce area with this tasty experience! Give each student a two-inch paper square, a four-inch paper square, and 16 Cheez-It crackers. To begin, have each child cover the smaller paper square with a layer of crackers and then count the squares to determine the area in square inches. Repeat the activity with the remaining paper square. Then invite students to munch on their crackers. Delicious!

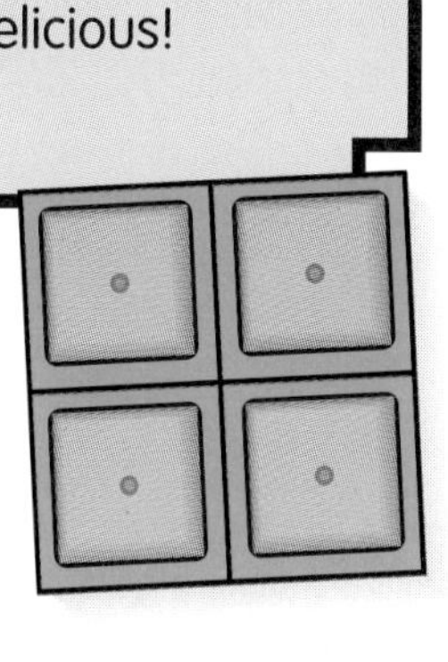

## Checkered Area

Check out this activity for reinforcing area and multiplication skills. In advance, cut a 2" x 4" paper rectangle, a 4" x 8" paper rectangle, and 32 one-inch paper squares for every two students. To begin, gather students around a checkers gameboard and remind them that the area of the gameboard can be determined by counting the squares. Guide students in counting the squares and record this number on the board. Next, have students count the squares along the length and the width of the board. Multiply the two numbers together and compare the two answers. Lead students to realize that the area of the board can also be found by multiplying its length by its width.

For practice, pair students and give each twosome the paper rectangles and squares. Instruct each pair to place the squares along the length and width of the smaller rectangle, as shown, and then determine the area of the rectangle by multiplying. Instruct the duo to check its work by covering the rectangle with squares and counting the total number. Then have each pair repeat the activity with the larger rectangle.

Find more student practice on pages 149–150.

Name ______________________ Date ______________

# All Around the Castle

Use a centimeter ruler to measure each side of each part of the castle.
Write the lengths.
Add to find the perimeter.

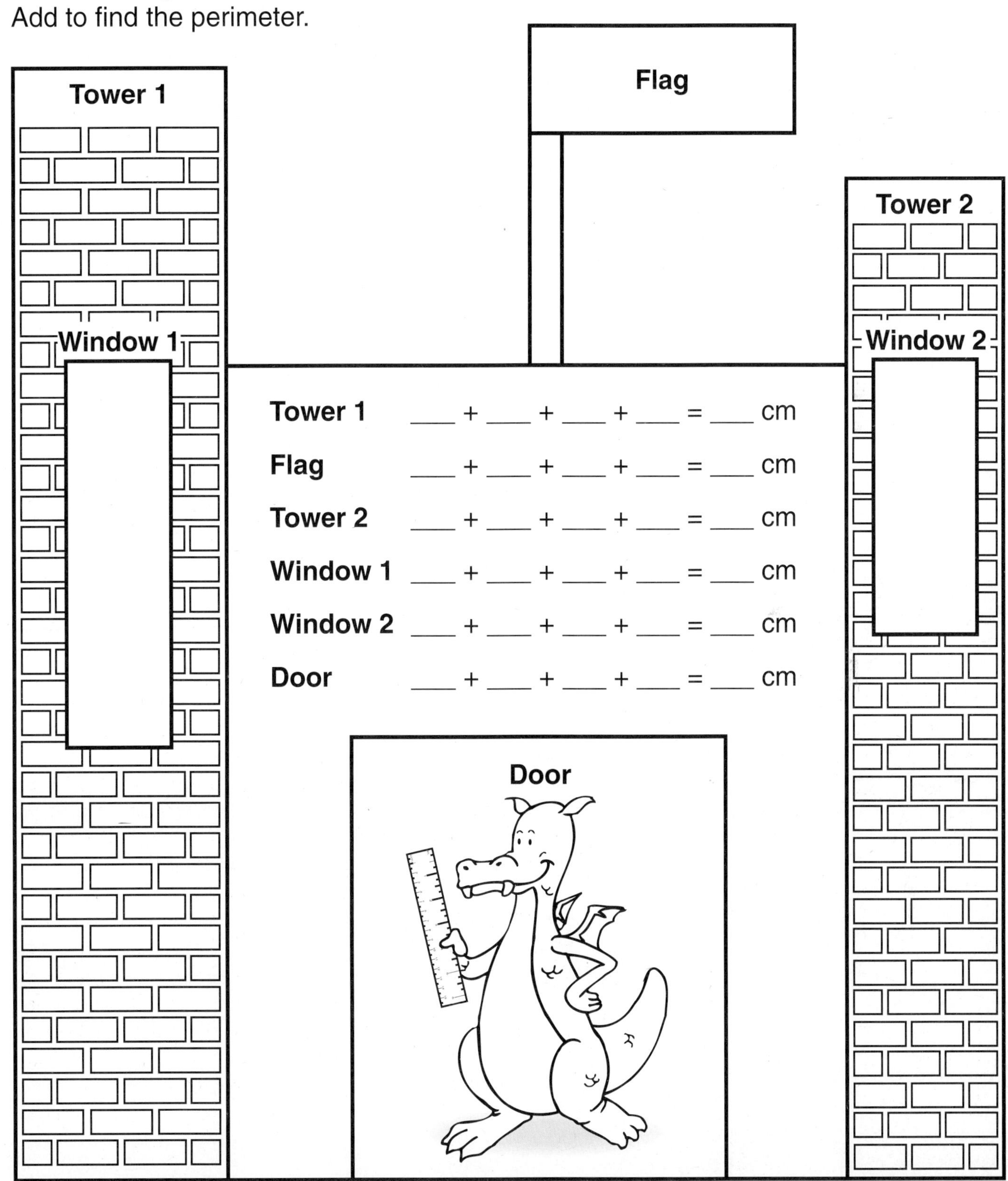


**Tower 1** ___ + ___ + ___ + ___ = ___ cm

**Flag** ___ + ___ + ___ + ___ = ___ cm

**Tower 2** ___ + ___ + ___ + ___ = ___ cm

**Window 1** ___ + ___ + ___ + ___ = ___ cm

**Window 2** ___ + ___ + ___ + ___ = ___ cm

**Door** ___ + ___ + ___ + ___ = ___ cm

Name ____________________ Date ____________

# A Quilting Bee

Write the length and width of each quilt block.
Find the area.
Write your answers on the chart.
Color the quilt by the code.

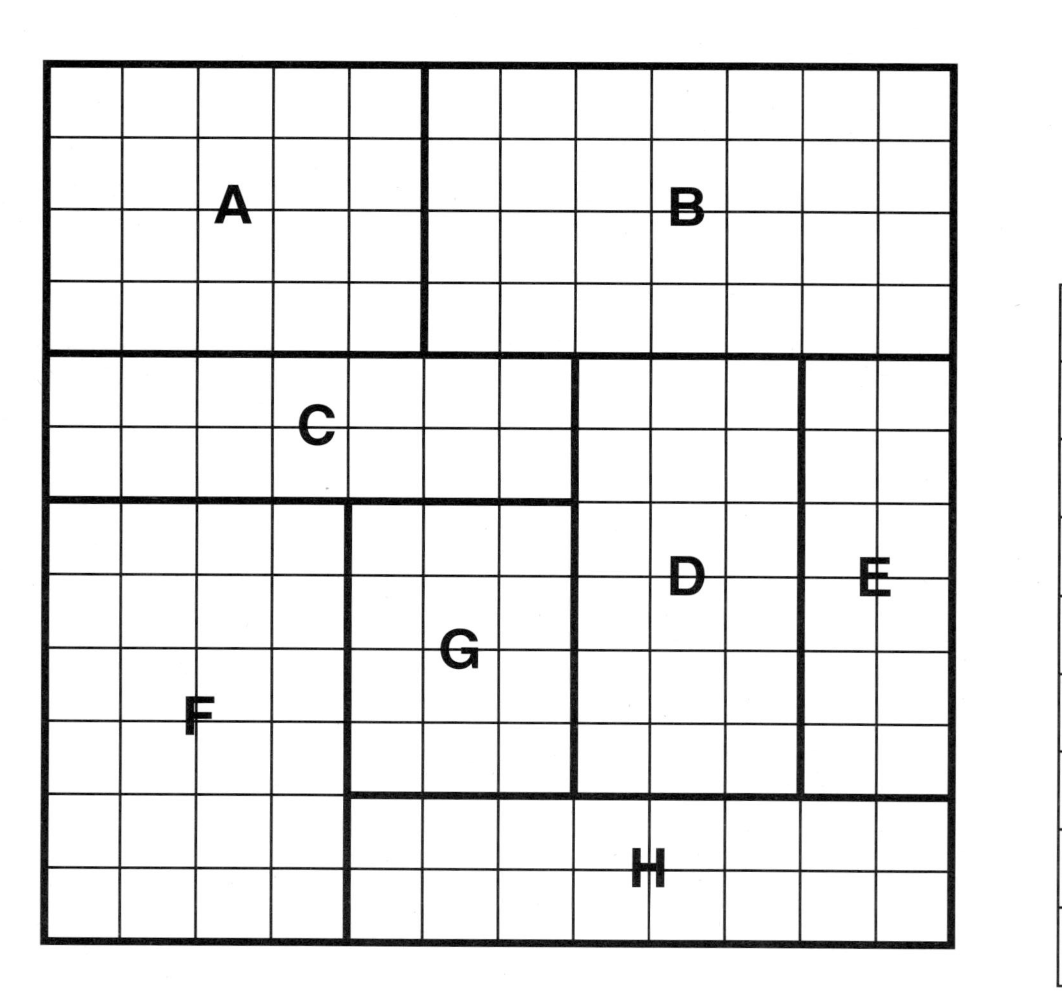


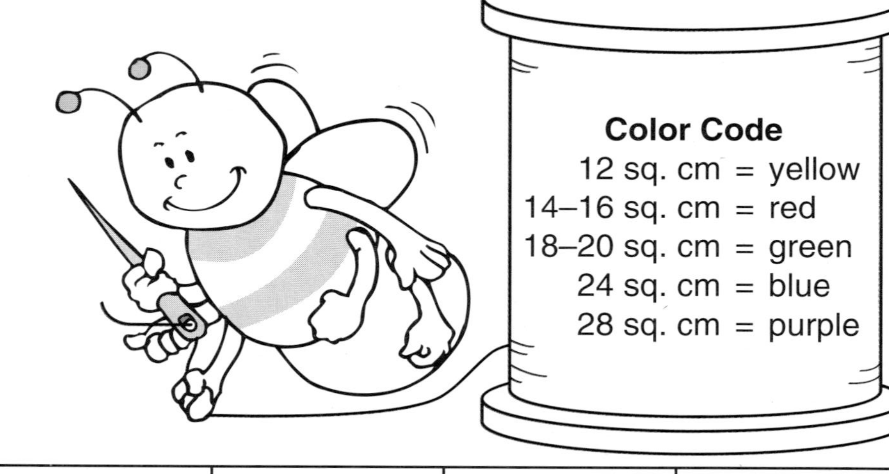


| Quilt Patch | Length | Width | Area |
|---|---|---|---|
| A | _____ cm | _____ cm | _____ sq. cm |
| B | _____ cm | _____ cm | _____ sq. cm |
| C | _____ cm | _____ cm | _____ sq. cm |
| D | _____ cm | _____ cm | _____ sq. cm |
| E | _____ cm | _____ cm | _____ sq. cm |
| F | _____ cm | _____ cm | _____ sq. cm |
| G | _____ cm | _____ cm | _____ sq. cm |
| H | _____ cm | _____ cm | _____ sq. cm |

## More or Less?

Give students a feel for a one-pound weight with this small-group activity. To prepare, obtain a balance scale, several objects to be weighed (some weighing a pound, some more than a pound, and some less than a pound), and a one-pound weight for students to use as a benchmark. To begin, circulate the one-pound weight among students. Then use the scale to balance two one-pound items against each other, preferably items of different sizes. Point out to students that the scales balance and confirm that each item weighs one pound. Lead students to understand that one pound of weight can come in a variety of shapes and sizes.

Next, have each student prepare a three-column chart labeled with the headings shown and ask him to list the remaining objects in the first column. Circulate the objects among students. After each child writes a guess as to whether each object weighs more or less than a pound, have volunteers help you weigh each object against the one-pound weight. Instruct each youngster to list on his paper whether the actual weight was more or less than a pound and discuss whether his guess was correct or incorrect.

| More or Less Than a Pound | | |
|---|---|---|
| Object | Guess | Actual |
| pencil | less | less |

## An "Egg-cellent" Weigh Station

| Name Joshua | | |
|---|---|---|
| Recording Sheet | | |
| Nonstandard Unit of Measure | Estimate | Actual |
| cubes | 6 | 11 |

Students will scramble to participate in this activity! Set up five or six stations in your classroom. In each station place a hard-boiled egg, a balance scale, and a different nonstandard unit of measure, such as crayons, cubes, pennies, or paper clips. Provide each child with a recording sheet similar to the one shown. Divide students into as many small groups as you have stations. At a station, have each student estimate how many units will equal the weight of the egg and then record it on his sheet. Have one student place the egg and a number of measuring units on the scale until it is balanced. Then have a different group member count the units out loud to determine the weight of the egg. Have each group member record this number on his paper. At your signal, have the groups rotate to different stations and repeat the activity, continuing in this manner until each group has visited each station. Encourage students to apply the knowledge they have learned from each previous weighing as they make their estimates. Then gather all the groups and compare the results. Now that's some Grade A practice with weight!

## Lightest to Heaviest

This center activity puts students' ability to recognize different weights to the test. In advance, collect six identical opaque margarine tubs and lids. Fill each tub with objects, making sure that the types of objects used are all of different weights, such as stones, gram weights, feathers, paper clips, sand, and macaroni. Cover each tub with its lid and tape it securely. Letter the bottom of the tubs from lightest to heaviest beginning with the letter *A*. To use the center, a student chooses two tubs. She holds one tub in each hand to determine the heavier tub. She continues comparing tubs, lining them up in order from lightest to heaviest. To check her order, she carefully flips the tubs over. If the letters *A, B, C, D, E, F,* and *G* appear in order, she has ordered them correctly.

## Ounces or Pounds?

Here's a fun activity for reinforcing units of weight. Create a set of picture cards by having each child cut from old magazines a picture of an object that should be weighed in ounces and a picture of an object that should be weighed in pounds. Have her glue each picture to a separate blank card and then write "ounces" or "pounds" on the back of each card to indicate the correct unit of measurement. Collect and shuffle the cards. Then give each child another blank card and have her write "ounces" on one side of the card and "pounds" on the other. To begin the activity, display a picture card and have each student hold up her card with "ounces" or "pounds" showing to indicate whether the object should be weighed in ounces or pounds. Scan students' cards for accuracy. Then continue the activity with the remaining cards.

**Find more student practice on pages 153–155.**

Name________________________________ Date________________

# A Cluttered Counter

Study each item on the counter.
If the item weighs more than 1 pound, color it red.
If the item weighs less than 1 pound, color it blue.

Remember: A box of spaghetti weighs 1 pound!

 = 1 pound

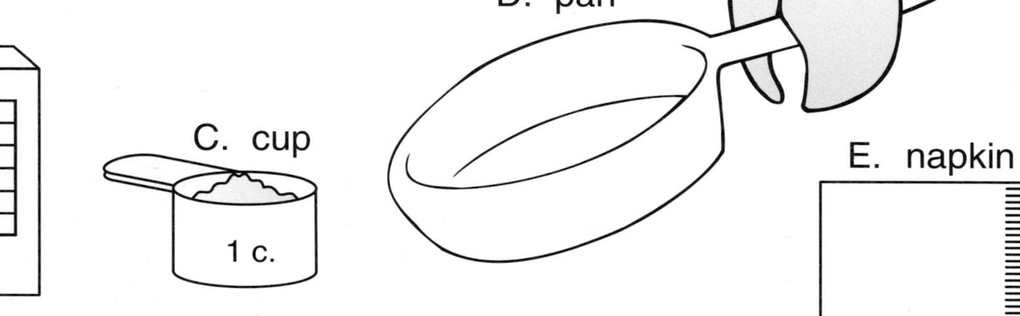

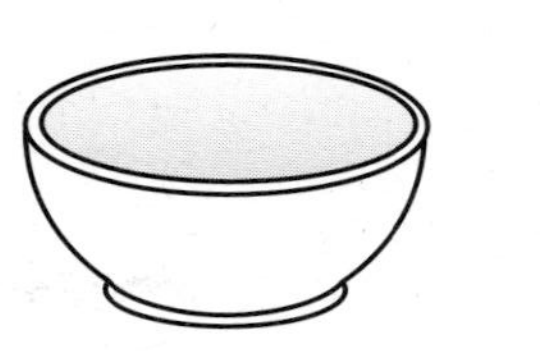

B. spoon

A. microwave

C. cup

1 c.

D. pan

E. napkin

F. blender

G. toaster

H. bowl

I. spatula

J. bowl of potatoes

K. fork

Name________________________ Date______________

# This Week's Specials

Think about which unit you would use to weigh each item.
Write "ounces" or "pounds" under each item.

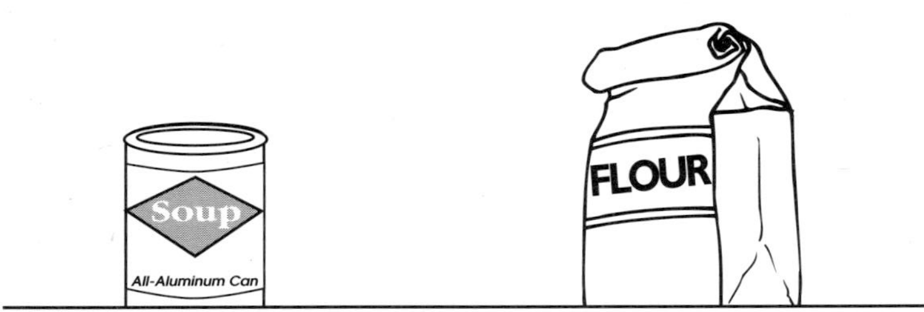

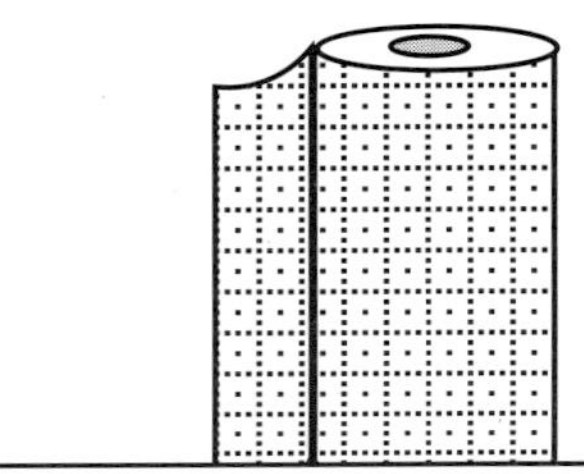

A. can of soup
12________

B. bag of flour
5________

C. shredded cheese
8________

D. freshly baked roll
5________

E. paper towels
10________

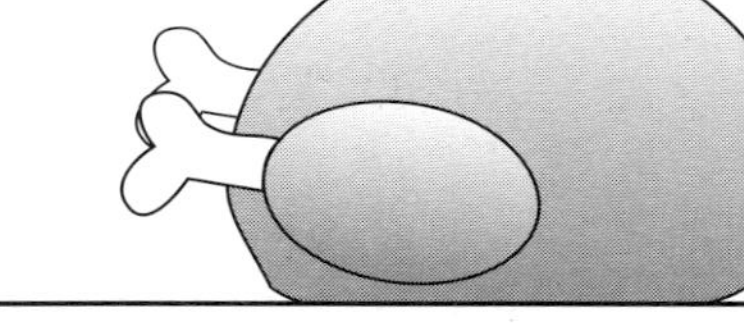

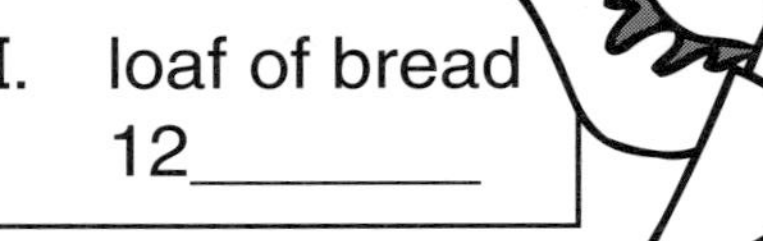

F. box of cereal
14________

G. peanut butter
18________

H. frozen turkey
10________

I. loaf of bread
12________

J. box of tissues
10________

K. bag of apples
5________

L. bag of dog food
7________

Name ______________________ Date ______________

# At the Garage

Cut out each box.
Glue the box in the matching column.

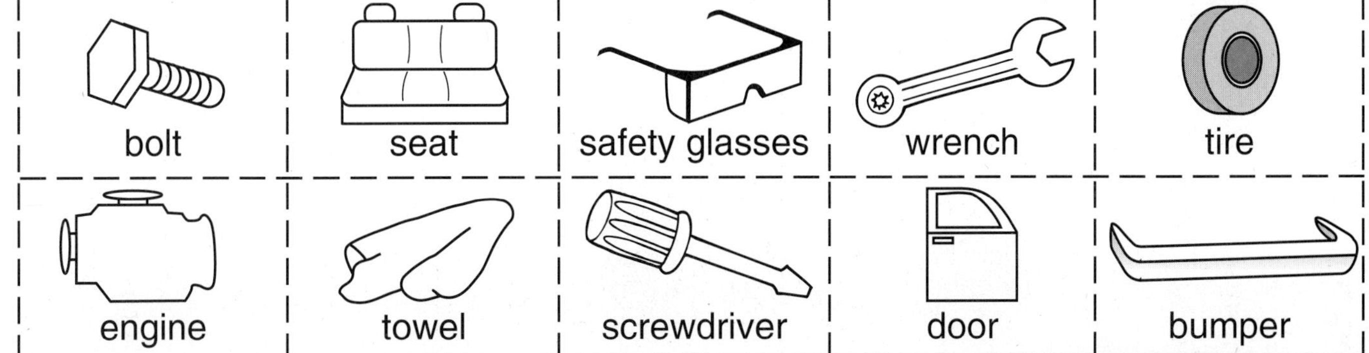

# Capacity

## A Cup Is a Cup

Pour on capacity learning with this comparison activity. Gather several plastic or metal one-cup measuring cups that are as varied in shape as possible. Place the cups in a sink or water table. Have one child in a pair measure in one of the cups exactly one cup of water. Next, instruct his partner to carefully pour the water into a different measuring cup. Have the partners take turns pouring the water into different cups until they have tested each cup. No matter the shape of the container, one cup of water is still one cup!

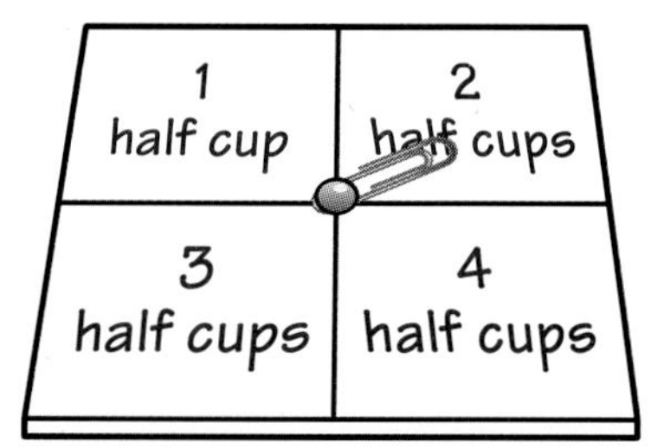


## Half a Cup

This activity on measuring by half cups is overflowing with practice! Provide each pair of students with a small bowl, a plastic measuring cup with half-cup increments clearly marked, a tray or cookie sheet, access to water, and a spinner similar to the one shown. Have each pair place its bowl on the tray (to catch spills). Instruct each child to spin the spinner, in turn, and then pour a matching number of half cups of water into the bowl. Play continues alternating between the two students until one child pours his water and makes the bowl overflow. This child is declared the winner.

# Does It Fit?

Fill your students with capacity explorations at this partner center. Gather six clean, empty containers in a variety of standard measures, such as two-cup storage containers, one-cup yogurt cups, half-cup pudding cups, and quarter-cup condiment cups. Label each of the containers with a different number from 1 to 6 and line them up at a center along with a die, a small plastic measuring cup, a bowl of uncooked rice, and paper. While at the center, a student rolls the die to determine which container to focus on. Next, he estimates the amount of rice the container will hold. To test his guess, he uses the measuring cup to measure that quantity of rice before pouring it into the container. If his estimate is accurate, he records it on his paper. If not, he adjusts his estimate to determine the container's capacity before recording his findings on his paper. Then his partner rolls and repeats the activity with a different container. Continue until all containers have been used. Now that's an idea that really measures up!

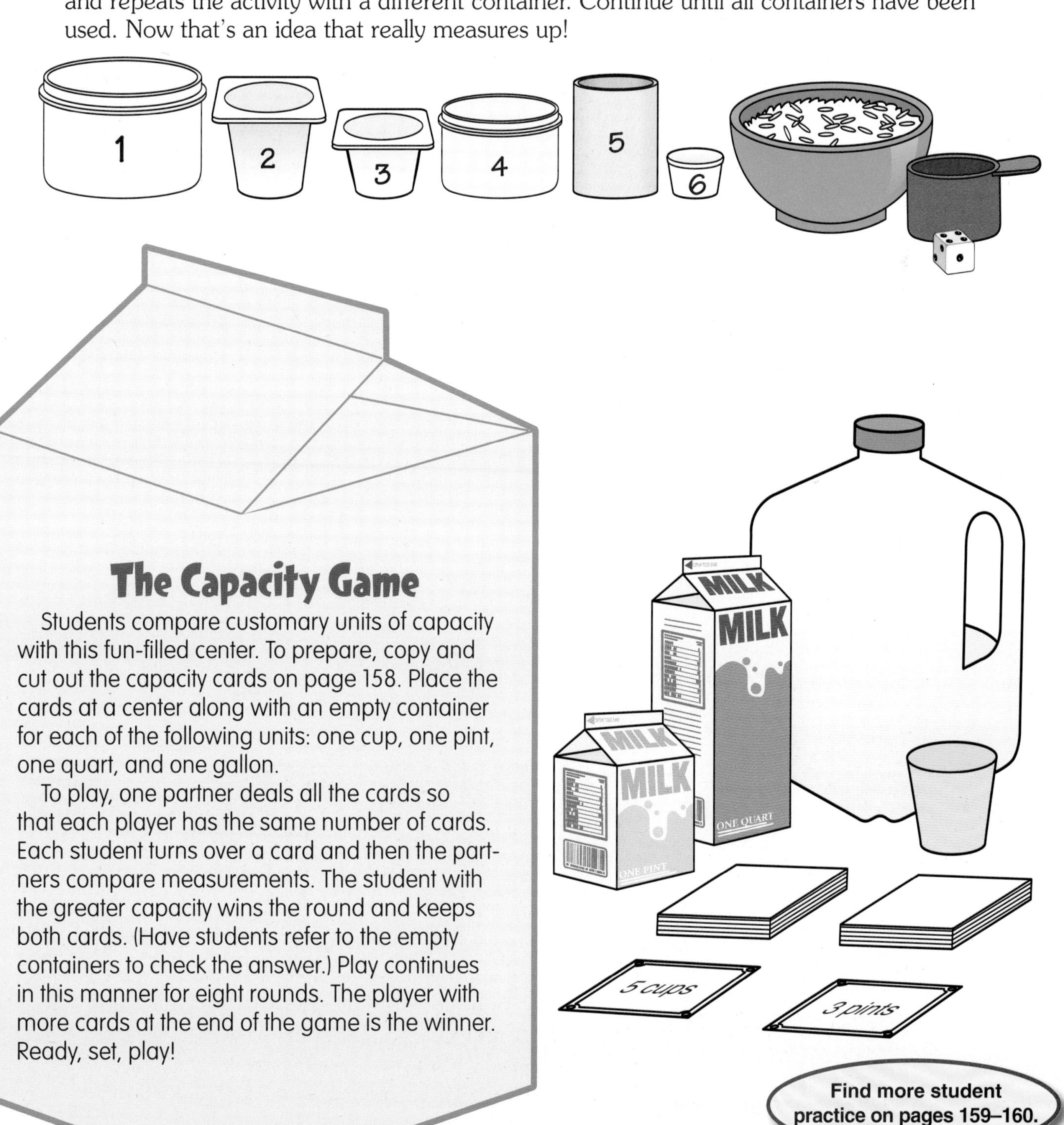


## The Capacity Game

Students compare customary units of capacity with this fun-filled center. To prepare, copy and cut out the capacity cards on page 158. Place the cards at a center along with an empty container for each of the following units: one cup, one pint, one quart, and one gallon.

To play, one partner deals all the cards so that each player has the same number of cards. Each student turns over a card and then the partners compare measurements. The student with the greater capacity wins the round and keeps both cards. (Have students refer to the empty containers to check the answer.) Play continues in this manner for eight rounds. The player with more cards at the end of the game is the winner. Ready, set, play!

**Find more student practice on pages 159–160.**

## Capacity Cards

Use with "The Capacity Game" on page 157.

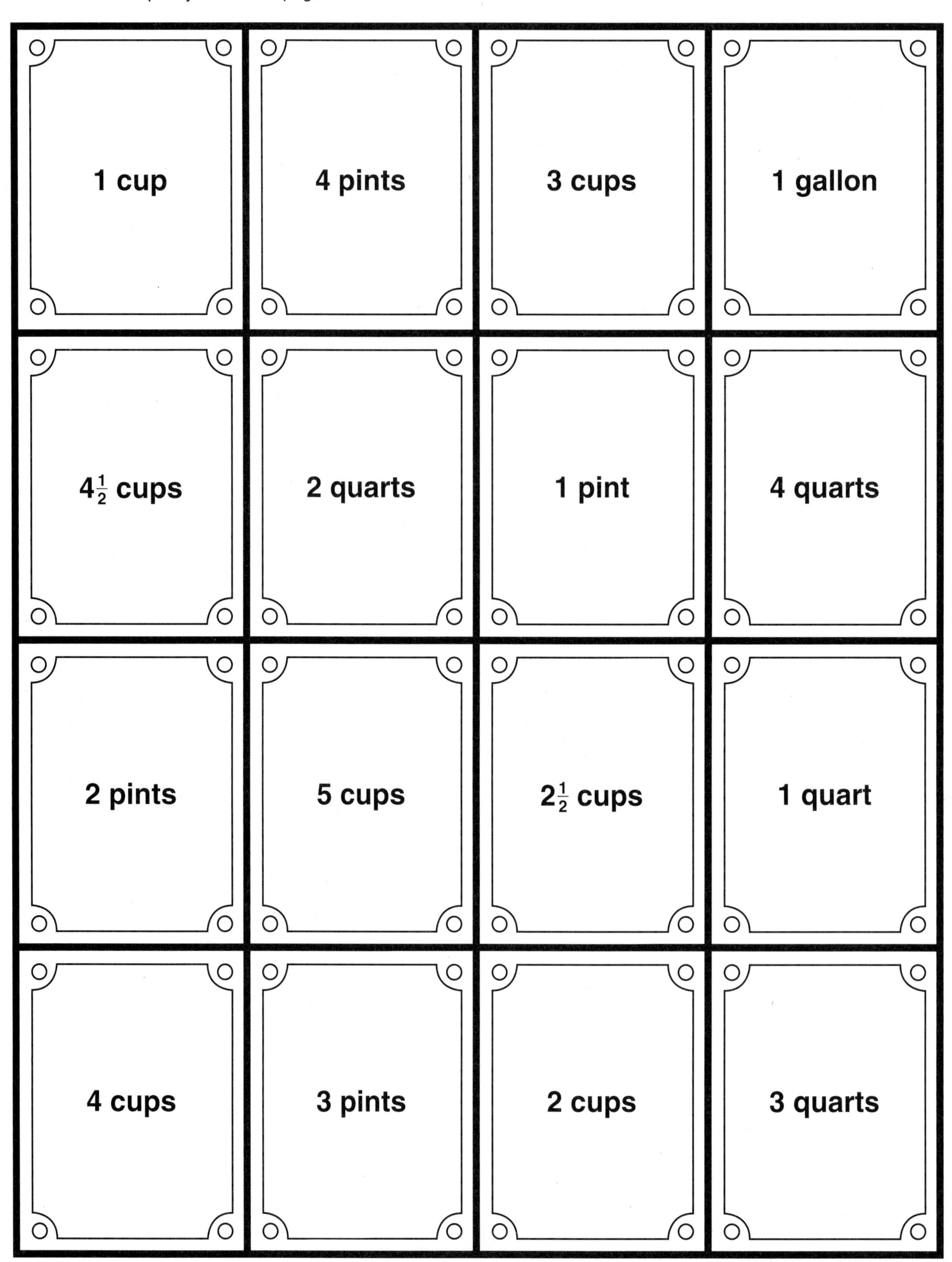

Name ______________________ Date ______________

# Plenty of Room

Choose the unit you would use to measure the liquid in each item.
Write "ml" for milliliter.
Write "L" for liter.

| 1. ______ | 2. ______ | 3. ______ | 4. ______ |
|---|---|---|---|
| 5. ______ | 6. ______ | 7. ______ | 8. ______ |
| 9. ______ | 10. ______ | 11. ______ | 12. ______ |

Home for Sale!

WELCOME

Name ______________________ Date ______________

# Lots to Paint

Circle the correct unit of measurement.
Color each paintbrush by the code.

cup pint quart gallon

**Color Code**
cup = red
pint = blue
quart = green
gallon = yellow

| Item | Unit | |
|---|---|---|
| 1. sink | pint or gallon | |
| 2. pool | gallon or quart | |
| 3. dog bowl | pint or gallon | |
| 4. mug | quart or cup | |
| 5. baby bottle | pint or gallon | |
| 6. teakettle | cup or quart | |
| 7. small bowl | pint or gallon | |
| 8. barrel | quart or gallon | |
| 9. drinking glass | quart or cup | |
| 10. large fish tank | pint or gallon | |
| 11. bathtub | gallon or quart | |
| 12. pitcher | pint or quart | |

# Temperature

## Picturing the Temperature

These picture prompts will encourage students to estimate approximate temperatures. Stock a center with several old magazines, drawing paper, scissors, glue, markers, and copies of either the Fahrenheit or the Celsius thermometer from page 162. To use the center, a child selects a magazine picture that implies a predictable temperature (such as a family playing in the snow or children at the beach). He cuts out the picture, glues it onto a sheet of provided paper, and colors the thermometer to represent a temperature that is appropriate for the activity shown in the picture. Then he glues the thermometer onto the page and writes about the temperature. Display the finished pages to use for a quick temperature review!

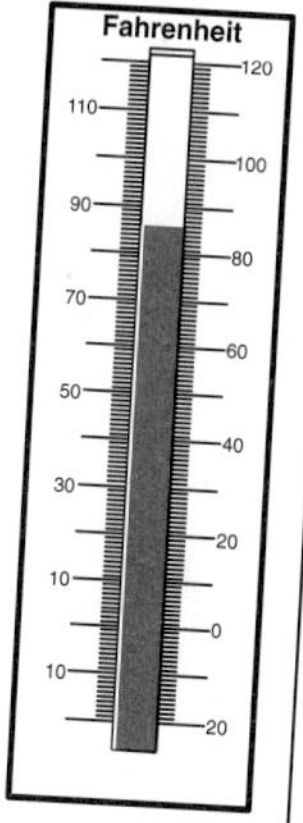


These kids are swimming. It's 86°F.

## Highs and Lows

This daily activity is just right for comparing and contrasting temperatures! Plus it can also lead to a geography lesson! In advance, post a U.S. map or, if desired, a map of a smaller area such as your state. Each day use a national newspaper or Web site to obtain the previous day's high and low temperatures in the United States (or your state). On the map, plot the locations that had the high and low temperatures and then tack a copy of a thermometer from page 162 near each location. Ask a different student volunteer to color each thermometer to show the designated high or low temperature. Then enlist students' help to determine the difference between the high and low temperatures, or have them compare how much colder or warmer the locations were than your local temperature listing. Wow! It was 22 degrees colder in Plattsburgh than it was here!

Canada
Plattsburgh
Vermont
New York
Utica
Schenectady
Albany
Binghamton
Connecticut
Pennsylvania
New York
High = 32° F
Low = 10° F
January 6, 2005

Find more student practice on page 161.

## Thermometers

Use with "Picturing the Temperature" and "Highs and Lows" on page 161.

### Fahrenheit

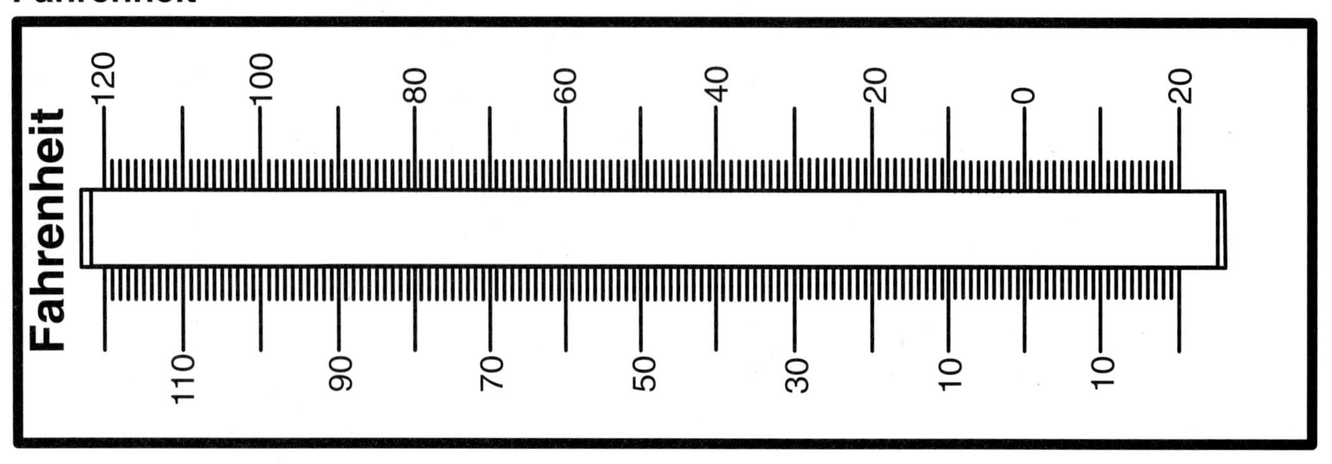


### Celsius

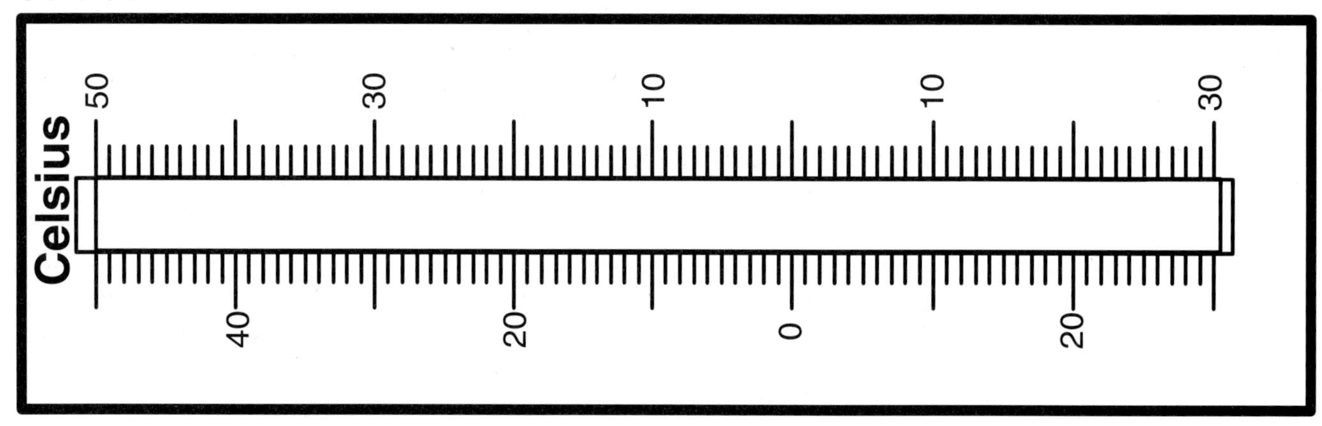


### Fahrenheit and Celsius

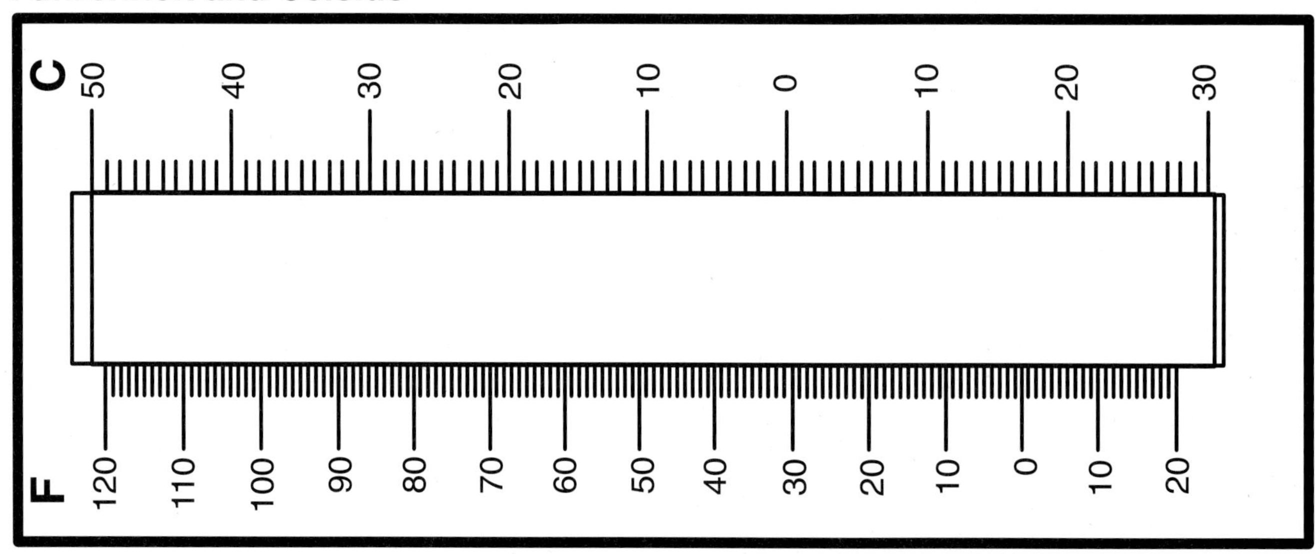

Name ____________________ Date ____________

# A Fine Day for...

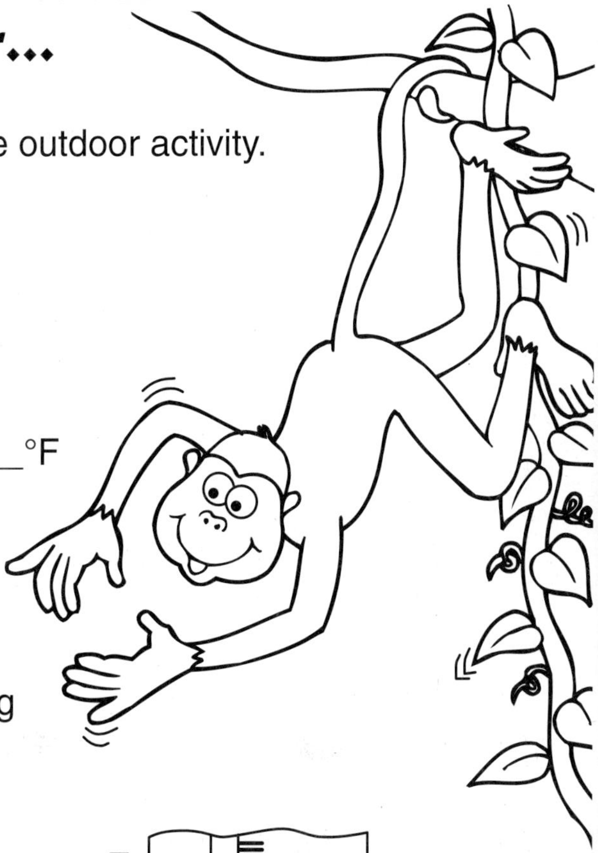

Write each temperature shown.
Then color the banana beside the most reasonable outdoor activity.

A. 40°F 30°F 20°F 10°F ____°F
jogging
building snowmen

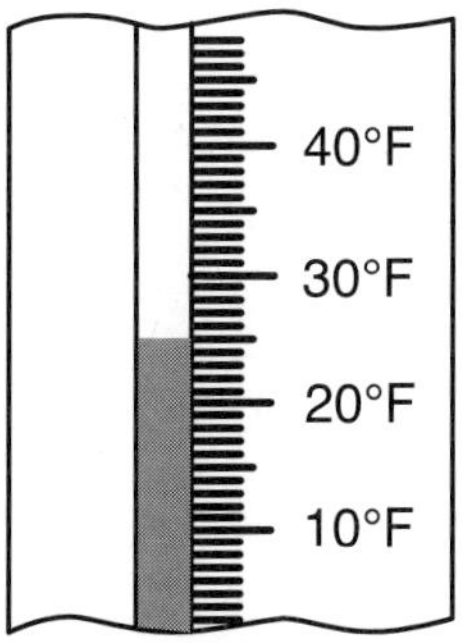

B. 80°F 70°F 60°F 50°F ____°F
bike riding
snow-skiing

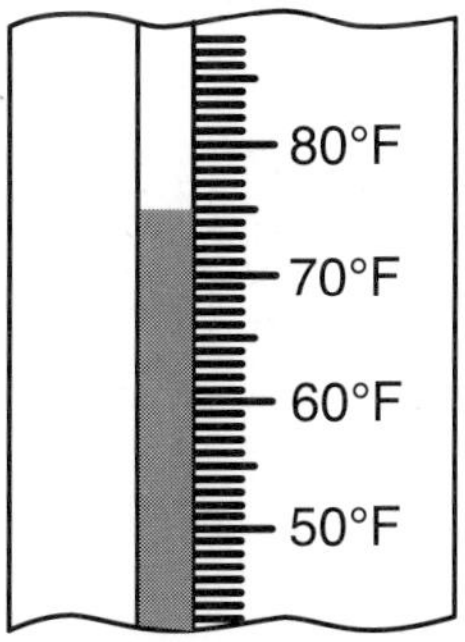

C. 70°F 60°F 50°F 40°F ____°F
gardening
swimming

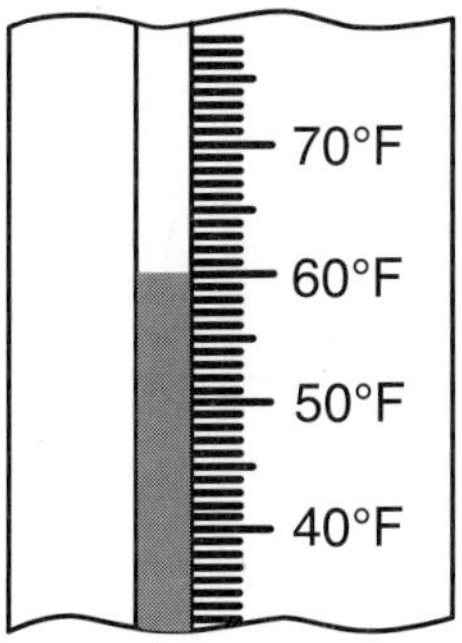

D. 100°F 90°F 80°F 70°F ____°F
raking leaves
surfing

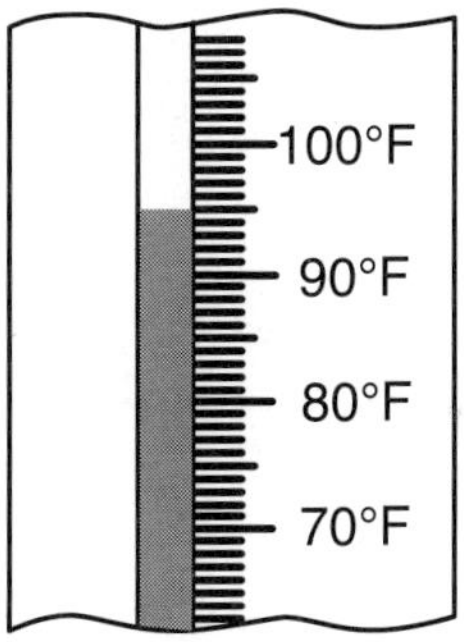

E. 40°F 30°F 20°F 10°F ____°F
water-skiing
shoveling snow

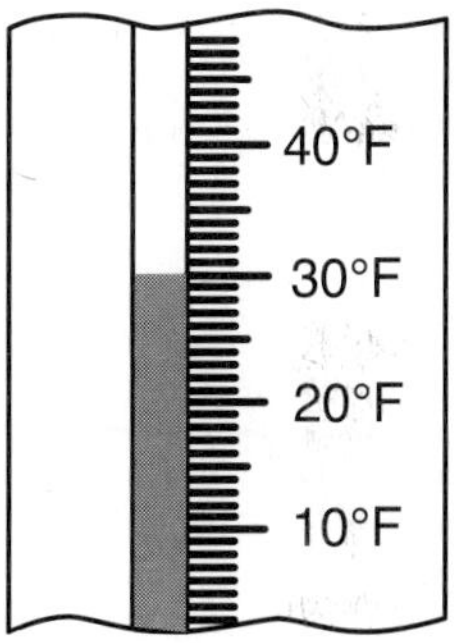

F. 90°F 80°F 70°F 60°F ____°F
sledding
sailing

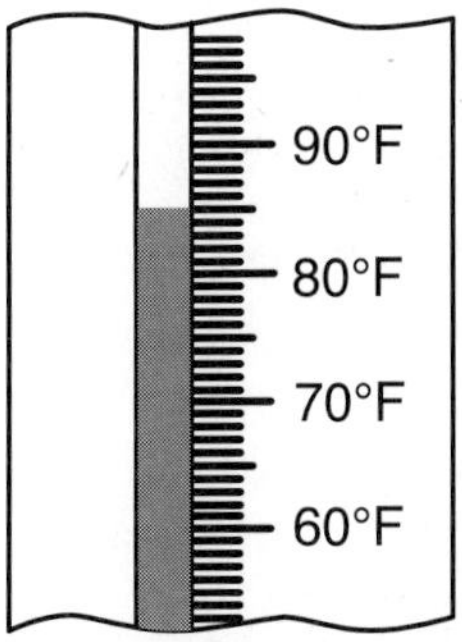

G. 80°F 70°F 60°F 50°F ____°F
hiking
snowboarding

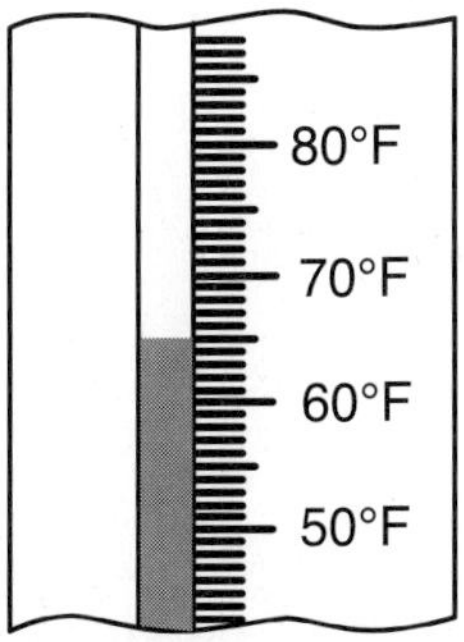

H. 40°F 30°F 20°F 10°F ____°F
roller-skating
ice-skating

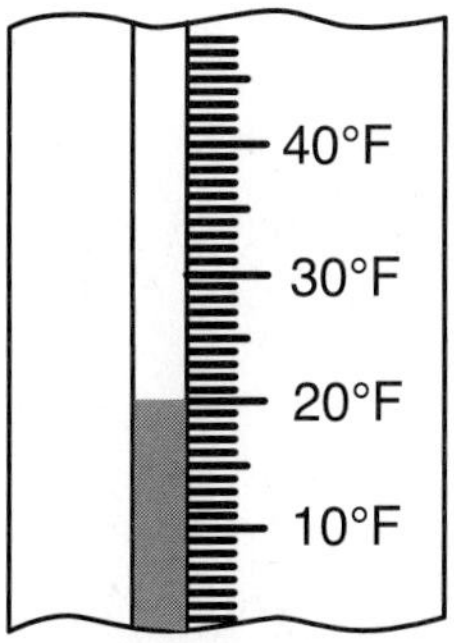

# Time

## The Long and Short of Clock Hands

Use this simple tip to help youngsters learn the difference between the hour and minute hands on a clock. Draw a short hour hand and a longer minute hand on your board. Then write the word *hour* above the shorter hand and the word *minute* above the longer hand. Explain to youngsters that the word *hour* is short just like the hour hand on a clock and that the word *minute* is longer to match the longer minute hand. Easy!

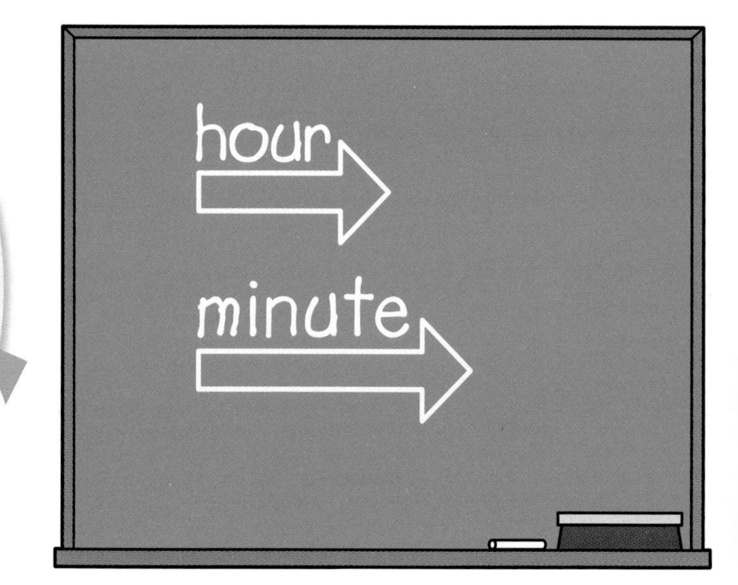


## Ticktock Time

Youngsters will be attracted to this activity, which has them showing off their time-telling skills. In advance, use a marker to draw on a 15-inch tagboard circle a clockface that includes numbers, minute markings, and a center dot to indicate placement of the clock hands. Also cut an hour hand and a longer minute hand from a contrasting color of tagboard. Laminate the clockface and hands; then cut them out. Attach magnetic tape to the back of each hand and the back of the clockface. Also make a set of time cards with times appropriate for your students. Place the cards, clockface, and clock hands at a center that has a magnetic surface. A student places the clock face on the magnetic surface. Next, he draws a card, reads its time, and uses the hands to display the correct time on the clockface. Then he announces the time that he has modeled on the clock. He repeats the process a predetermined number of times before he readies the center for the next child. Ticktock!

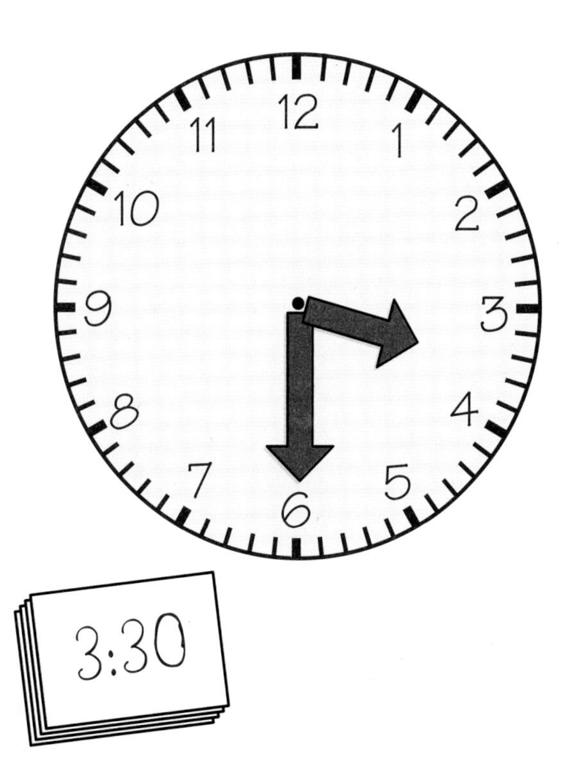

## Checking Time

Reinforce telling time with this simple time check activity. Have each child write her name on a sticky note and attach it to her desk. Then, at five different times during the day, spontaneously declare, "Time check!" Each child looks at an analog clock and writes the time on her note. (Be sure to keep a running record of the times yourself.) At the end of the day, announce the times students should have listed. Repeat the activity each day to give students additional timekeeping practice. Time check!

Abby

8:15 1:25
9:30 2:45
10:10

## Magazine Minutes or Hours

You'll know at a glance who is able to estimate time to minutes or hours after students complete this collage activity. Have each child fold a 12" x 18" sheet of construction paper in half, unfold it, and label one half "Minutes" and the other half "Hours." Next, provide each child with an old magazine. Have her search the magazine for pictures that depict activities that take minutes to complete and others that take hours to complete. Have her cut out the pictures and then glue them to the appropriate section of her paper. After the projects are finished, have each student share her collage with the class and explain why she chose to glue her pictures on her paper the way she did. Doesn't time fly when you're having fun?

## Elapsed Time From Start to Finish

Youngsters build an understanding of elapsed time while learning something about school personnel, family, or neighbors with this timely idea. Have each child ask an adult to name a daily activity in which she engages and the approximate start and stop time of the activity. Then instruct each child to write a story problem, similar to the one shown, using the information collected and focusing on determining the elapsed time of the task. Have him draw two clocks, one with the starting time of the activity and the other clock with the ending time. Next, instruct him to calculate the time it takes the person to complete the task and write it on the back of his paper. Have each child swap papers with a neighbor and then check each other's work. Ready, set, go!

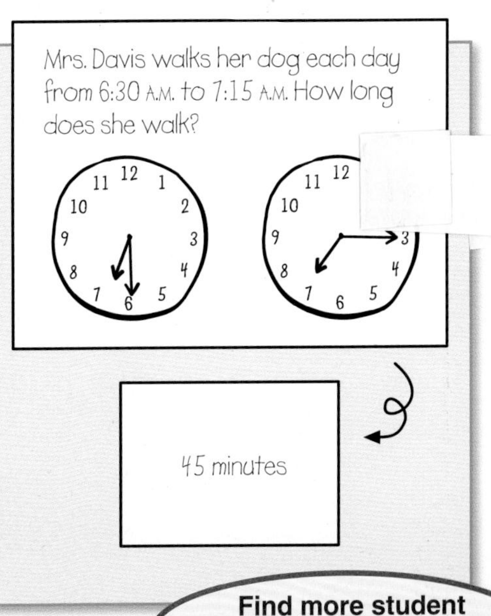


Find more student practice on pages 166–169.

Name______________________ Date______________

# Picnic Time

Draw clock hands to match each time.

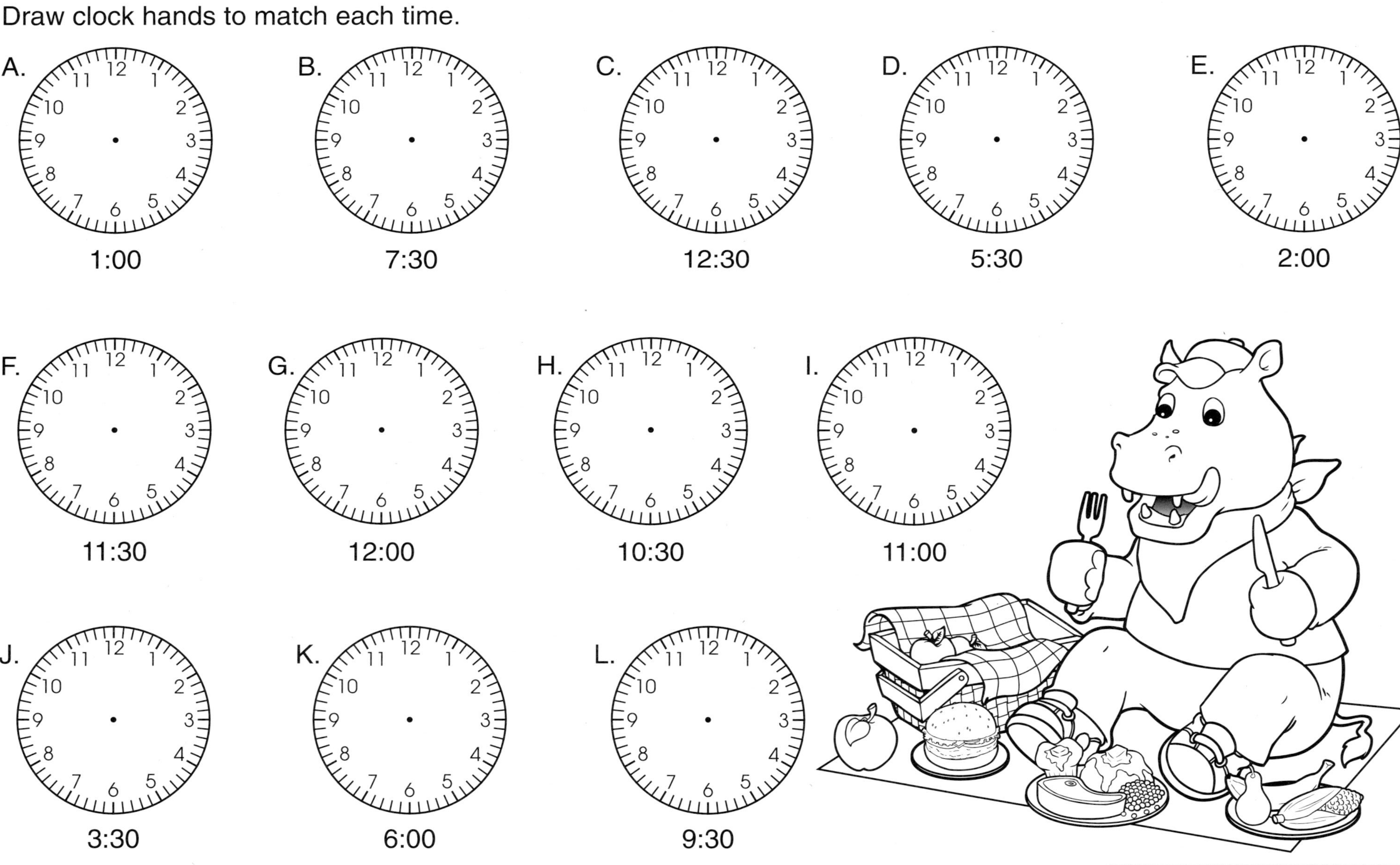




Time: telling time to the half hour

Name____________________ Date______________

# Time for More Lotion!

For each clock, write the time shown.
Think about the words you would use to read the times.
Color the clocks by the code.

| **Color Code** | |
|---|---|
| o'clock = red | a quarter past = yellow |
| half past = blue | a quarter to = green |

A. 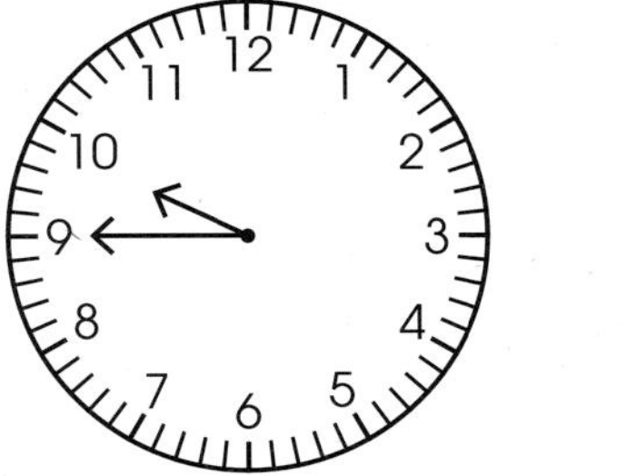
______

B. 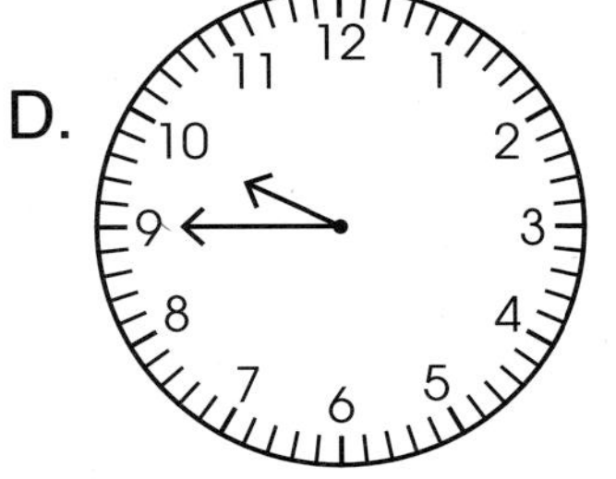
______

C. 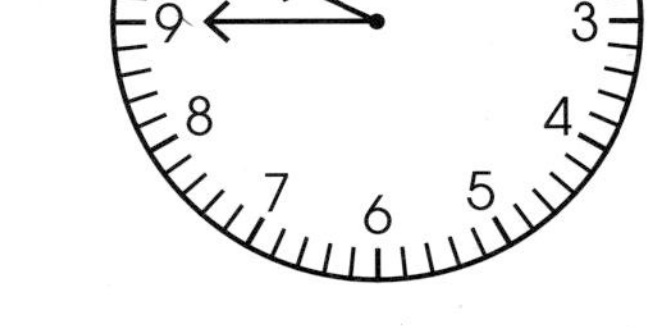
______

D. 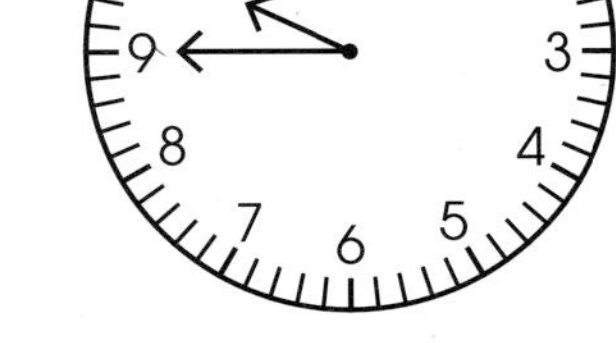
______

E. 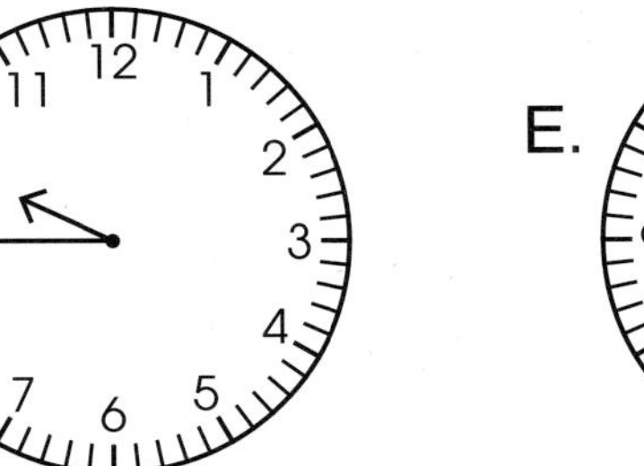 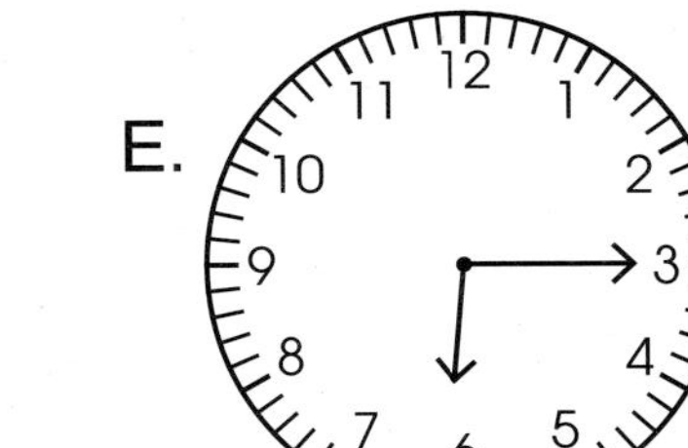
______

F. 
______

G. 
______

H.  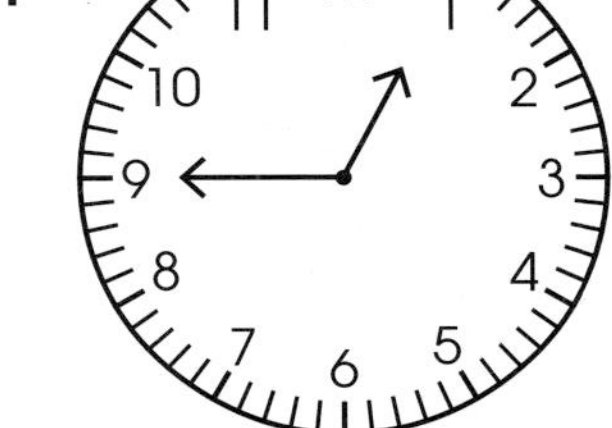
______

I. 
______

J. 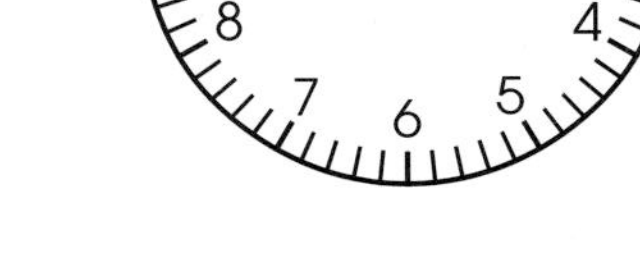
______

K. 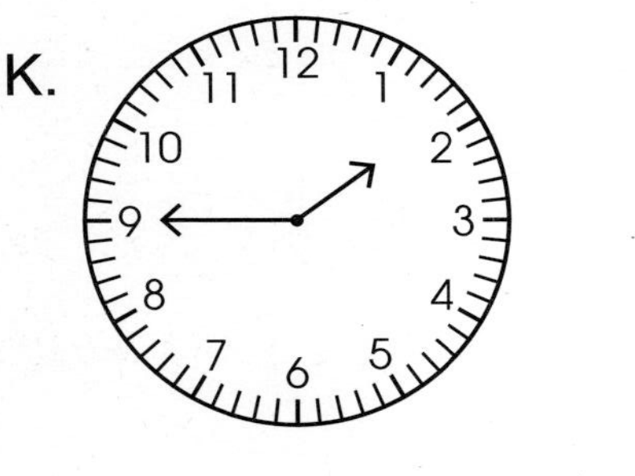 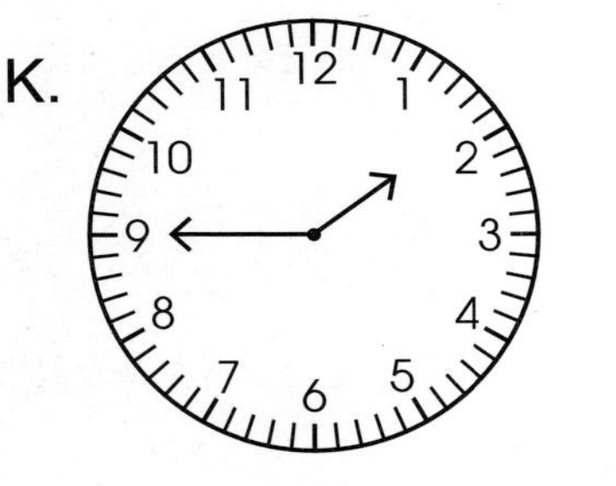 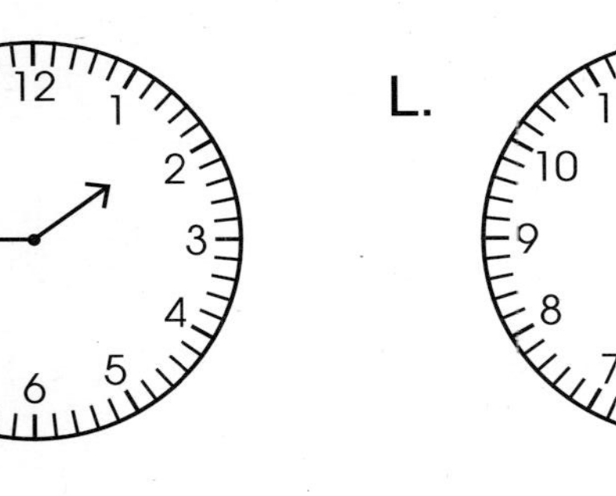
______

L.  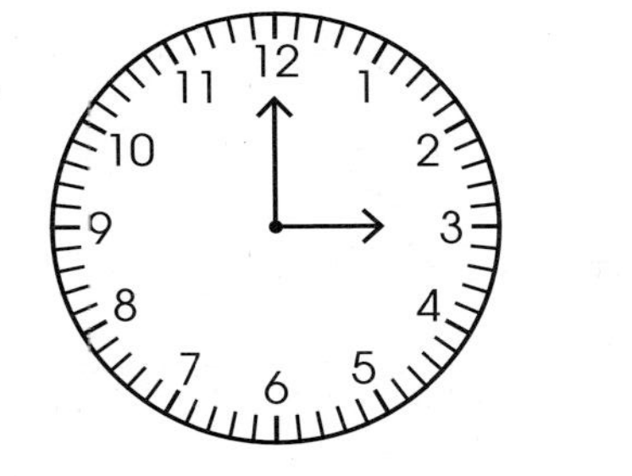
______

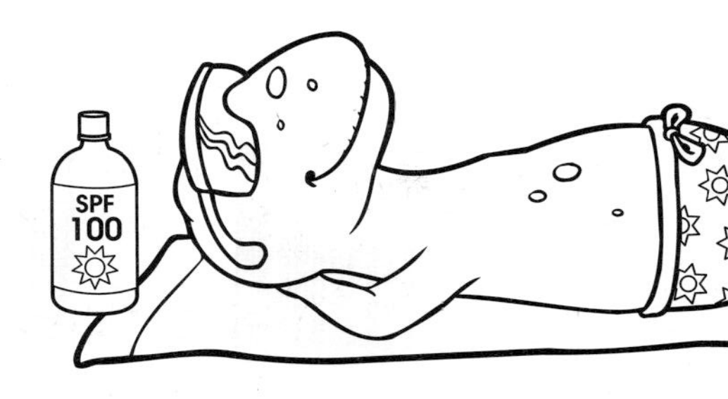

7:20 P.M. We arrived back at the dock.

6:40 A.M. We set sail.

3:25 P.M. We played cards.

11:55 A.M. We ate lunch.

6:05 A.M. We arrived at the dock.

5:50 P.M. We ate fish and chips for dinner.

8:10 A.M. We caught a fish.

12:35 P.M. We passed Feather Island.

Name________________ Date________

# Arriving on Time!

Captain Quack tells about his day at sea.
Cut out each event.
Glue each event beside the matching clock.

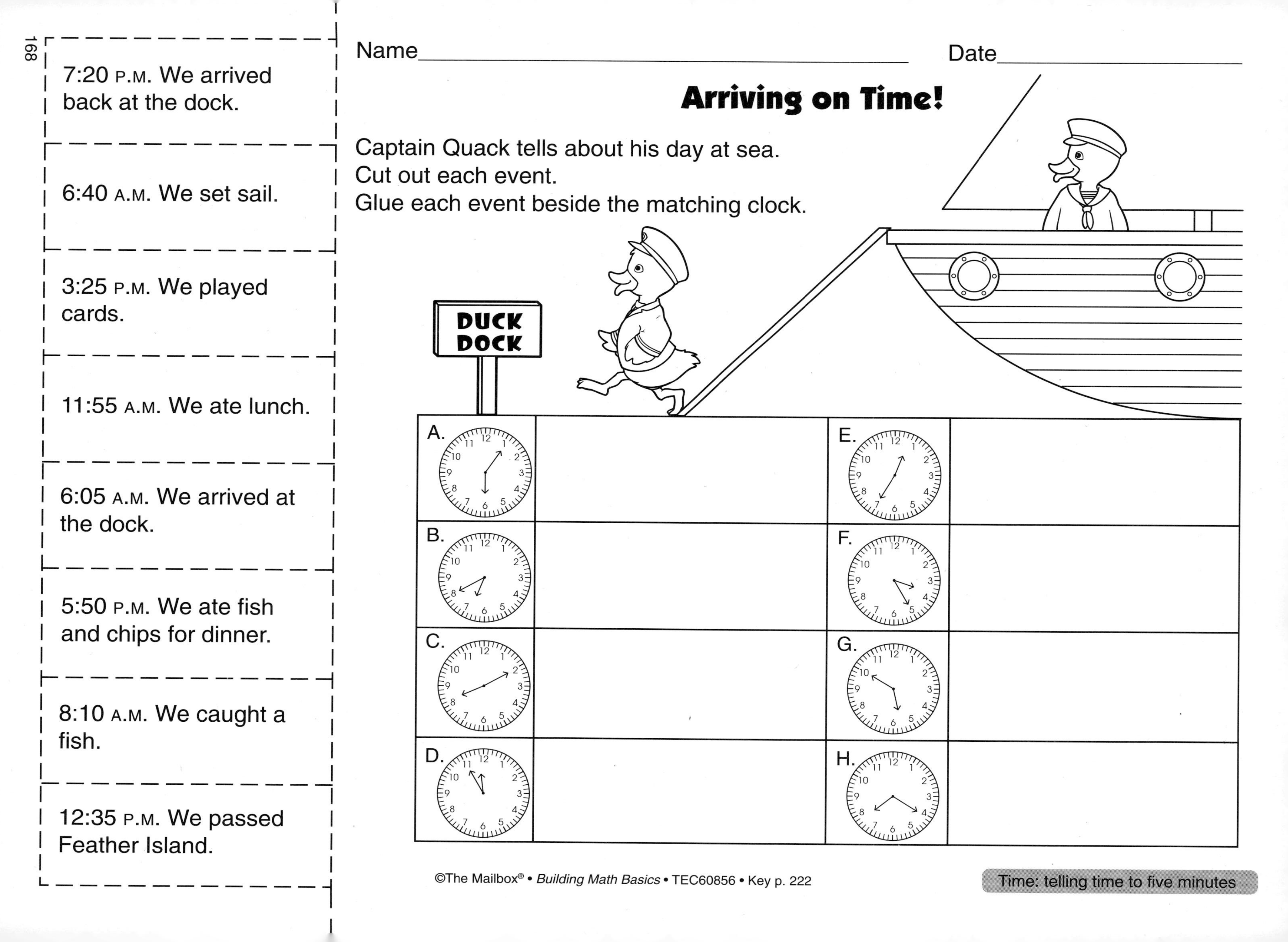

Name ______________________ Date ______________________

# Time to Crow!

Write the time below each clock.

A. ____________ B. ____________ C. ____________ D. ____________

E. ____________ F. ____________ G. ____________ H. ____________

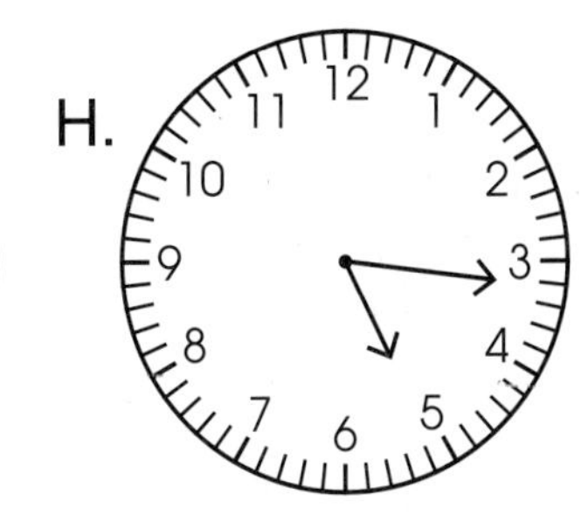

I. ____________ J. ____________

**My Schedule**

5:00–8:00 Morning wake-up calls

8:01–12:00 Singing lessons

12:01–4:59 Feather cleaning

Underline the times Roy is doing morning wake-up calls.

Circle the times Roy has singing lessons.

Draw a box around the times Roy cleans his feathers.

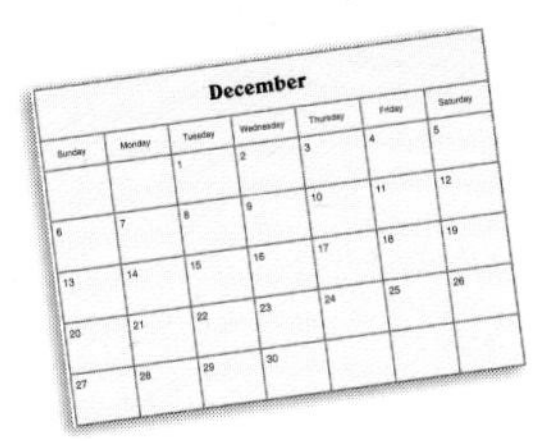

# Calendar

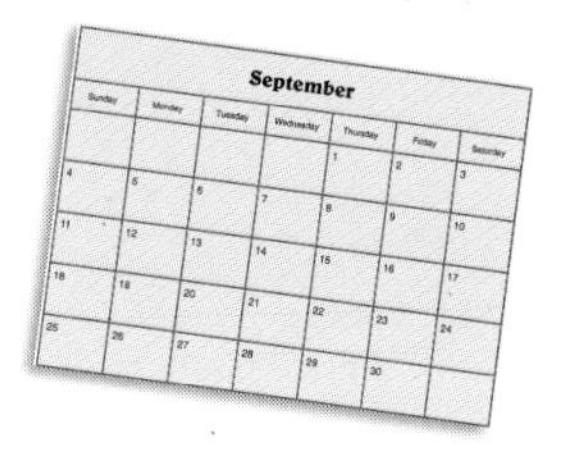

## Personal Calendar

Reinforce a variety of math skills with this monthly calendar activity. On the first school day of each month, give each student a blank calendar. Have each child write the current month and number the days on the calendar. Then challenge students to complete tasks similar to those listed below to reinforce skills as desired. Finally, have each student create a key to help identify the items on his calendar. Encourage students to take their calendars and keys home and post them in a prominent place for a month-long reference of special events. How useful!

Sample Tasks

- Trace the even numbered dates with a green marker.
- Trace the odd numbered dates with a red marker.
- Use a blue marker to circle the dates when we are not in school.
- Highlight days with early dismissal.
- Mark each birthday with a cake and the student's name.
- Mark each holiday with a symbol, such as a man in a hard hat for Labor Day.

September

| Sunday | Monday | Tuesday | Wednesday | Thursday | Friday | Saturday |
|---|---|---|---|---|---|---|
| | | | | 1 | 2 | 3 |
| 4 | 5 | 6 | 7 | 8 | 9 | 10 |
| 11 | 12 | 13 | 14 Jordan | 15 | 16 | 17 |
| 18 | 19 | 20 | 21 | 22 | 23 Emma | 24 |
| 25 | 26 | 27 | 28 | 29 | 30 | |

= Labor Day

= birthday

September

| Sunday | Monday | Tuesday | Wednesday | Thursday | Friday | Saturday |
|---|---|---|---|---|---|---|
| | | | | 1 | 2 | |
| | 5 | 6 | 7 | 8 | 9 | 10 |
| 11 | 12 | | 14 | | 16 | 17 |
| 18 | 19 | 20 | 21 | | 23 | 24 |
| | 26 | 27 | 28 | 29 | 30 | |

## Mystery Date

Who can guess the mystery date? Your calendar-savvy students, that's who! Mentally select a date within the current month, such as a holiday or the day of a special event. Direct students' attention to a calendar and announce that you've selected a mystery date. Give a clue to eliminate dates, such as "The mystery date does not occur on Friday." Have a volunteer cover the eliminated date(s) with a sticky note. Continue naming clues and having students cover dates until only one date remains: the mystery date! Then reveal the significance of that date as students applaud their search efforts.

# Special Events

This center activity provides calendar-reading practice and introduces students to some unusual special events. On a current monthly calendar, label several days with special events that occur during that month. (Use a resource such as *Chase's Calendar of Events.*) Then create clue cards that relate to the calendar similar to those shown. Program the back of each card with the date and the event for self-checking. Place the cards and the calendar at a center. As a student visits the center, he selects a card, reads the clue, locates the appropriate date on the calendar, and checks his answer. Wow, reading the calendar has never been so informative!

**November**

| Sunday | Monday | Tuesday | Wednesday | Thursday | Friday | Saturday |
|---|---|---|---|---|---|---|
| | | 1 National Authors' Day | 2 | 3 Sandwich Day | 4 | 5 |
| 6 | 7 | 8 General Election Day | 9 | 10 | 11 Veterans Day | 12 |
| 13 | 14 | 15 | 16 | 17 | 18 Mickey Mouse's Birthday | 19 |
| 20 | 21 World Hello Day | 22 | 23 | 24 Thanksgiving Day | 25 | 26 |
| 27 | 28 | 29 Electronic Greetings Day | 30 | | | |

This event happens on the Tuesday after the first Monday.

November 8
General Election Day

This event takes place on a Friday. It's a birthday for a mouse friend.

November 18
Mickey Mouse's Birthday

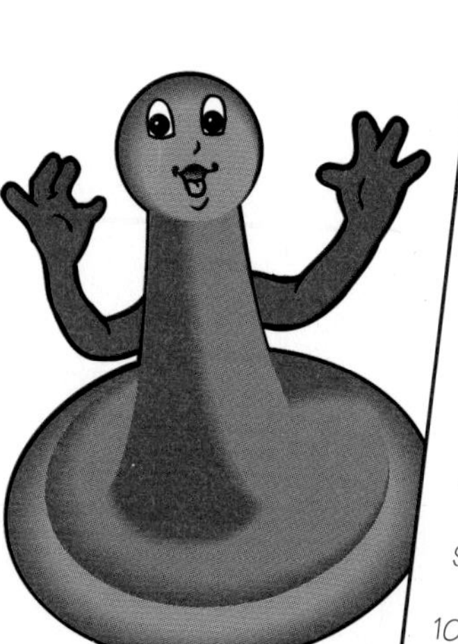

**February Calendar Fun**

1. Mark the first Wednesday of the month.
2. Find Valentine's Day on your calendar. Mark that date.
3. Our 100th Day celebration is on February 8. Mark that date.
4. Mark the second Thursday of the month.
5. Mark the date that is the last day of the month.
6. Mark the date that is one week after February 17.
7. Mark the date that is one day before Jake's birthday.
8. Find the first Saturday of the month. Mark that date.
9. Put a marker on the last Friday of the month.
10. Mark the day of the class pizza party.

**February**

| Sunday | Monday | Tuesday | Wednesday | Thursday | Friday | Saturday |
|---|---|---|---|---|---|---|
| | | | 1 | 2 | 3 | 4 |
| 5 | 6 | 7 | 8 100th Day Celebration | 9 | 10 Class Pizza Party | 11 |
| 12 | 13 | 14 | 15 | 16 | 17 | 18 |
| 19 | 20 | 21 | 22 Jake's Birthday | 23 | 24 | 25 |
| 26 | 27 | 28 | | | | |

# Mark That Date!

Reinforce a variety of math skills with this whole-group calendar game. Create a list of ten directions, similar to the ones shown, that relate to the current month. To begin, give each child a copy of a calendar of the current month and ten game markers. Read aloud the first direction and have each child place a marker on his calendar to indicate the date. Continue in this manner with the remaining directions. To have students check their work, place a calendar transparency on the overhead and reveal the answers by placing markers on the appropriate calendar dates. Then, for additional practice, have students clear their calendars and repeat the activity with a different set of directions.

Find more student practice on pages 172–174.

Name ______________________________ Date ______________________

# Off to the Zoo

When is the class trip to the zoo?
To find out, read each clue. Write the answer on the line.
Cross off the matching date(s) on the calendar.

| May | | | | | | |
|---|---|---|---|---|---|---|
| Sunday | Monday | Tuesday | Wednesday | Thursday | Friday | Saturday |
| 1 | 2 | 3 | 4 | 5 | 6 | 7 |
| 8 | 9 | 10 | 11 | 12 | 13 | 14 |
| 15 | 16 | 17 | 18 | 19 | 20 | 21 |
| 22 | 23 | 24 | 25 | 26 | 27 | 28 |
| 29 | 30 | 31 | | | | |

1. The trip will not be on a Saturday or Sunday. What are those dates?
   ______________________________________________
2. The zoo is closed on Mondays. What dates are Mondays?
   ______________
3. The trip is not the second Wednesday of May. What is the date?
   ________
4. Tickets for the zoo are sold out on May 6, 13, 20, and 27. On what day of the week are these dates? ______________
5. The class picnic is the third Wednesday in May. What is the date?
   ________
6. The seals do not perform on Tuesdays and Thursdays. What dates are Tuesdays and Thursdays? ______________________________
7. The zoo is closed on May 25. What day of the week is that? ____________

**The trip will be on ____________________.**

Name________________________ Date________________

# Underwater Parade of Months

Cut.
Glue the months in order.

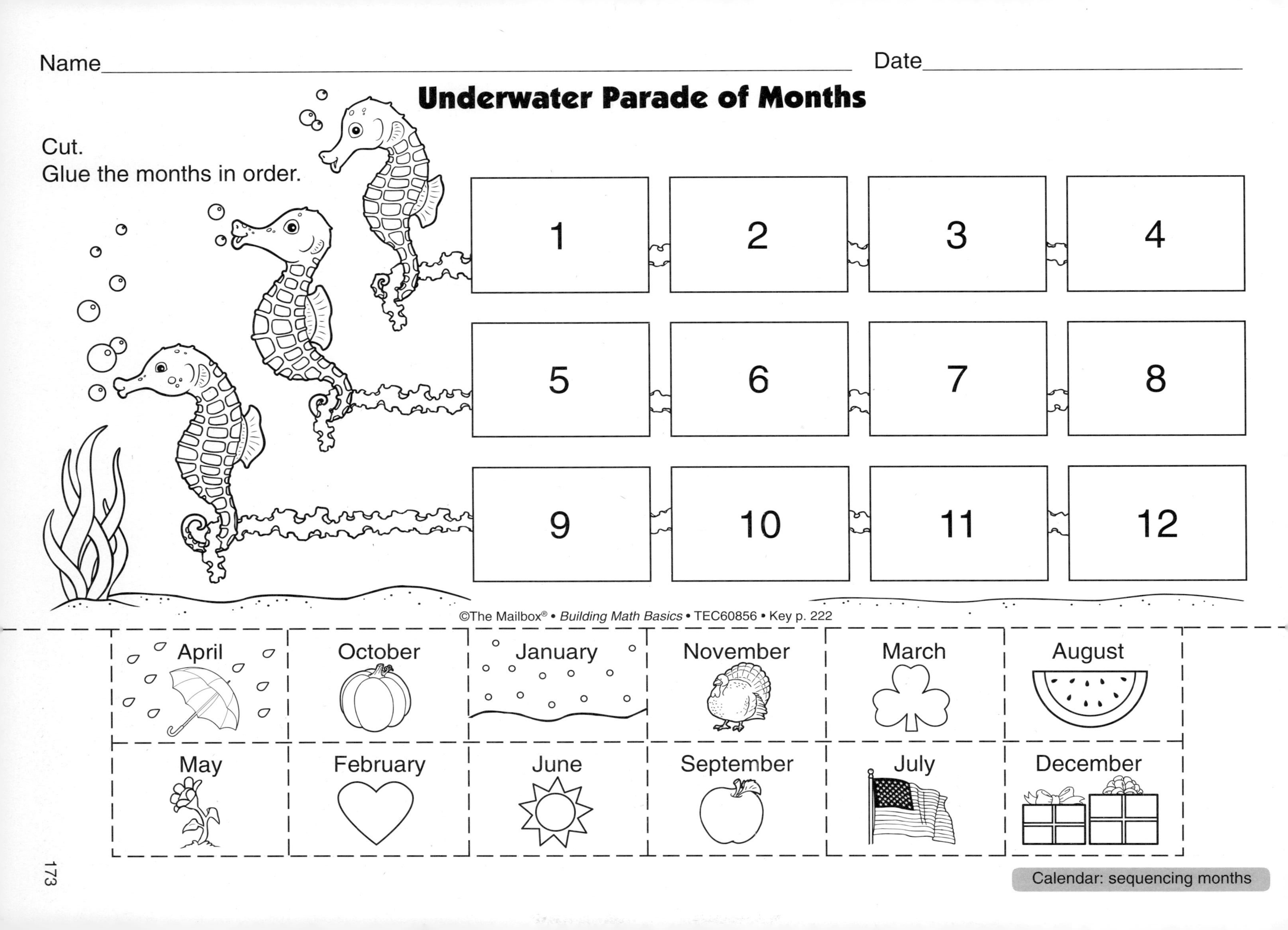




| April | October | January | November | March | August |
|---|---|---|---|---|---|
| May | February | June | September | July | December |

Calendar: sequencing months

Name ______________________________ Date ____________________

# Through the Year

Use the calendar to answer each question.

**January**

| Su | M | T | W | Th | F | Sa |
|---|---|---|---|---|---|---|
| | | | | | | 1 |
| 2 | 3 | 4 | 5 | 6 | 7 | 8 |
| 9 | 10 | 11 | 12 | 13 | 14 | 15 |
| 16 | 17 | 18 | 19 | 20 | 21 | 22 |
| 23 | 24 | 25 | 26 | 27 | 28 | 29 |
| 30 | 31 | | | | | |

**February**

| Su | M | T | W | Th | F | Sa |
|---|---|---|---|---|---|---|
| | | 1 | 2 | 3 | 4 | 5 |
| 6 | 7 | 8 | 9 | 10 | 11 | 12 |
| 13 | 14 | 15 | 16 | 17 | 18 | 19 |
| 20 | 21 | 22 | 23 | 24 | 25 | 26 |
| 27 | 28 | | | | | |

**March**

| Su | M | T | W | Th | F | Sa |
|---|---|---|---|---|---|---|
| | | 1 | 2 | 3 | 4 | 5 |
| 6 | 7 | 8 | 9 | 10 | 11 | 12 |
| 13 | 14 | 15 | 16 | 17 | 18 | 19 |
| 20 | 21 | 22 | 23 | 24 | 25 | 26 |
| 27 | 28 | 29 | 30 | 31 | | |

**April**

| Su | M | T | W | Th | F | Sa |
|---|---|---|---|---|---|---|
| | | | | | 1 | 2 |
| 3 | 4 | 5 | 6 | 7 | 8 | 9 |
| 10 | 11 | 12 | 13 | 14 | 15 | 16 |
| 17 | 18 | 19 | 20 | 21 | 22 | 23 |
| 24 | 25 | 26 | 27 | 28 | 29 | 30 |

**May**

| Su | M | T | W | Th | F | Sa |
|---|---|---|---|---|---|---|
| 1 | 2 | 3 | 4 | 5 | 6 | 7 |
| 8 | 9 | 10 | 11 | 12 | 13 | 14 |
| 15 | 16 | 17 | 18 | 19 | 20 | 21 |
| 22 | 23 | 24 | 25 | 26 | 27 | 28 |
| 29 | 30 | 31 | | | | |

**June**

| Su | M | T | W | Th | F | Sa |
|---|---|---|---|---|---|---|
| | | | 1 | 2 | 3 | 4 |
| 5 | 6 | 7 | 8 | 9 | 10 | 11 |
| 12 | 13 | 14 | 15 | 16 | 17 | 18 |
| 19 | 20 | 21 | 22 | 23 | 24 | 25 |
| 26 | 27 | 28 | 29 | 30 | | |

**July**

| Su | M | T | W | Th | F | Sa |
|---|---|---|---|---|---|---|
| | | | | | 1 | 2 |
| 3 | 4 | 5 | 6 | 7 | 8 | 9 |
| 10 | 11 | 12 | 13 | 14 | 15 | 16 |
| 17 | 18 | 19 | 20 | 21 | 22 | 23 |
| 24 | 25 | 26 | 27 | 28 | 29 | 30 |
| 31 | | | | | | |

**August**

| Su | M | T | W | Th | F | Sa |
|---|---|---|---|---|---|---|
| | 1 | 2 | 3 | 4 | 5 | 6 |
| 7 | 8 | 9 | 10 | 11 | 12 | 13 |
| 14 | 15 | 16 | 17 | 18 | 19 | 20 |
| 21 | 22 | 23 | 24 | 25 | 26 | 27 |
| 28 | 29 | 30 | 31 | | | |

**September**

| Su | M | T | W | Th | F | Sa |
|---|---|---|---|---|---|---|
| | | | | 1 | 2 | 3 |
| 4 | 5 | 6 | 7 | 8 | 9 | 10 |
| 11 | 12 | 13 | 14 | 15 | 16 | 17 |
| 18 | 19 | 20 | 21 | 22 | 23 | 24 |
| 25 | 26 | 27 | 28 | 29 | 30 | |

**October**

| Su | M | T | W | Th | F | Sa |
|---|---|---|---|---|---|---|
| | | | | | | 1 |
| 2 | 3 | 4 | 5 | 6 | 7 | 8 |
| 9 | 10 | 11 | 12 | 13 | 14 | 15 |
| 16 | 17 | 18 | 19 | 20 | 21 | 22 |
| 23 | 24 | 25 | 26 | 27 | 28 | 29 |
| 30 | 31 | | | | | |

**November**

| Su | M | T | W | Th | F | Sa |
|---|---|---|---|---|---|---|
| | | 1 | 2 | 3 | 4 | 5 |
| 6 | 7 | 8 | 9 | 10 | 11 | 12 |
| 13 | 14 | 15 | 16 | 17 | 18 | 19 |
| 20 | 21 | 22 | 23 | 24 | 25 | 26 |
| 27 | 28 | 29 | 30 | | | |

**December**

| Su | M | T | W | Th | F | Sa |
|---|---|---|---|---|---|---|
| | | | | 1 | 2 | 3 |
| 4 | 5 | 6 | 7 | 8 | 9 | 10 |
| 11 | 12 | 13 | 14 | 15 | 16 | 17 |
| 18 | 19 | 20 | 21 | 22 | 23 | 24 |
| 25 | 26 | 27 | 28 | 29 | 30 | 31 |

1. How many days are in September? ____________
2. Which is the fifth month of the year? ____________
3. How many Thursdays are in November? ____________
4. Which month has the fewest days? ____________________________
5. What date follows October 31? _______________________
6. What is the date one week before May 14? ____________________
7. What is the date one week after June 17? _____________________
8. How many months have 31 days? ____________
9. What month follows April? _________________
10. What day is August 2? _____________________

# Money

## It's Mine!

This "cent-sational" small-group game prompts plenty of coin-counting practice. From old catalogs and store circulars, cut out pictures of several kid-pleasing items. Glue each picture on a construction paper card and add a price. Place the cards facedown in a stack and place a plastic dish near the stack for the bank. Gather a small group of students and give each child an equal set of coins that includes pennies, nickels, dimes, and quarters. To play, a child draws a card, reads the price, and counts out that amount of money from his set of coins. If he has enough, he puts the money in the bank, keeps the card, and announces, "It's mine!" (If he needs to get change from the bank, he may do so.) If a child does not have enough change to buy the item, he returns the card to the bottom of the draw pile and his turn is over. The winner is the player who has the least amount of money left after all of the cards are gone or no one can buy the remaining cards. If desired, students may also count to see which player was able to purchase the most items for his money. Now that's a bargain!

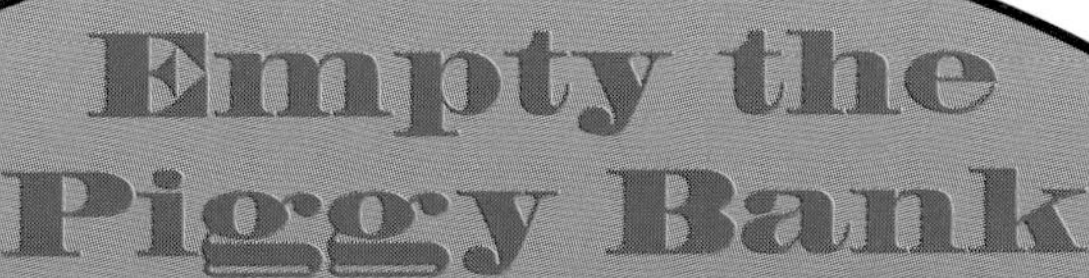


Students go hog-wild for this game, which practices coin recognition and counting. Prepare a piggy bank cutout for each child. Also make a spinner like the one shown for each group of four students. Divide the class into small groups. Give each group a spinner and a sharpened pencil and large paper clip to use with the spinner. Then give each group member a piggy bank cutout and one of each of the following coins to place on her cutout: penny, nickel, dime, quarter, and half dollar. (For an especially challenging round, give students two of each coin.) To play, each player, in turn, spins the spinner and removes from her bank the coin that matches the value she spun. She moves this coin to a personal discard pile. If a student spins the pig, she puts all of her coins from her discard pile back into her bank. At that time, each player must announce the value of the coins in her bank. Then play resumes with the next player. The first player to empty her bank wins!

# Let's Count the Ways...

Cash in with this independent activity, which will soon be a favorite. Prepare a chart similar to the one shown and place it on a table along with coin stamps and a stamp pad, coin stickers, or paper coin cutouts. Challenge each student to determine a different coin combination that equals $1.00 and create it on the chart. When each child has had a turn, gather students and enlist their help in checking each row and adding additional rows if other combinations are possible. Lead students to recognize how many different combinations were made. If desired, post a new chart the following day with a new monetary value.

How many ways can you make $1.00 ?

| |
|---|
| Jaylen |
| Mia |
| Ben |
| Davis |
| Lydia |
| Sarah |
| Jackson |

## Number Card Coin Count

Shake up an interest in counting money with this small-group game. Label 18 small blank cards with the numerals 1 to 9 (two cards for each). Place the cards in a paper lunch bag and fold down the top of the resulting money bag. Give each child an assortment of coins and scrap paper and a pencil for keeping score. Player 1 shakes the money bag, reaches inside it without looking, and removes two cards. Next, she uses the cards to create a money value in cents. (For example, cards labeled "6" and "3" can be 63 cents or 36 cents.) Then each player uses her coins to show that value. When all the players are finished, each player takes a turn counting her coins aloud for verification. A point is rewarded to each player who has a unique coin combination. Player 1 returns the two cards to the bag and refolds the top of the bag. Then the remaining players take turns in a similar manner. The first player to reach five points is declared the winner. Count 'em up!

5 2

Find more student practice on pages 177–181.

Name________________________ Date______________

# "Toad-ally" Tasty!

Cut out the candy cards below.
Glue each one in the row with the matching amount.

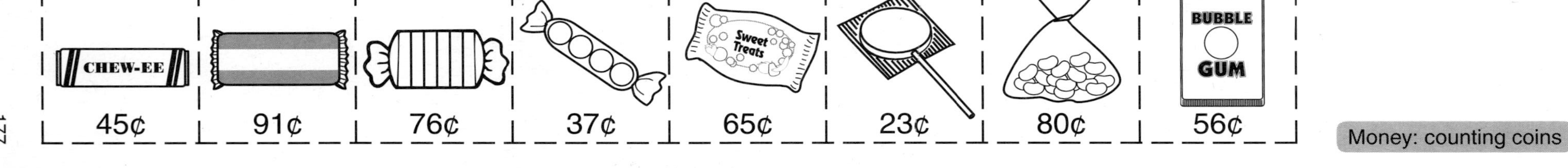


Money: counting coins

Name ______________________ Date ______________

# Pocket Change

Cut out the coin cards below.
Use exactly three coins to match each amount.
Glue the cards in place.

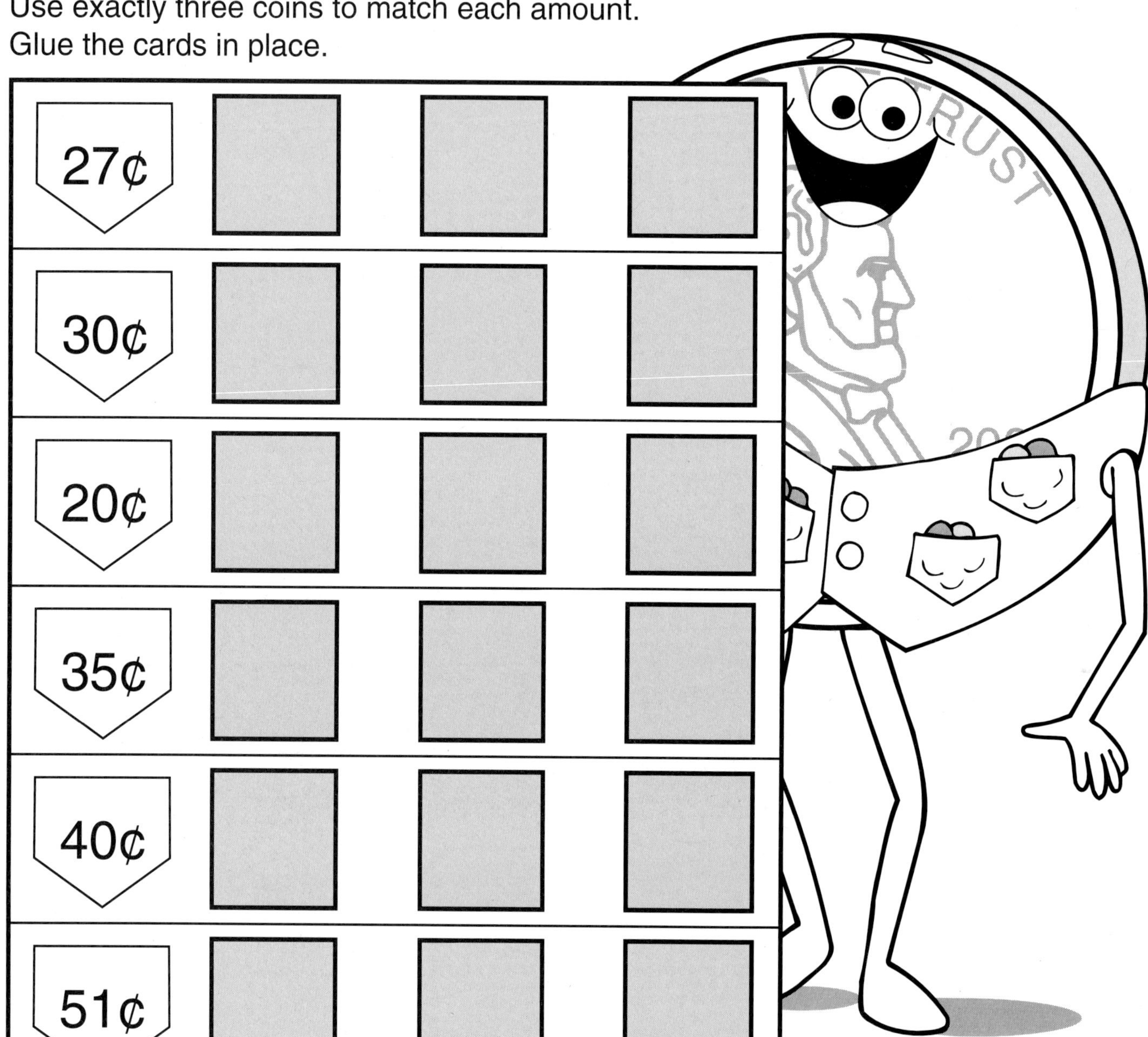

Name ____________________ Date ____________________

# Shopping Sheep

Color the coins to show the price.
Write the amount of the coins that are not colored in the circle above each bag.

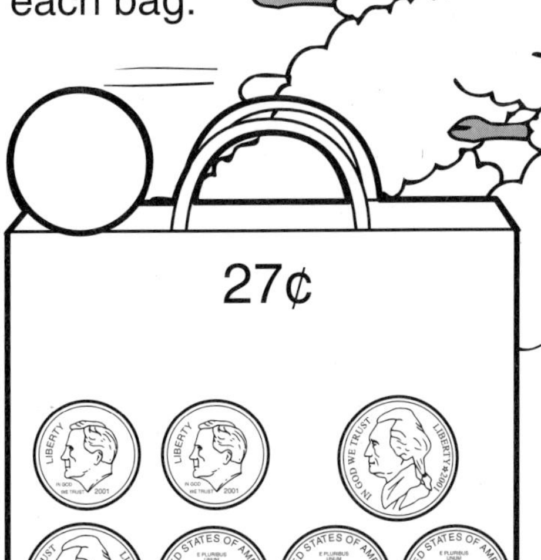


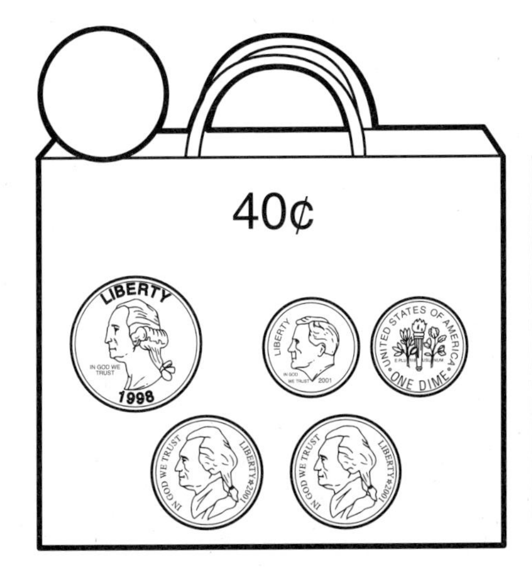

Name ______________________ Date ____________

# Money in the Bank

Write the total amount of each set of one-dollar bills and coins.
On the bank slip, write the name beside the matching amount.

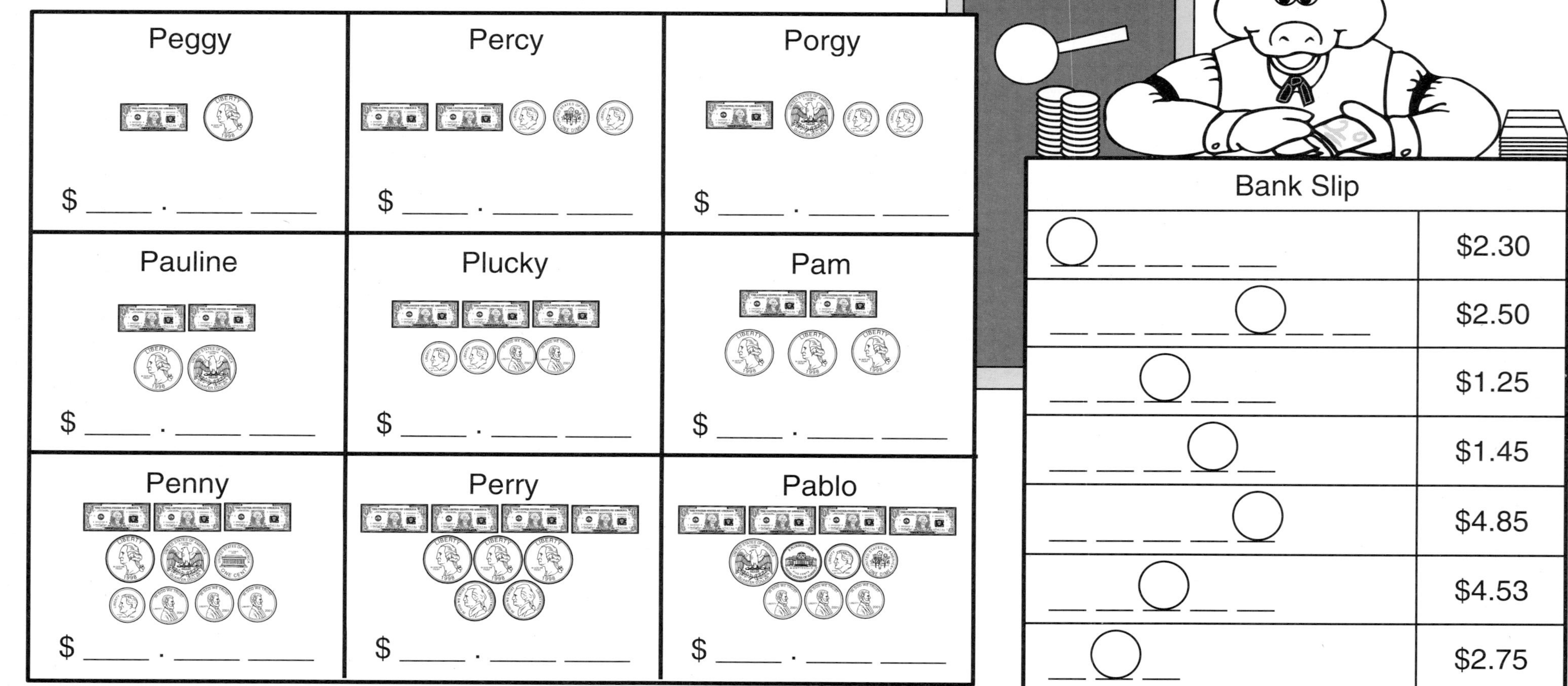

**Why did the bank teller squeal?**

To solve the riddle, write the circled letters in order in the spaces below.

Because he worked in a ___ ___ ___ ___ ___ ___ ___ ___ ___!

Name ______________________________ Date ____________________

# Garage Sale Goodies

Estimate the total price of the two toys.
Write your estimate in the box.
Write the prices.
Add.

A.

$1.25 $ ___.___ ___

$1.80 + ___.___ ___

$ ___.00 $ ___.___ ___

B.

$2.65 $ ___.___ ___

$0.25 + ___.___ ___

$ ___.00 $ ___.___ ___

C.

$1.10 $ ___.___ ___

$3.20 + ___.___ ___

$ ___.00 $ ___.___ ___

D.

$1.10 $ ___.___ ___

$1.75 + ___.___ ___

$ ___.00 $ ___.___ ___

E.

$1.75 $ ___.___ ___

$1.00 + ___.___ ___

$ ___.00 $ ___.___ ___

F.

$2.15 $ ___.___ ___

$2.65 + ___.___ ___

$ ___.00 $ ___.___ ___

## Just Clip It!

Here's a timesaving way to provide daily graphing practice! Designate a bulletin board for this purpose. Each day, use sentence strips to post a graph-related question and possible responses as shown. Mount a pushpin below each response. Set a class supply of large paper clips nearby. As students arrive, have each child read the question and carefully attach a paper clip below his chosen response. When all students have clipped their preferences, have them compare the resulting paper clip chains to interpret the results!

## On the Go

Set graphing skills in motion with a transportation pictograph! Ask students to think about which way they would prefer to travel if they were taking a dream vacation. Give examples, such as by airplane, car, bus, boat, or train. Have each child write her name on a provided square and then draw her preferred mode of transportation. Collect the squares and loosely tape them to the board in a random arrangement.

Next, wonder aloud how many students preferred a particular mode of transportation. Count the corresponding pictures, intentionally counting some twice. When students notice your mistake, suggest that sorting the pictures into categories would help ensure accurate counting. Organize the pictures into columns and add labels. Then pose questions to help students read the resulting graph.

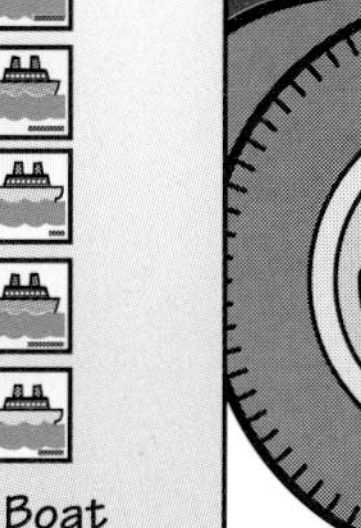

# Class Favorites

Help students see how their classmates' preferences stack up with this student-generated graphing idea. Program a sentence strip "What is your favorite _______?" Cut out several shorter lengths for possible responses. Then have each child personalize a blank card. (Tell him to make sure that his name is visible from a distance.) Laminate the strips and cards for durability. Mount a strip of magnetic tape to the back of each one. Place the cards in a container.

At the end of the school day, randomly select a name card. Have that child use a wipe-off marker to complete the graph question as desired and program the blank cards with possible responses. Tell him to arrange the pieces on the board to prepare for a graph like the one shown. When students arrive in the morning, direct each child to display his name card above his response. Discuss the results. Then use the display to take attendance!

What is your favorite sport ?

Janice

Tyler | Suzy | Allye

Mike | LaJuan | Fred

Nia | Moby | Jamison

Basketball | Baseball | Soccer

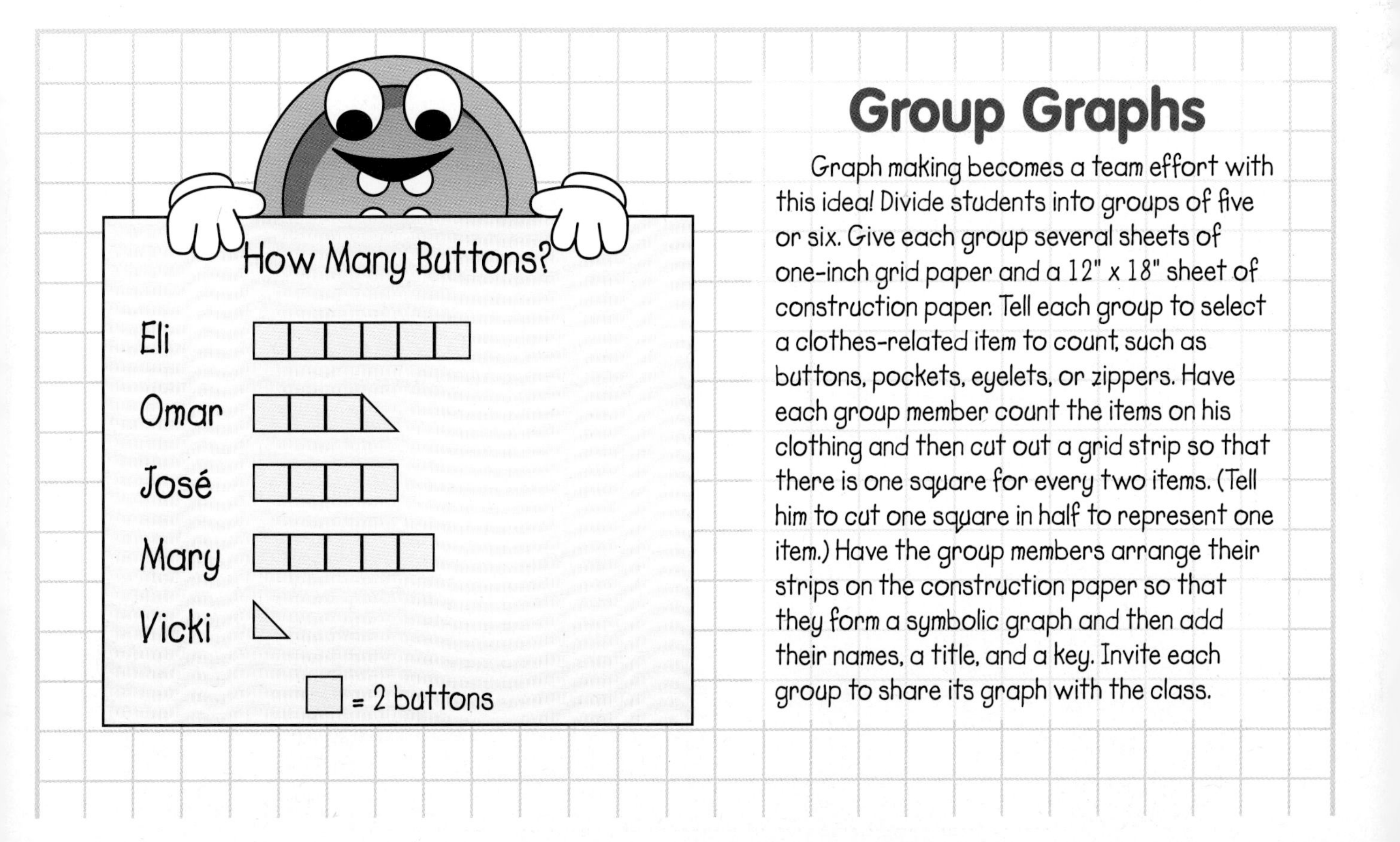


## Group Graphs

Graph making becomes a team effort with this idea! Divide students into groups of five or six. Give each group several sheets of one-inch grid paper and a 12" x 18" sheet of construction paper. Tell each group to select a clothes-related item to count, such as buttons, pockets, eyelets, or zippers. Have each group member count the items on his clothing and then cut out a grid strip so that there is one square for every two items. (Tell him to cut one square in half to represent one item.) Have the group members arrange their strips on the construction paper so that they form a symbolic graph and then add their names, a title, and a key. Invite each group to share its graph with the class.

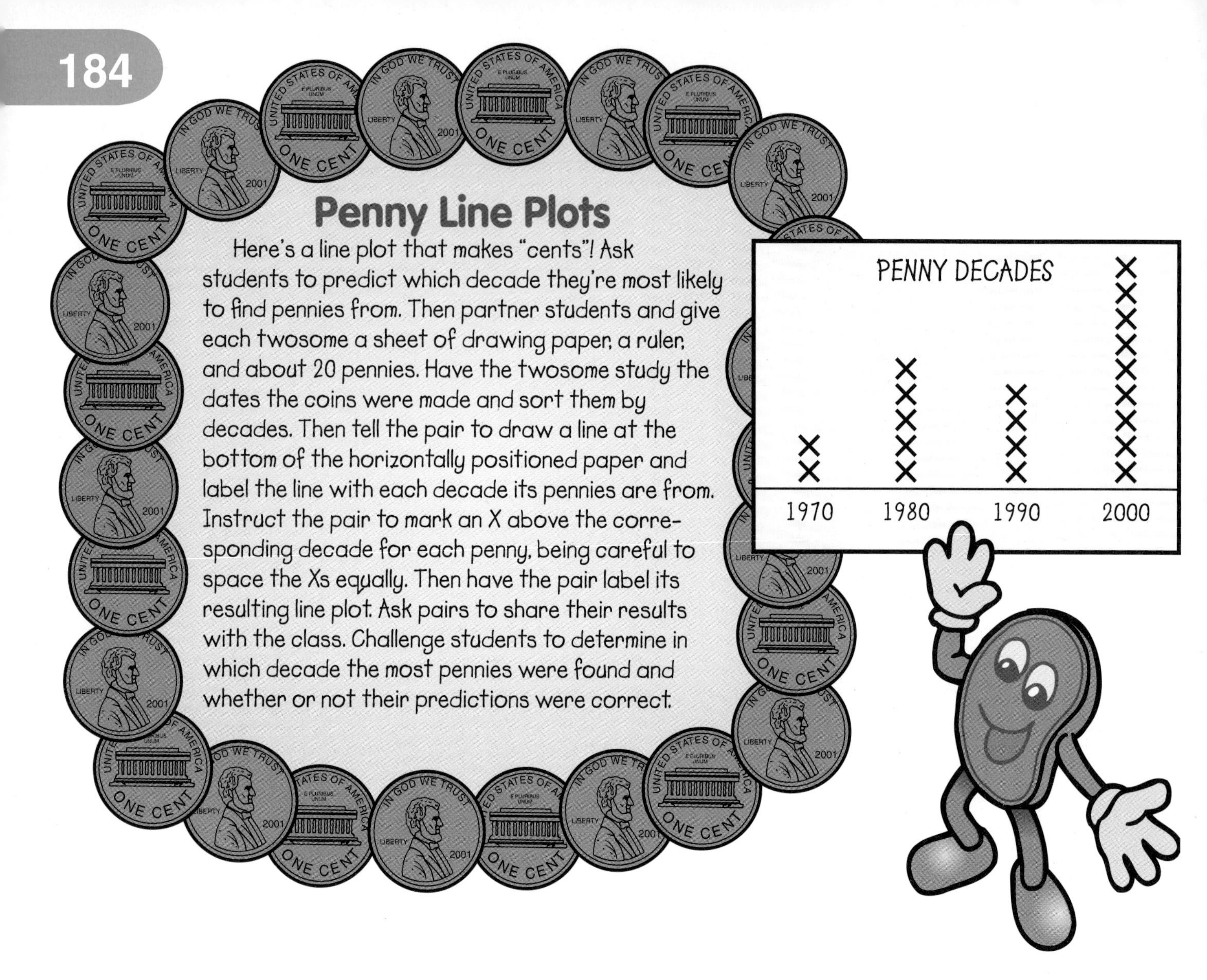


## Penny Line Plots

Here's a line plot that makes "cents"! Ask students to predict which decade they're most likely to find pennies from. Then partner students and give each twosome a sheet of drawing paper, a ruler, and about 20 pennies. Have the twosome study the dates the coins were made and sort them by decades. Then tell the pair to draw a line at the bottom of the horizontally positioned paper and label the line with each decade its pennies are from. Instruct the pair to mark an X above the corresponding decade for each penny, being careful to space the Xs equally. Then have the pair label its resulting line plot. Ask pairs to share their results with the class. Challenge students to determine in which decade the most pennies were found and whether or not their predictions were correct.

## Math by Any Other Name

Here's a letter-perfect idea to help students determine how the lengths of their names compare. Give each student a sheet of lined paper. Ask the class to help determine the least number of letters in any classmate's first name as well as the greatest number of letters. Then have each child draw a line along the bottom of his paper and number it using this range. (See the example.) Next, ask each child, in turn, to announce the number of letters in his name. Tell each student to mark an X above the corresponding number. Then have students study the graph and interpret the results. Gee, what about last names?

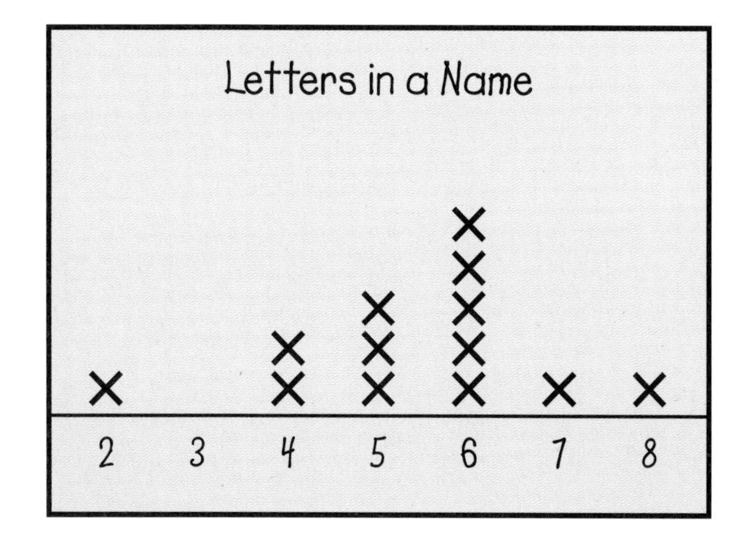

## Absence Results

When are the most students absent—toward the beginning, middle, or end of the week? Find out with this line graph activity! Explain to students that a line graph reveals how data changes over time. Prepare a sheet of chart paper for a line graph, as shown, including the days for a predetermined time period. (Plan to collect data for at least two weeks.) After you take attendance each day, have a student volunteer plot a point to indicate the number of absent students. When all the data has been collected, use a ruler to connect the points; then prompt a discussion about any trends the graph reveals. Ask questions such as "On which day were students absent the most?" or "Do more students tend to be absent at the beginning or end of the week?" Invite students to speculate about students' attendance trends using the graph to support their reasoning.

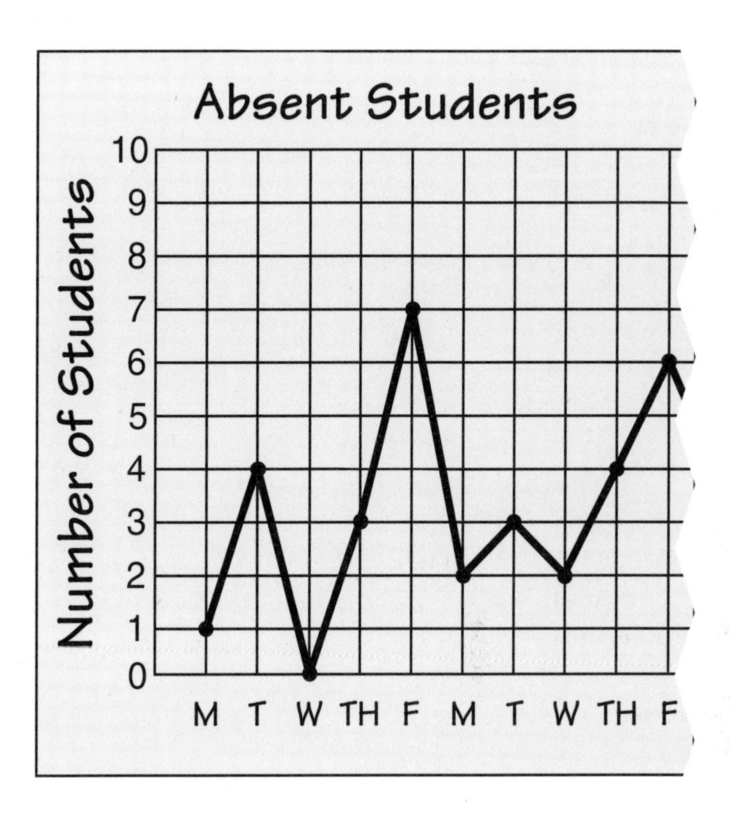

## Fancy Footwork

When it comes to understanding circle graphs, this idea is a "shoe-in"! You will need a length of clothesline for this activity. To begin, have students stand in a circle so that students wearing similar types of shoes (such as athletic shoes or sandals) are standing side by side. Stand in the middle of the circle. Use the clothesline to visually divide the groups into sections. Identify the resulting data display as a circle graph. Then ask students to compare the display with a bar graph.

## Sweet Results

Try this tasty way to make a circle graph! Enlarge the pattern below and make student copies. Give each child a circle and ten M&M's candies. Have her place each candy on an individual section so that like-colored candies are grouped together. Then tell her to color the sections accordingly. Next, ask each student to write three statements about her candy colors using her circle graph. As an added challenge, have each student write the fractional amount or percentage of each color. Then invite students to eat their candy. Yum!

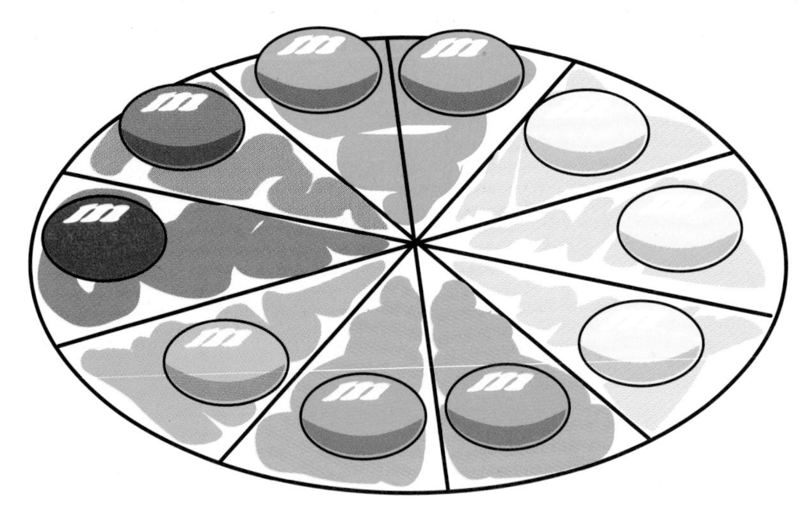

Find more student practice on pages 187–191.

### Circle Graph Pattern

Use with "Sweet Results" on this page.

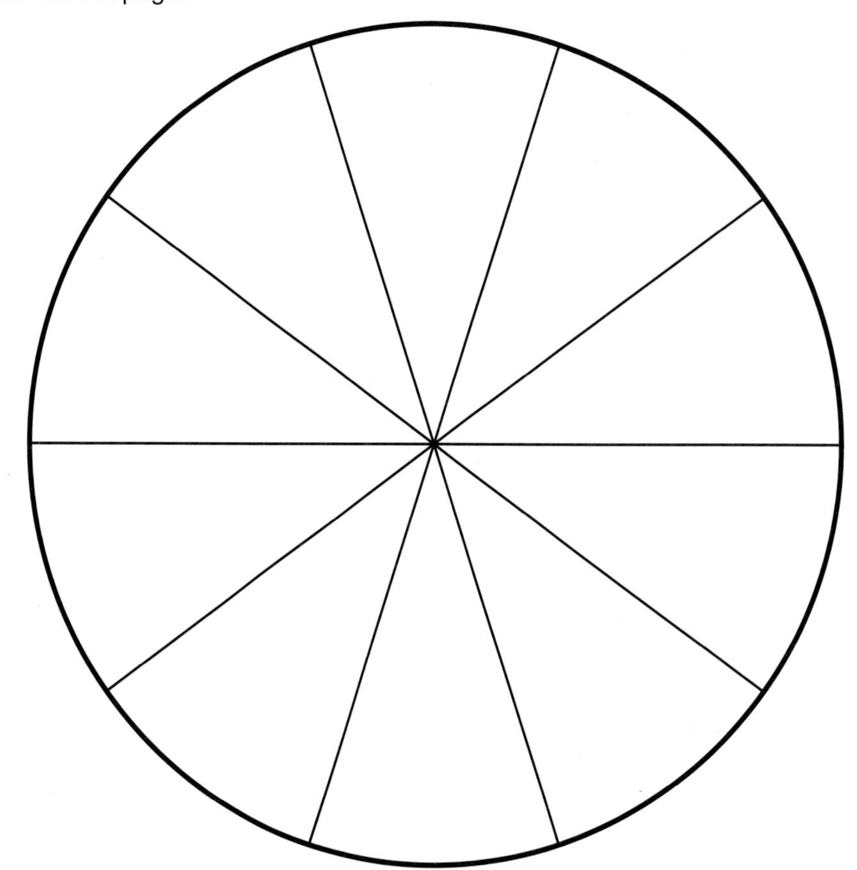

Name ______________________ Date ______________

# The Cheese Shop

Use the graph to answer the questions.

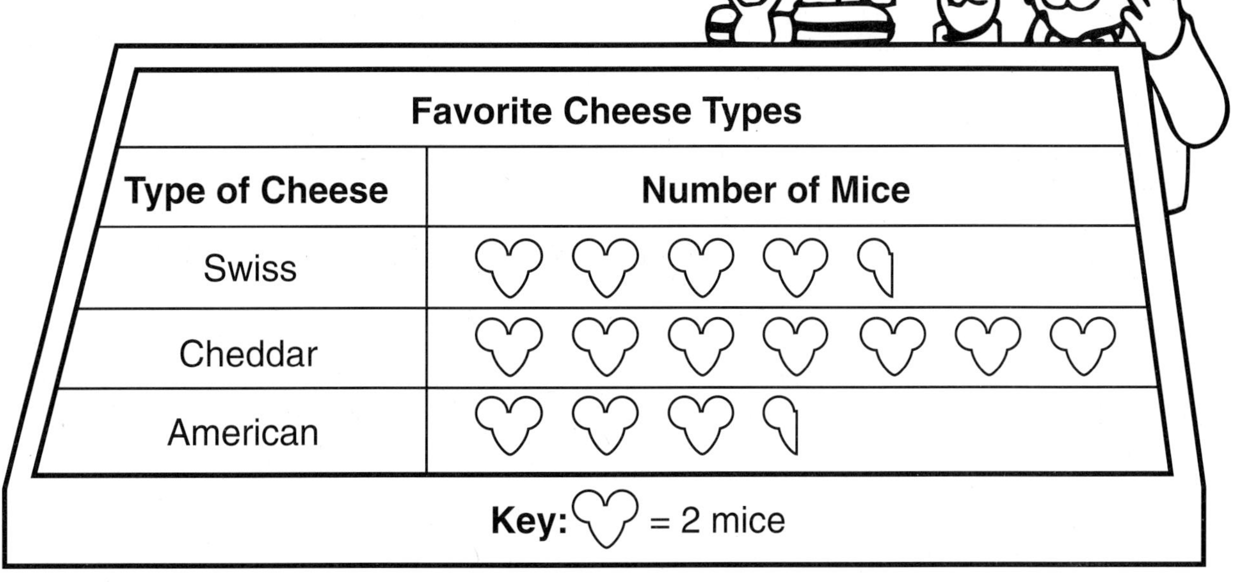

| Favorite Cheese Types | |
|---|---|
| **Type of Cheese** | **Number of Mice** |
| Swiss | |
| Cheddar | |
| American | |

**Key:** = 2 mice

1. How many mice like American cheese best? __________
2. Which cheese do the mice like the most? ______________________

   The least? ______________________
3. How many more mice like cheddar cheese than Swiss cheese? __________

Study the table.
Use the key to complete the graph.

| Cheese Sold | |
|---|---|
| **Type of Cheese** | **Number of Slices** |
| Swiss | 7 |
| Cheddar | 12 |
| American | 14 |

| Cheese Sold | |
|---|---|
| **Type of Cheese** | **Number of Slices** |
| Swiss | |
| Cheddar | |
| American | |

**Key:** △ = 2 slices

Name ______________________ Date ______________

# Sweet Sale

Study the list.
For each name, mark an X above the matching number on the line plot.

**Number of Candy Bars Sold**

Alex—4
Becky—2
Denise—7
Frank—4
Irv—4
Jason—7
Kelly—3
Mike—5
Nia—3
Ollie—4
Pablo—5
Raj—4
Sue—1
Tasha—3

Read each statement below.
Study the line plot.
If the statement is true, color the candy bar.

5 students sold 4 candy bars.

7 students sold 2 candy bars.

14 students sold candy bars.

Some students sold 6 candy bars.

The number of candy bars sold the most was 4.

Name ____________________ Date ____________________

# Moe's Animal Crackers

Make a tally mark on the table for each type of cracker.
Write each total.

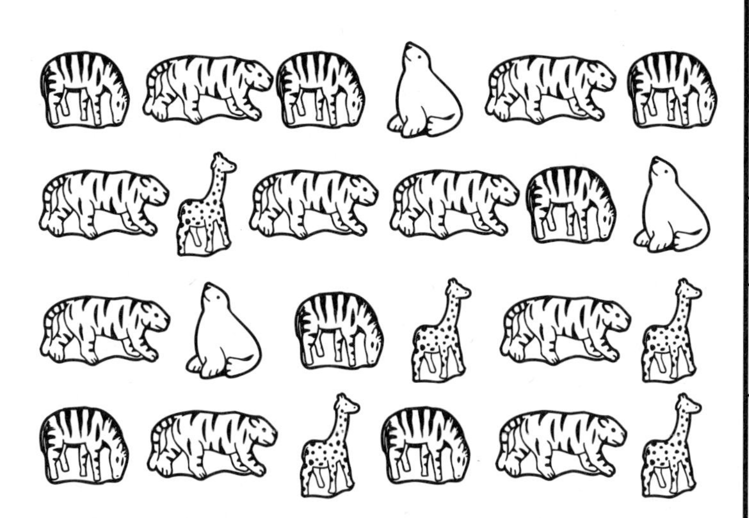

**How Many Crackers Were Eaten?**

| Type of Cracker | Tally Marks | Total |
|---|---|---|
| | | |
| | | |
| | | |
| | | |

Use the table to complete the graph below.

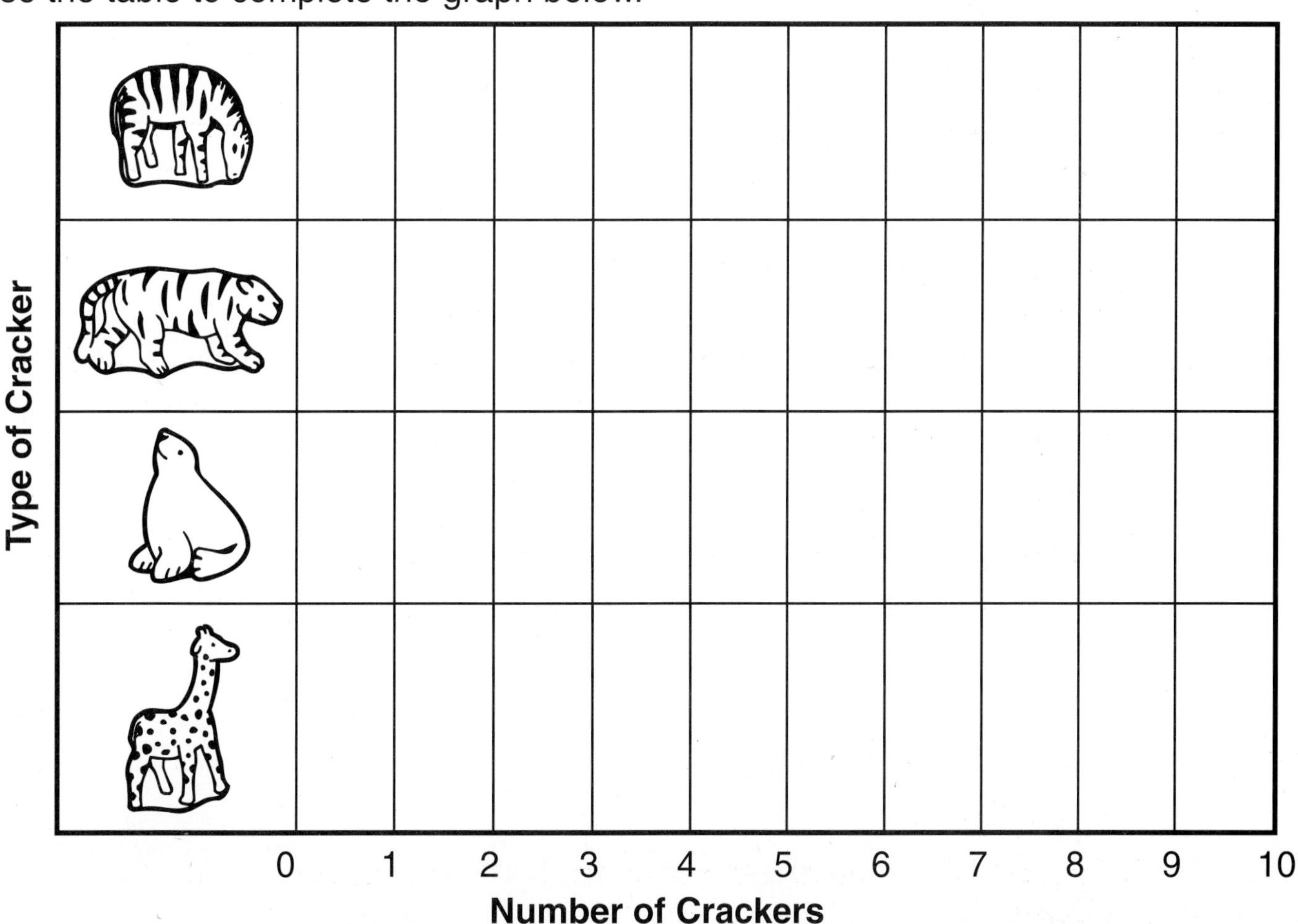

Name ______________________ Date ______________

# Easy As Pie!

Flo took a survey. She asked these questions:

- What is your favorite kind of pie?
- What do you like to drink with pie?

Make a tally mark to show each response.
Write each total.

**Responses**
apple, coffee
peach, coffee
cherry, milk
apple, tea
apple, milk
peach, coffee
cherry, milk
apple, coffee
peach, tea
cherry, milk

| Answer | Tally Marks | Total |
|---|---|---|
| apple | | |
| peach | | |
| cherry | | |

| Answer | Tally Marks | Total |
|---|---|---|
| coffee | | |
| tea | | |
| milk | | |

Complete each key.
Color each circle graph to show the results.

**Favorite Kind of Pie**

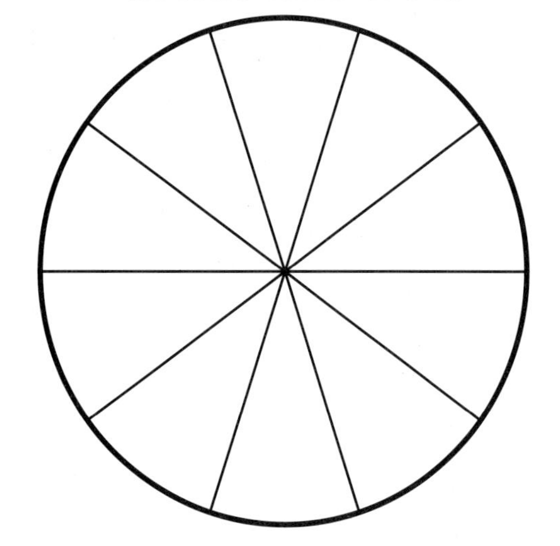

**Key**
☐ apple ☐ peach ☐ cherry

**Favorite Drink With Pie**

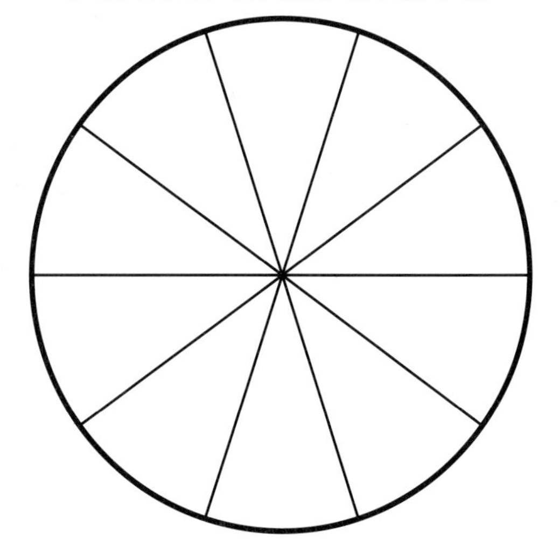

**Key**
☐ coffee ☐ tea ☐ milk

Name ______________________ Date ______________

# Open for Business!

GRAND OPENING

Look at the table.
Mark a dot to show the number of customers for each day.
Connect the dots.

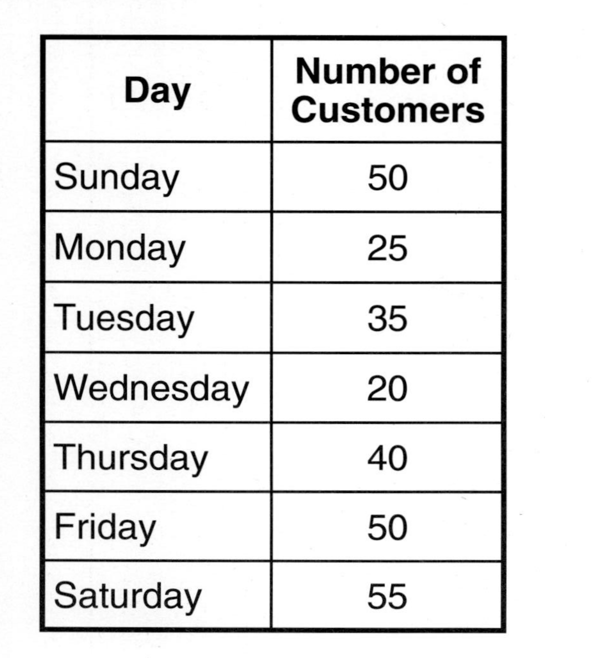

| Day | Number of Customers |
|---|---|
| Sunday | 50 |
| Monday | 25 |
| Tuesday | 35 |
| Wednesday | 20 |
| Thursday | 40 |
| Friday | 50 |
| Saturday | 55 |

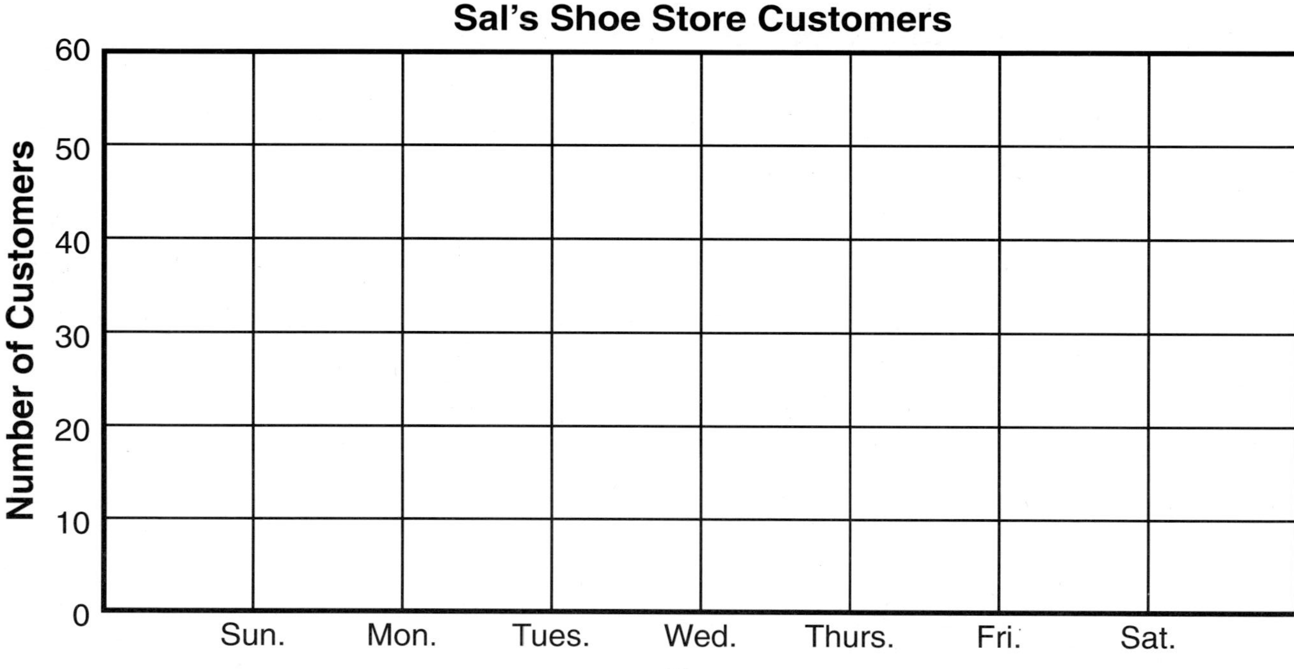


Use the graph to answer the questions.

1. On which day did the least number of customers visit the store?
   ______________
2. How many customers came on Wednesday and Thursday combined? ________
3. The number of customers went down between Sunday and Monday.
   Between which other days did the number of customers go down?
   ______________ and ______________
4. Why might there be more customers on Saturday than any other day?
   Write your answer on the back of this sheet.

# Probability

## What Are the Odds?

Students are sure to gain a better understanding of probability terms with this activity! Give each student about ten red, ten yellow, and ten blue cubes. List the colors in a column on the board. In another column, list the words *certain, likely, unlikely,* and *impossible*. Next, draw a check after one word in each list, such as *red* and *unlikely*. Challenge each child to form a group of cubes to represent the checked words (for example, a group from which it is unlikely that a red cube will be selected). Invite students to explain their selections. Then erase the check marks and repeat the activity with a different pairing of words.

## People Predictions

Personalize probability with this kid-pleasing idea! In advance, program a set of blank cards with students' names and then place the cards in a large bowl. To begin, identify an attribute that applies to at least one student and pose a probability-related question about it. (See the sample questions.) As a class, count the number of students to which the attribute applies. Also verify how many students are in the class. Then ask students to predict whether it is certain, very likely, likely, unlikely, or impossible that one of these students' names will be randomly drawn from the bowl in ten tries. Next, ask a student to draw a card from the bowl, read the name aloud, and mark a tally on a table like the one shown. When she returns the card to the bowl, have her mix up the cards. Repeat this process for a total of ten times. Then have students interpret the tally table to determine whether their predictions and results match. Repeat the activity several times using a variety of attributes.

Alexis

Jerome

Questions

**What are the chances of picking a student who...**

is wearing tennis shoes?
has green eyes?
is a girl?
has more than four pencils today?
has a sister?
has a dog?
brought lunch to school today?

| Attribute | Tally Marks |
|---|---|
| wearing tennis shoes | 𝍸 II |
| not wearing tennis shoes | III |

# What's for Lunch?

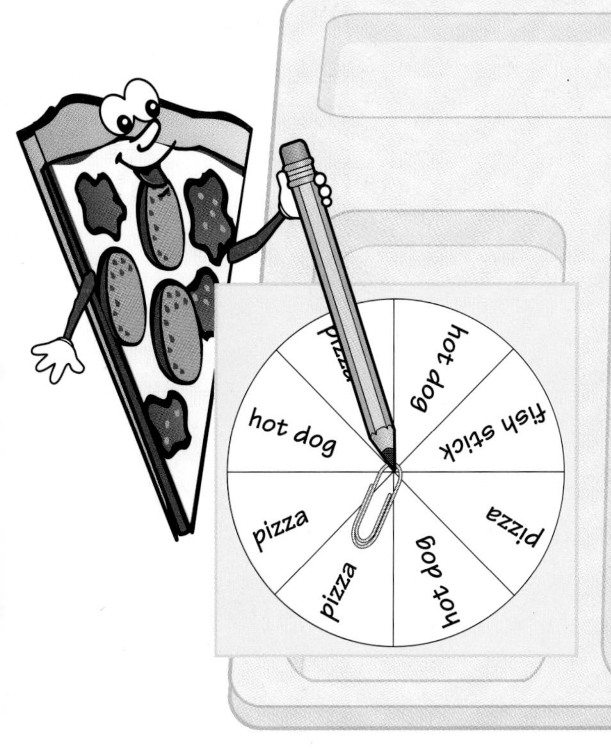


Serve up probability practice with these school lunch spinners! Give each student a copy of the spinner on page 194, a paper clip, and a pencil. Tell students to pretend that their spinners will determine what their school lunches will be for the next 20 days. Have each student program the eight sections of his spinner using only three lunch selections—his two favorites and his least favorite. Challenge him to label his spinner so that his favorite lunch choices are likely to be spun.

Next, direct each child to use his paper clip and a pencil to spin the spinner 20 times, keeping track of his lunch results on a recording sheet. Invite students to share how many times in the next 20 days they would be eating a favorite lunch.

# Fishy Chances

Reel in a fresh way to explore probability! Give each child a napkin, a copy of the recording sheet on page 194, and a cup containing an odd number of Goldfish crackers in two different colors. Ask the child to count each color of cracker and then describe on the recording sheet the probability of randomly selecting it. For example, if a child has 15 total crackers, three of which are orange, the chances of selecting orange are three out of 15. Next, have her predict which color of cracker will be chosen more often and draw an X in the appropriate box in the third column. To complete her sheet, a student takes one cracker at random, makes a mark on her chart, and then returns the cracker to the cup and gently shake the cup. Have her repeat this process until she has 20 tally marks. Tell her to write the total and analyze her data. Then invite her to compare her findings with a classmate's. Guide students to understand that their results are different because they each started with different amounts of each type of cracker. Finally, invite each student to eat her catch of the day!

Name Jennifer

| Color | What are the chances? | Which is more likely to be chosen? | Tally Marks | Total |
|---|---|---|---|---|
| orange | 3 out of 15 | | I | ___ |
| brown | 12 out of 15 | X | IIII | ___ |

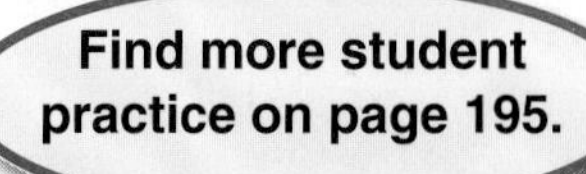
Find more student practice on page 195.

## Spinner

Use with "What's for Lunch?" on page 193.

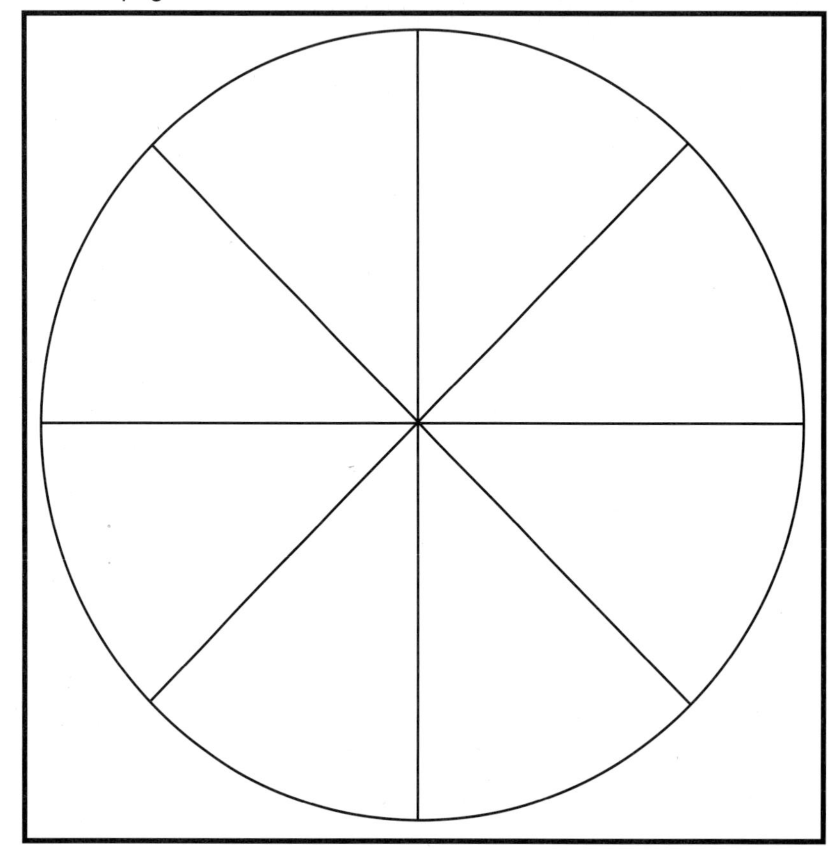

## Recording Sheet

Use with "Fishy Chances" on page 193.

Name ____________________

| Color | What are the chances? | Which is more likely to be chosen? | Tally Marks | Total |
|---|---|---|---|---|
| | ____ out of ____ | | | ____ |
| | ____ out of ____ | | | ____ |

Name ______________________ Date ______________

# Monkey Business

Balloon Animals 50¢

1. Make a **prediction.** Color the balloon cards below. If you put the cards in a bag and then pulled one out, which color would

   **most** likely be drawn? ____________

   **least** likely be drawn? ____________

2. **Follow the steps** to complete a probability experiment.
   A. Cut out the balloon cards below.
   B. Put the cards in the paper bag.
   C. Close your eyes and draw a card. Color one of the squares to show which color you pulled out. Put the card back in the bag.
   D. Repeat the steps above until all of the squares are colored.

| | | | | | | | | | | | |
|---|---|---|---|---|---|---|---|---|---|---|---|
| | | | | | | | | | | | |
| | | | | | | | | | | | |

3. Find your **results.**

   How many times did you pull out a blue balloon? ________

   How many times did you pull out a yellow balloon? ________

   How many times did you pull out a red balloon? ________

4. Are your results similar to your prediction? ______ Why or why not? ______

   ______________________________________________



| | | | | | |
|---|---|---|---|---|---|
| blue | blue | blue | blue | blue | blue |
| blue | yellow | yellow | yellow | yellow | red |

**Note to the teacher:** Give each child a copy of this page, crayons, scissors, and a paper lunch bag.

# Sorting

## So Many "Pasta-bilities"

Pasta makes an inexpensive and interesting manipulative for sorting practice! Purchase several boxes in different shapes and colors. Assign students to small groups, and give each group a small amount of pasta. Challenge students to examine it and then decide how to sort it. Display each pasta assortment and have each group explain its method for sorting. It's "im-pasta-ble" to run out of options!

## All-About-Me Manipulative

This is just the "sort" of learning center that students will want to visit again and again! Give each child a strip of tagboard. Ask her to write her name on it and to decorate it. Laminate the strips and then adhere magnetic strips to the back of each one. To create the center, display the names on a magnetic surface. As a student visits the center, challenge her to sort the names in a variety of ways, such as by the number of letters, vowels, consonants, vowel sounds, and so forth. This center will be popular all year long!

## Anything Boxes

Turn your junk drawer into a learning experience! Give small groups of students "anything boxes" containing assorted items such as keys, buttons, small lids, and seashells. Direct students to examine and discuss the items. Then each group chooses criteria and sorts the objects. The students write their method for sorting on an index card and place it facedown near the objects. Once groups are done, invite each of them to examine each assortment. Once the group thinks it knows the method for sorting, it flips the index card over and checks its answer. Who knew so much learning could come from odds and ends?

Find more student practice on page 197–198.

Name ______________________ Date ______________

# Fishy Families

Cut apart the fish cards below.
Glue each card on the matching tank.

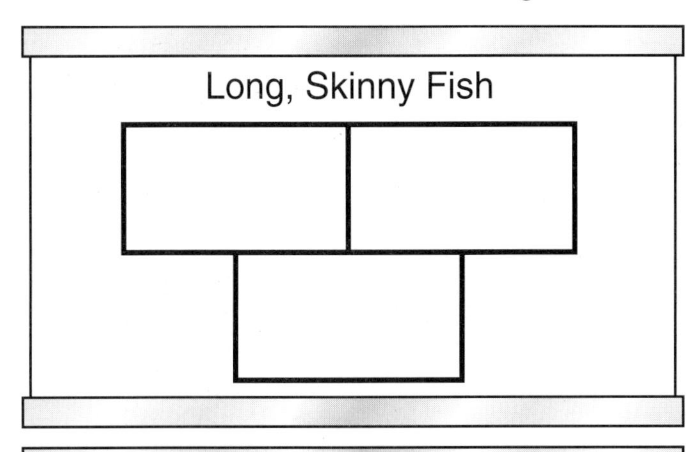


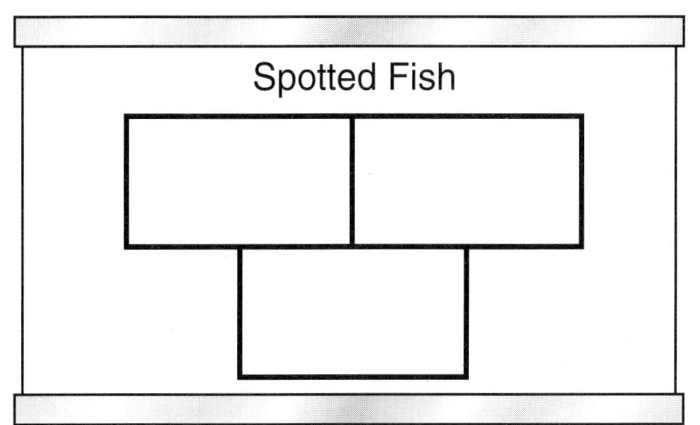


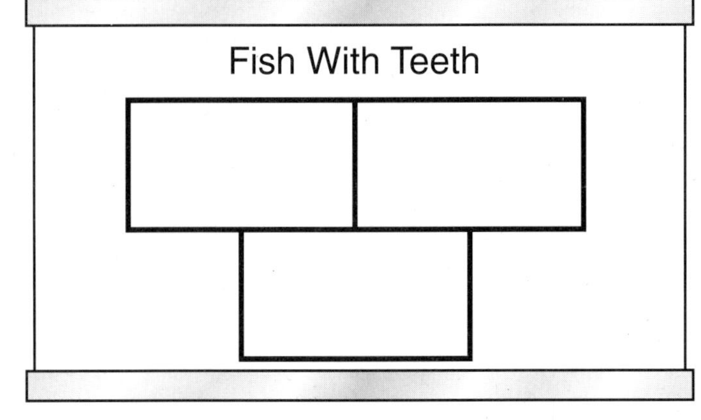


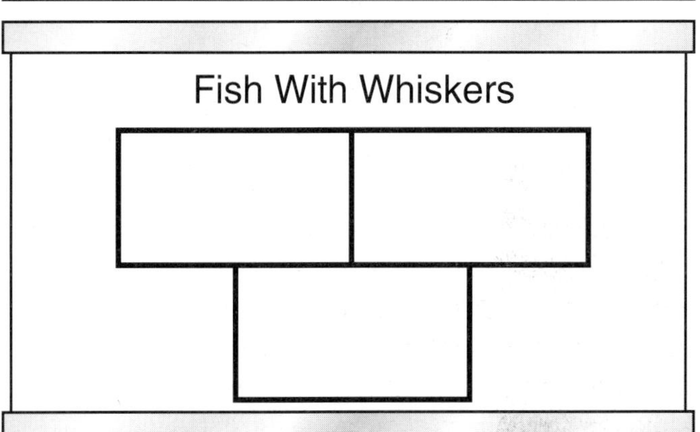


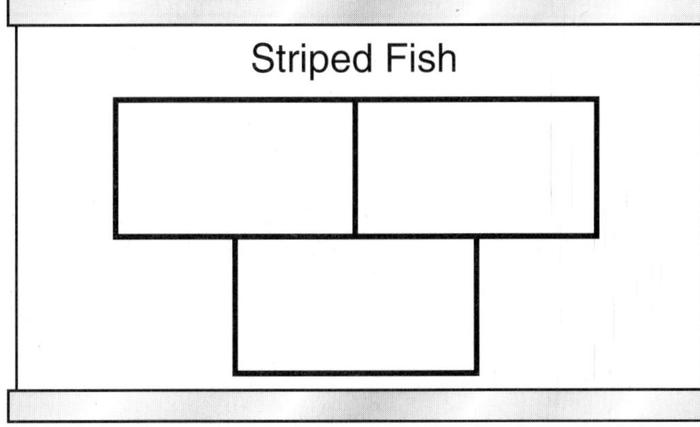


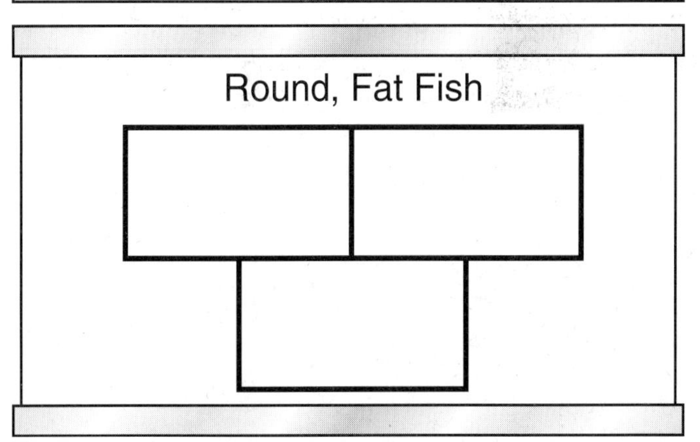

Name ______________________________ Date ____________________

# Miss Mouse Takes a Vacation

Read Miss Mouse's list.
Write the name of each object on the suitcase that is the best match.
Write four objects on each suitcase.

**Things to Pack**

- beach towel
- shoes
- socks
- comb
- toothbrush
- raft
- soap
- suntan lotion
- pants
- toothpaste
- shirts
- pool toy

# Classifying

## Follow the Rule

Introduce the skill of classifying with this quick pattern block activity. State the criteria for students to sort by, such as red blocks with four sides. Then display a set of blocks. Have students examine them and determine whether they fit the criteria. Repeat with more complicated parameters. Next, write individual criteria, such as the color of the block, number of sides, shape, and so forth, on separate index cards. Give pattern blocks to students. Explain that a child will pull two cards from the stack and read them to the class. Then have students arrange their pattern blocks according to the chosen guidelines. Students will classify this activity as fun!

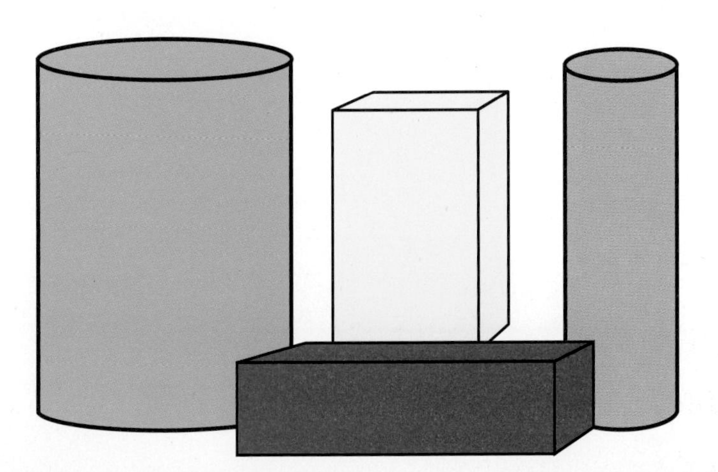


## Classification Station

Once students get the hang of classifying, why stick with just pattern blocks? Challenge students to find other items that can be used for classifying practice. For example, empty food containers, blocks, toy cars, candy bars, or even students themselves work well for classifying practice. Appoint one child to be in charge of the classification station each day. She brings in a set of items, places them at the learning center, and then writes criteria for classifying them. Students will be eager to visit the center to see—and classify—each day's assortment of items!

Find more student practice on pages 200–201.

Name ______________________________ Date ______________________

# Out-of-This-World Creatures!

Read each definition.
Circle the creatures that match the definition.

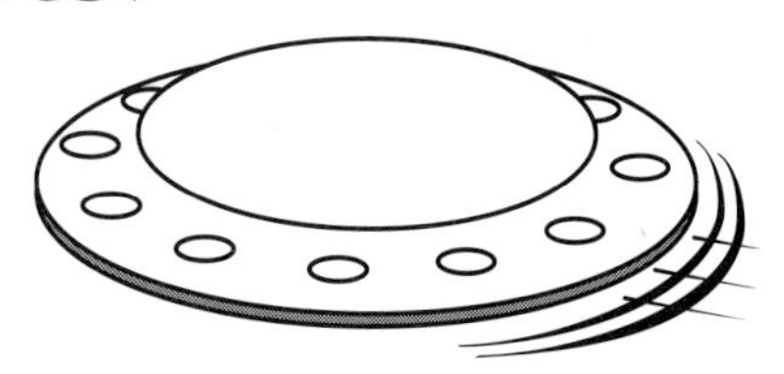

1. Acks have three eyes.

2. Otts have four feet.

3. Cluds have two ears and spots on their bellies.

4. Dints have one antenna and three feet.

5. Ells have stripes on their tails and two eyes.

6. Febs have long noses and short arms.

Name________________________ Date________________

# Now Playing...the Stripes!

Read the definitions.

- The instruments in the string family have strings.
- The instruments in the percussion family make music when someone hits them.
- The instruments in the brass family make music when someone blows into them.

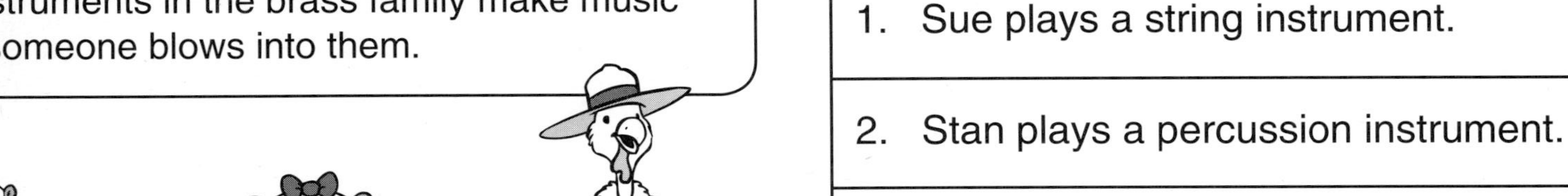

Stan Sara Tommy

Sally Simon Sue

Look at the band members.
Circle the correct letter for each sentence.

| | True | False |
|---|---|---|
| 1. Sue plays a string instrument. | C | R |
| 2. Stan plays a percussion instrument. | T | M |
| 3. Simon plays a brass instrument. | S | W |
| 4. Tommy plays a percussion instrument. | T | P |
| 5. Sally plays a brass instrument. | J | D |
| 6. Sara plays a percussion instrument. | U | B |

**Why did the skunks let the turkey join the band?**

To solve the riddle, write the circled letters above on the numbered lines below.

He had the __ R __ __ S __ I __ K __ !
(lines numbered: 5, 6, 2, 4, 1, 3)



Classifying: according to criteria

# Geometric Patterns

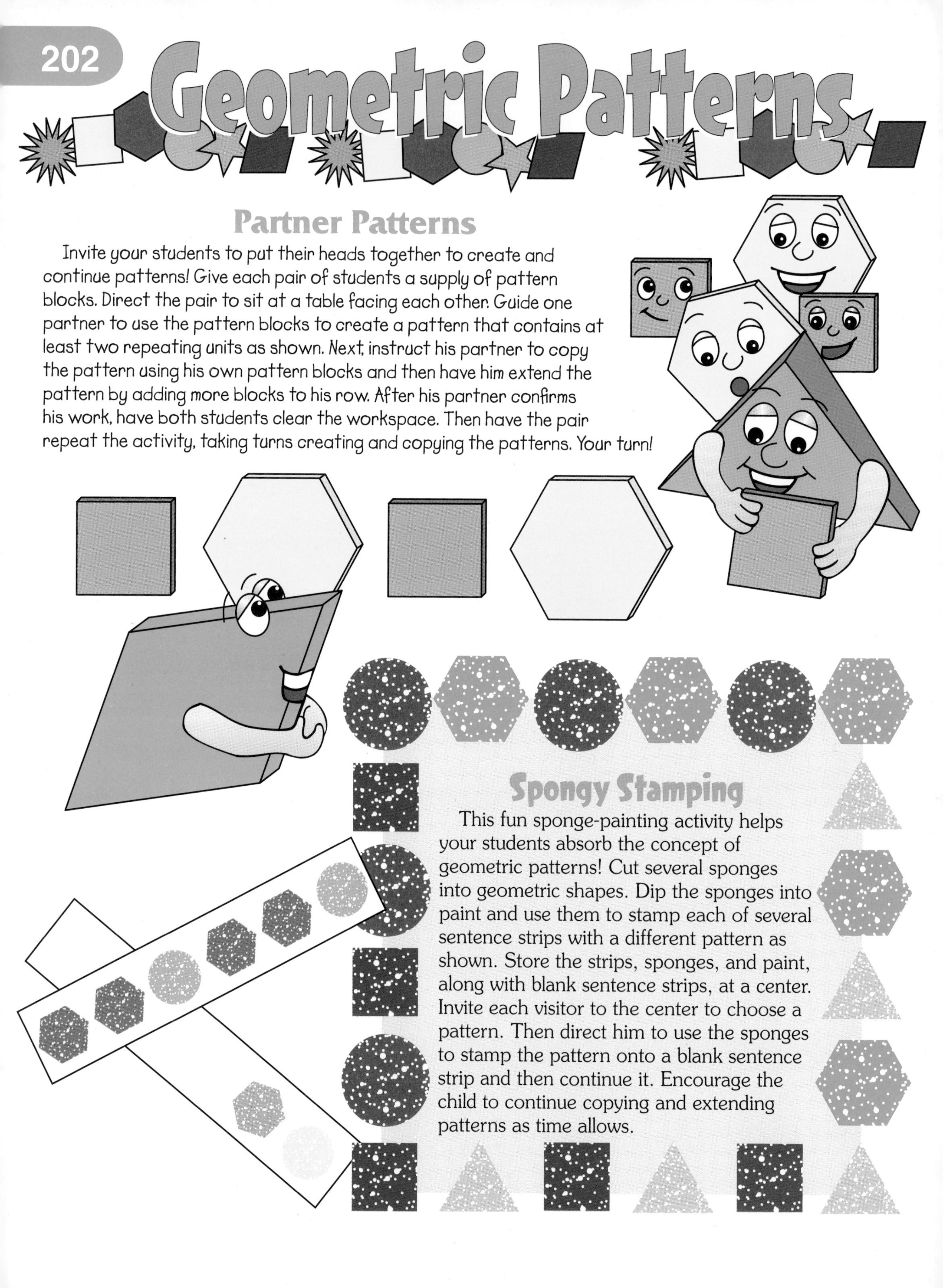

## Partner Patterns

Invite your students to put their heads together to create and continue patterns! Give each pair of students a supply of pattern blocks. Direct the pair to sit at a table facing each other. Guide one partner to use the pattern blocks to create a pattern that contains at least two repeating units as shown. Next, instruct his partner to copy the pattern using his own pattern blocks and then have him extend the pattern by adding more blocks to his row. After his partner confirms his work, have both students clear the workspace. Then have the pair repeat the activity, taking turns creating and copying the patterns. Your turn!

## Spongy Stamping

This fun sponge-painting activity helps your students absorb the concept of geometric patterns! Cut several sponges into geometric shapes. Dip the sponges into paint and use them to stamp each of several sentence strips with a different pattern as shown. Store the strips, sponges, and paint, along with blank sentence strips, at a center. Invite each visitor to the center to choose a pattern. Then direct him to use the sponges to stamp the pattern onto a blank sentence strip and then continue it. Encourage the child to continue copying and extending patterns as time allows.

## The Missing Piece

This partner activity really enhances students' understanding of geometric patterns! Give each pair a file folder and a supply of pattern blocks. Direct one partner to stand the file folder between herself and her partner to create a shield. Have her create a pattern with her blocks. Once she has finished the pattern, have her remove one piece from the pattern. Next, direct her to place the piece, along with three different unused pieces, in front of her partner. Then instruct her to remove the folder and have her partner study the pattern along with the four pieces in front of him. Have the partner choose the piece that belongs in the blank space and place it in the pattern. After his partner confirms his work, invite the partners to switch roles and repeat the activity.

## Pattern or Not?

For a tasty lesson, invite students to create these crunchy patterns! To begin, give each child a paper napkin and a large handful of Sociables crackers. Have the child unfold the napkin and use it to cover her desktop. To begin, specify two shapes for each student to place side by side on her napkin. Continue specifying shapes to be added until there are eight crackers in the row. (Make extra crackers available to students in case they do not have the named crackers.) Instruct students to study the row and give a thumbs-up if the row is a pattern or a thumbs-down if the row is not a pattern. Invite a student who gives the correct signal to explain why the row is or is not a pattern. Then have students clear their desktops and repeat the activity as time allows. If desired, have student volunteers take your place calling out cracker arrangements. Once the activity is complete, invite students to munch on their geometric pieces!

Find more student practice on pages 204–205.

Name ______________________ Date ______________________

# Crunchy Cookies

Two cookies on each cookie tray are switched.
Find the pattern and then circle the switched cookies.
Cut out the cookie cards below and glue them in
place to make the correct pattern.

Name ____________________ Date ____________

# Bugs and Blossoms

Cut apart the shape cards below.
Start with 1 across.
Then complete each row and column in order by gluing each card in place.

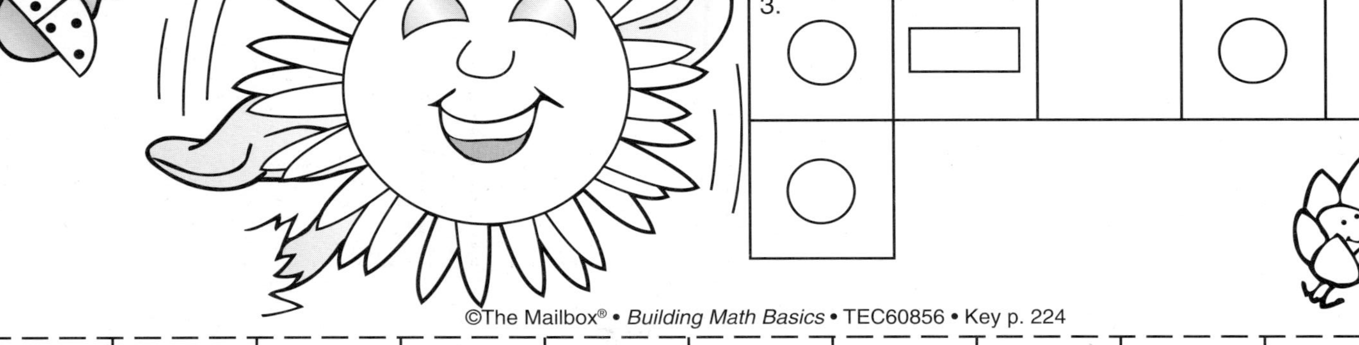

# Number Patterns

## Dangle a Caret

Could your students use a little help figuring out number patterns? Turn to a tool that's typically used for writing and editing—the caret, which shows where words should be added. Explain that when a student wants to determine a pattern, she writes a caret between the first and second numbers. Then she determines the relationship between the numbers and writes it above the caret. Next, she writes a caret between the second and third numbers, determines the difference, and writes it above the caret. The student continues until the pattern emerges. What a helpful hint!

+2 +2 +2 +2
2, ^ 4, ^ 6, ^ 8, ^ 10

−2 −3 −4 −5
21, ^ 19, ^ 16, ^ 12, ^ 7

## What a Snap!

Add a little rhythm to number patterns with this skip-counting activity! Choose a number to skip-count by, such as five. Have students sit in a circle. Show the children a clap-and-slap pattern by slapping your knees twice, clapping your hands twice, and then snapping with the right hand and snapping with the left. Explain that on the snaps, a student will say the next number in the pattern. Go around the circle until each child has had a turn. What very snappy number patterns!

## From 1 to 1,000

Invite students to join the 1,000 Club, and make practice with number patterns fun! Have students sit in a circle. Choose a number to skip-count by, such as by 30. Instruct students to go around the circle and say the next number in the pattern. Invite the student who reaches 1,000 to become the first member of the 1,000 Club. He gets to sit in the middle of the circle and choose the next number to skip-count by, based on your suggestions. Then students skip-count by that number until a new club member is chosen. Number patterns will be no problem!

5

11

13

## Pick Your Own Pattern

Challenge students to create their own number patterns! Here's how. To prepare, write each number from 1 to 9 on an index card. Place the cards number side down in a basket. A student randomly selects three cards. Challenge him to study the cards and determine how to add and subtract to link the numbers. Then have each student write the pattern and extend it for two more cycles. After you check the pattern, the student recopies it and passes it to another student to solve. These customized patterns are sure to be popular!

5, 11, 13, 19, 21, 27, 29, 35, 37

Find more student practice on page 208–209.

Name____________________ Date __________

# Patterns at the Pond

Read each rule.
Glue a lily pad over the mistake in each pattern.
Write the correct number on the lily pad.

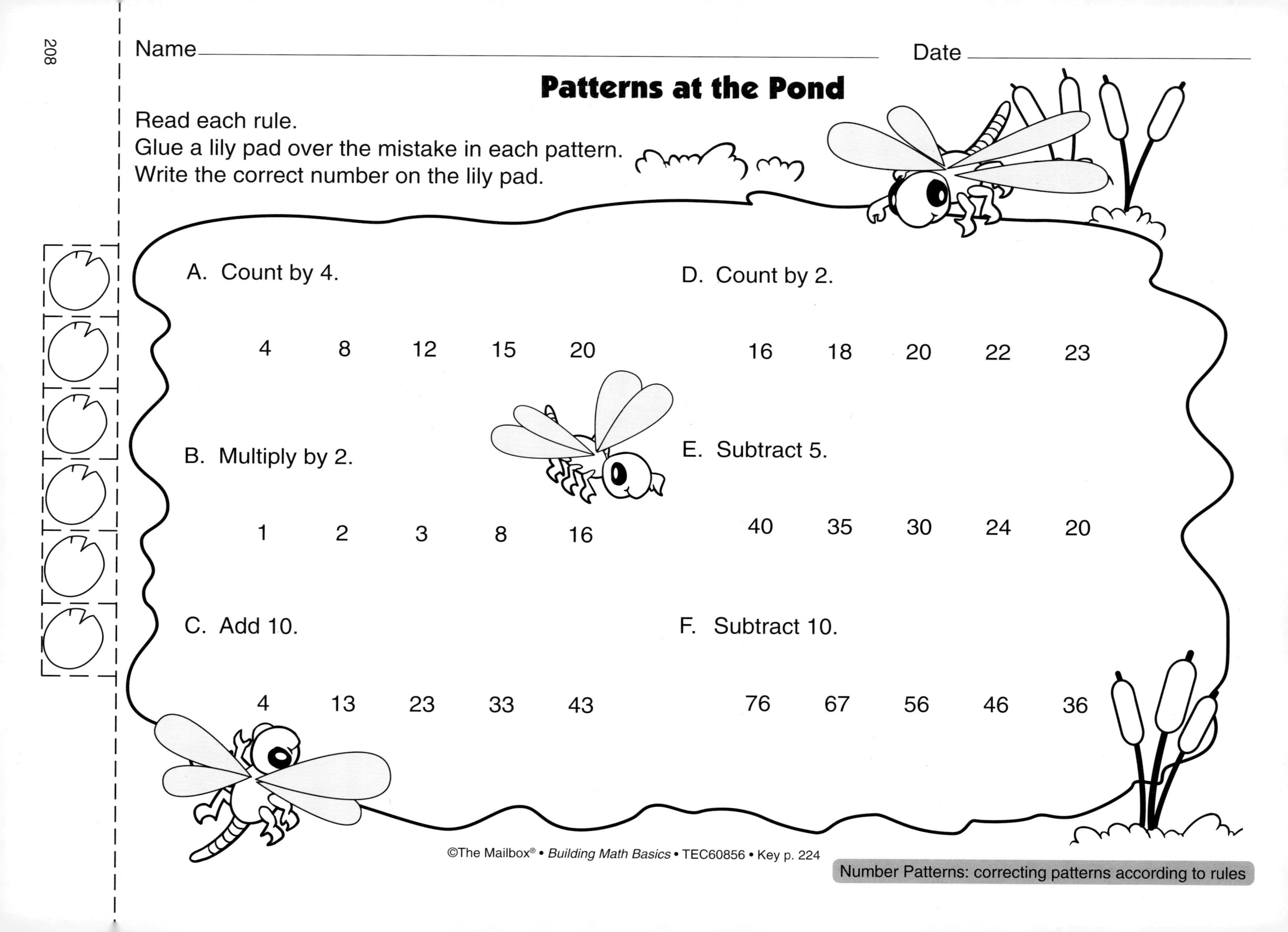

A. Count by 4.

4 8 12 15 20

B. Multiply by 2.

1 2 3 8 16

C. Add 10.

4 13 23 33 43

D. Count by 2.

16 18 20 22 23

E. Subtract 5.

40 35 30 24 20

F. Subtract 10.

76 67 56 46 36

Key p. 224

Name ______________________ Date ______________

# Sky-High Panda

Write the rule for each number pattern.
Use the rules to help you.
Add three more numbers to each pattern.

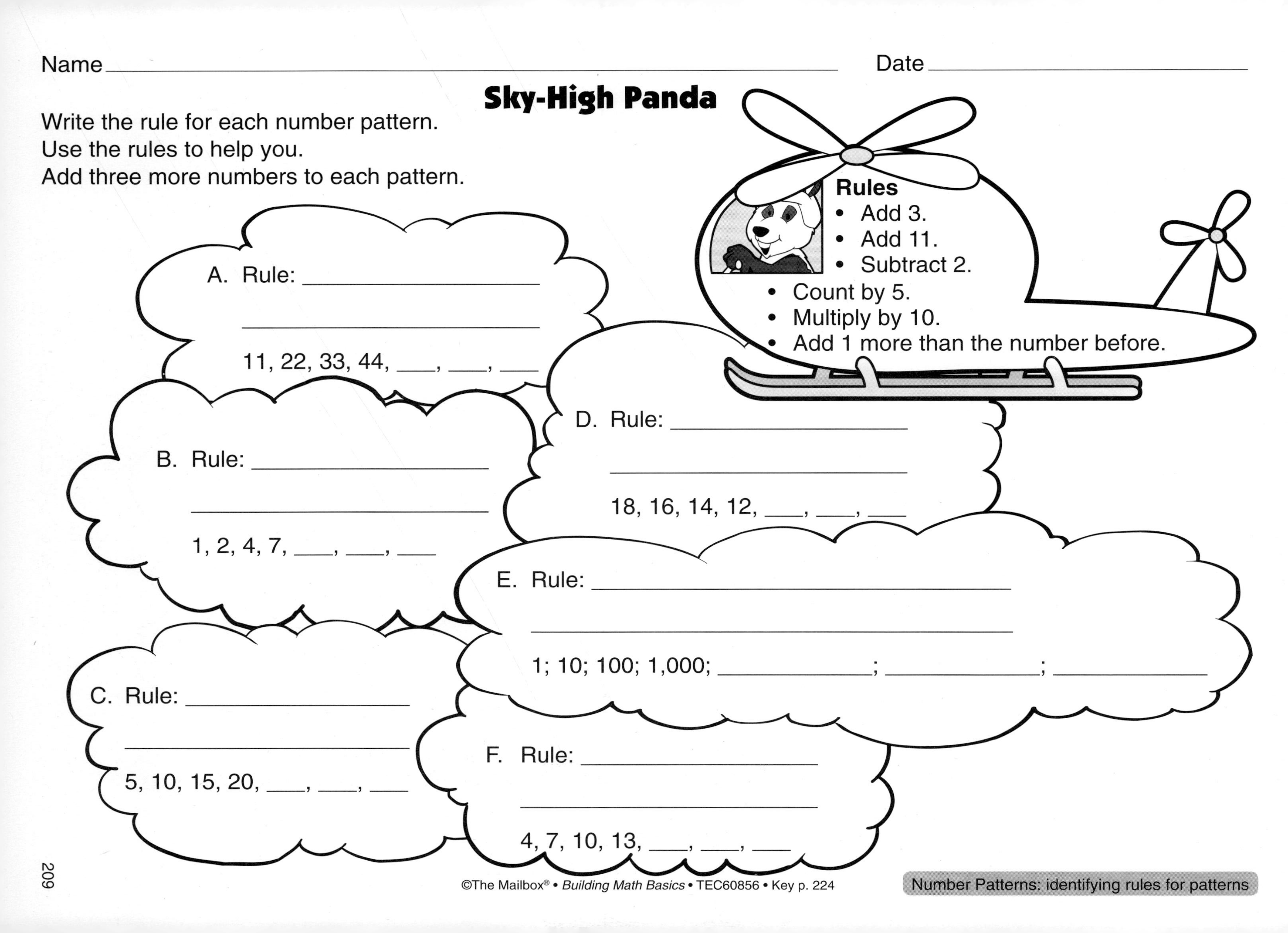


A. Rule: ______________________

______________________

11, 22, 33, 44, ___, ___, ___

B. Rule: ______________________

______________________

1, 2, 4, 7, ___, ___, ___

C. Rule: ______________________

______________________

5, 10, 15, 20, ___, ___, ___

D. Rule: ______________________

______________________

18, 16, 14, 12, ___, ___, ___

E. Rule: ______________________

______________________

1; 10; 100; 1,000; ________; ________; ________

F. Rule: ______________________

______________________

4, 7, 10, 13, ___, ___, ___



Number Patterns: identifying rules for patterns

# Logical Reasoning

## Clever Climb

Take students' logic skills to new heights with this partner game! Give each pair of students a 12-inch tagboard strip and 11 paper clips. Explain that starting at the bottom of the strip each student, in turn, will "climb" up the strip by clipping one, two, or three paper clips to the strip. The winner is the player who forces his partner to attach the last paper clip. After a few rounds, invite students to share their winning strategies. If desired, reverse the game and challenge students to try to be the first person to attach the final paper clip. This game is tops!

## Right in Line

Here's a picture-perfect way to practice reasoning skills! Cut a piece of tagboard to 36" x 12". Holding it horizontally, clip five paper clips across the top; then write the numbers 1–5 across the bottom. Post it at the front of the classroom. Next, have each child write her name at the bottom of a vertically positioned 6" x 9" piece of white construction paper. Then ask her to draw her self-portrait on the page.

Divide students into groups of up to five. Each group places its pictures in a row and writes logic clues to help the class figure out the order of the illustrations. Check the clues. Next, one group clips its pictures to the tagboard in order, with the illustrations hidden. Give the rest of the class small squares of paper to represent the portraits. Have students label each of their squares with a group member's name and then use the squares to figure out the problem. To begin, one group member reads the group's first clue aloud. (If desired, the student could also write the clue on the board.) Class members arrange their paper squares based on the clue. Repeat until all the clues have been read. When a student guesses the order correctly, unveil the portraits. What a lineup!

# Got Your Number

Count on this idea to show students the benefits of logic boxes! Direct each child to write his name and a three-digit number on a slip of paper. Then ask four students to stand at the front of the room, show each other their numbers, and then pass you their slips of paper. On the board, list their numbers across the top of a 4 row x 4 column chart and their names down the side. Do not reveal which student has which number. Have students create a similar chart on drawing paper.

Challenge the class to figure out which student chose each number by asking number-related yes-or-no questions. For example, a student might ask, "Chris, is your number larger than Katie's number?" or "Nina, does your number have a three in the tens place?" If a question has two possible answers, direct the student to rephrase the question so that it has only one answer. To give the answer, the child who was asked the question writes a check mark for "yes" or an X for "no" in the appropriate space on the chart. Students copy that information onto their charts. Model how the chart helps to determine the answers. Very logical!

| | 132 | 371 | 512 | 809 |
|---|---|---|---|---|
| Nina | | | | |
| Chris | | | | |
| Katie | | | | |
| Rex | | | | |

## Shaping Up

Shape up students' thinking skills with this brain-tickling activity! Write the shape equation shown on the board. First, have students list all of the addends of 8 *(1 + 7, 2 + 6, 3 + 5, 4 + 4)*. Next, direct students to study the second part of the problem and determine the addends that have a difference of four *(6 – 2)*. Then challenge students to determine the value of each shape *(triangle = 6, circle = 2)*. Guide students through several examples until they feel comfortable.

To extend the activity, have each student cut two different shapes from construction paper. The child writes a number on the back of each shape. Direct her to place the shapes atop a plain sheet of paper (number side down) and write an addition sentence and a subtraction sentence like the ones shown. Then pair students and challenge them to solve one another's problems. Students can also trade shapes and write new number sentences. The activity is sure to shape up to be lots of fun!

Find more student practice on pages 212–213.

Name ______________________________ Date ____________________

# Center Court

Read the statements and complete each chart.
Write a ✓ if the statement is true.
Write an X if the statement is not true.

How many points did each player score?

- Rita scored the most points.
- Jim scored one more point than Kate.

| | 25 | 32 | 33 | 50 |
|---|---|---|---|---|
| Al | | | | |
| Jim | | | | |
| Kate | | | | |
| Rita | | | | |

Which jersey number belongs to each player?

- One of the digits in Jim's number is 4.
- Rita's number has a 5 in the tens place.
- Al's number is smaller than Jim's number.

| | 27 | 53 | 74 | 95 |
|---|---|---|---|---|
| Al | | | | |
| Jim | | | | |
| Kate | | | | |
| Rita | | | | |

Name ____________________ Date ____________

# Hooray for Weekends!

Read the clues and complete each logic chart.
Write a ✓ if the statement is true and an X if it is not true.

- Fran wears special footwear and practices her hobby in the park.
- Harry needs a bat to practice his hobby.
- Jerry's hobby requires paper.
- Polly's hobby requires paper and is not drawing.
- Frank needs an oven to practice his hobby.

| | baseball | drawing | in-line skating | reading | baking |
|---|---|---|---|---|---|
| Harry | | | | | |
| Fran | | | | | |
| Jerry | | | | | |
| Polly | | | | | |
| Frank | | | | | |

- John's sport is played on a court.
- Patty goes to the pool to practice her sport.
- Debbie's sport is not basketball.
- Chuck needs a racket for his sport.
- Tina goes to the stables to practice her sport.

| | running | basketball | tennis | swimming | horseback riding |
|---|---|---|---|---|---|
| Chuck | | | | | |
| Patty | | | | | |
| John | | | | | |
| Tina | | | | | |
| Debbie | | | | | |



 • Key p. 224

# Answer Keys

**Page 7**

1. 754
2. 891
3. 132
4. 679
5. 468
6. 583
7. 335
8. 246

**Page 8**

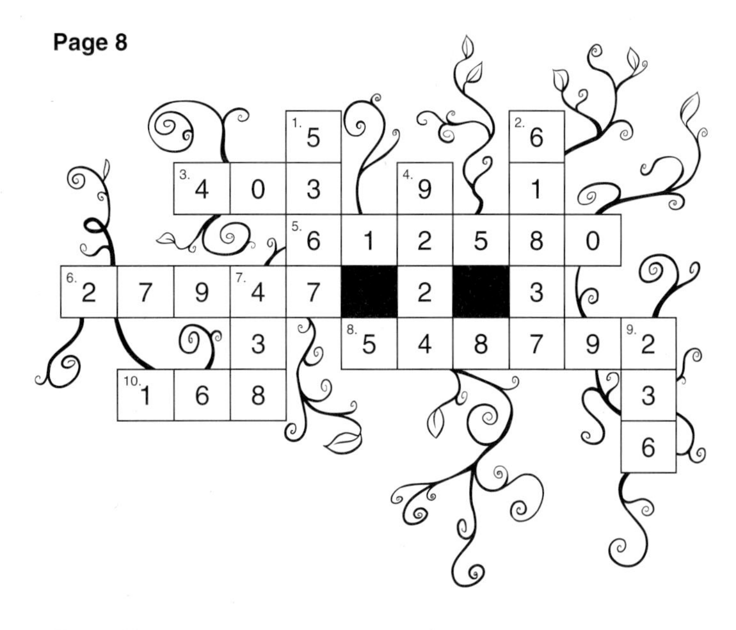


**Page 14**

Order of answers on each pin may vary.

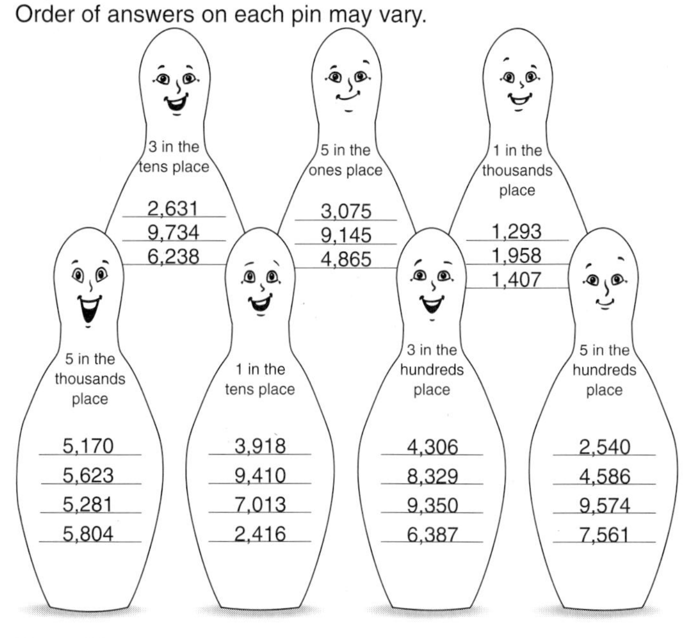


**Page 11**

1. 14, 15, 16, 17, 18, 19, 20
2. 136, 137, 138, 139, 140, 141, 142
3. 261, 262, 263, 264, 265, 266, 267
4. 398, 399, 400, 401, 402, 403, 404
5. 524, 525, 526, 527, 528, 529, 530
6. 675, 676, 677, 678, 679, 680, 681
7. 739, 740, 741, 742, 743, 744, 745
8. 852, 853, 854, 855, 856, 857, 858
9. 895, 896, 897, 898, 899, 900, 901
10. 917, 918, 919, 920, 921, 922, 923

Because he OVERSWEPT!

**Page 12**

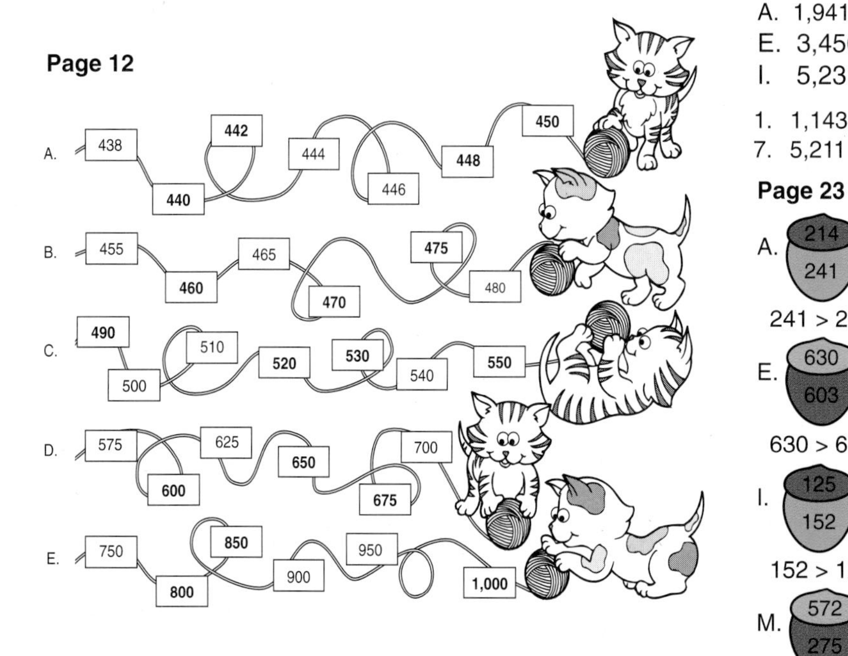


**Page 15**

a. 401
b. 6,239
c. 1,345
d. 28,650
e. 698
f. 5,978
g. 42,180
h. 72,346

**Page 19**

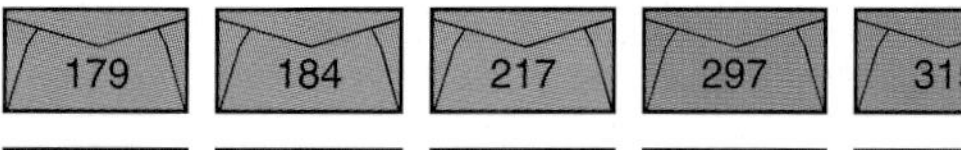

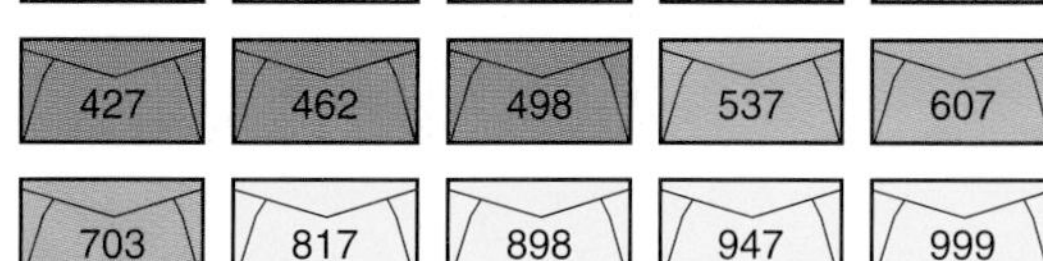


**Page 20**

| | | | |
|---|---|---|---|
| A. 1,941 | B. 2,049 | C. 1,431 | D. 6,729 |
| E. 3,456 | F. 9,800 | G. 7,567 | H. 4,987 |
| I. 5,231 | J. 8,901 | K. 9,786 | L. 3,856 |

| | | | | | |
|---|---|---|---|---|---|
| 1. 1,143 | 2. 1,546 | 3. 2,021 | 4. 3,269 | 5. 3,290 | 6. 4,657 |
| 7. 5,211 | 8. 6,042 | 9. 7,489 | 10. 8,747 | 11. 8,900 | 12. 9,546 |

**Page 23**

A. 214 241

241 > 214

B. 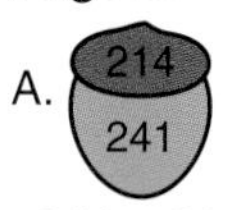


304 < 340

C. 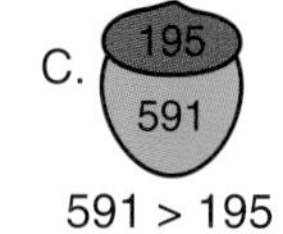


591 > 195

D. 682 826

682 < 826

E. 630 603

630 > 603

F. 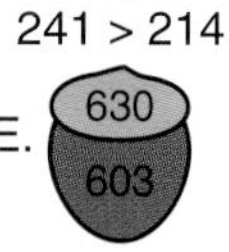


947 > 794

G. 289 298

298 > 289

H. 551 515

515 < 551

I. 125 152

152 > 125

J. 724 722

722 < 724

K. 

918 > 891

L. 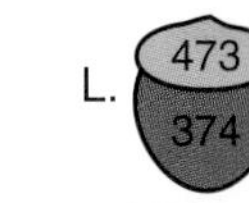


374 < 473

M. 

275 > 572

N. 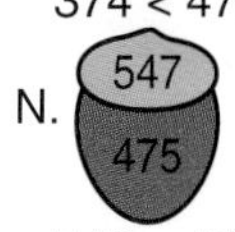


547 > 475

## Page 27

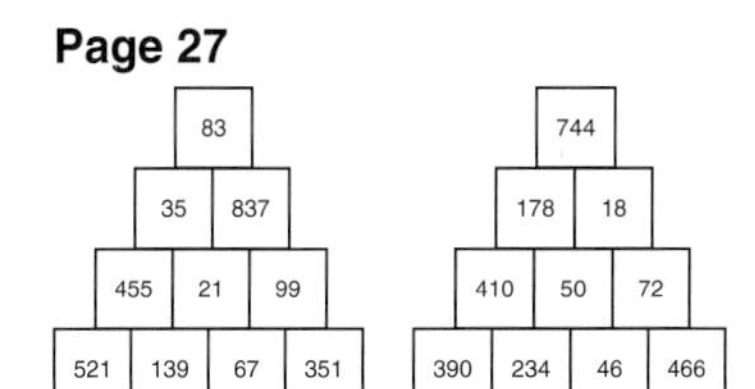


## Page 28

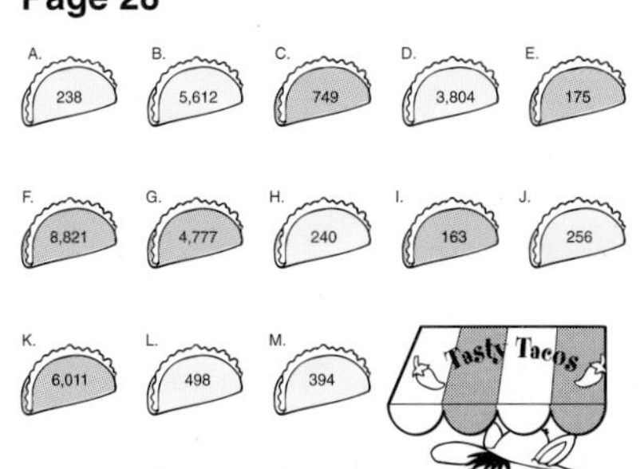


## Page 32

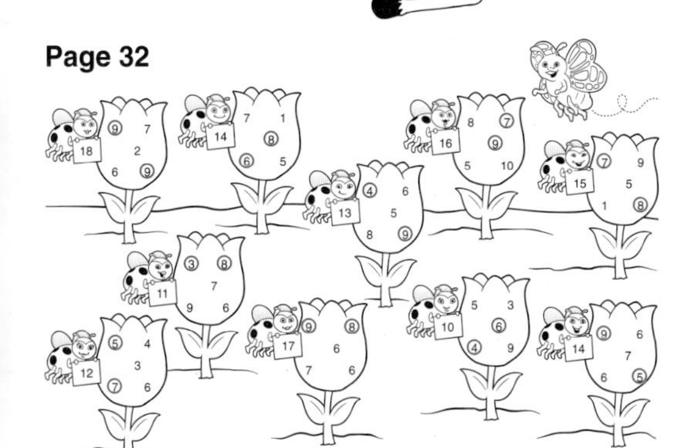


## Page 33

Order of answers may vary.

A.
9 + 4 = 13
6 + 7 = 13
8 + 5 = 13

B.
4 + 6 = 10
5 + 5 = 10
9 + 1 = 10

C.
6 + 3 = 9
1 + 8 = 9
5 + 4 = 9

D.
7 + 7 = 14
6 + 8 = 14
5 + 9 = 14

E.
8 + 3 = 11
5 + 6 = 11
9 + 2 = 11

F.
4 + 8 = 12
5 + 7 = 12
6 + 6 = 12

## Page 34

Addition facts for leftover sums will vary.

## Page 35

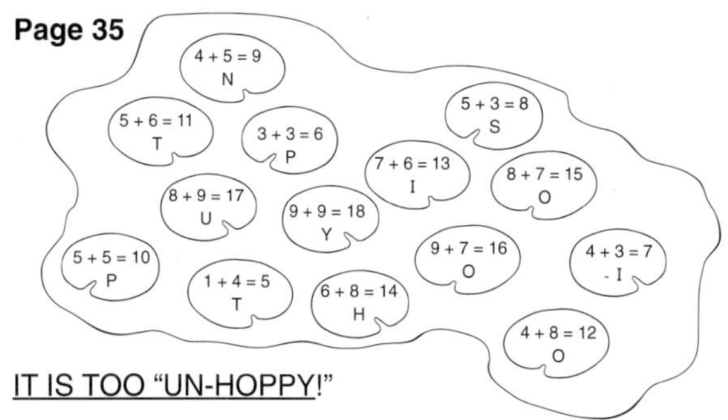


IT IS TOO "UN-HOPPY!"

## Page 36

| | | | | | |
|---|---|---|---|---|---|
| 6 | 6 | 12 | 7 | 8 | 15 |
| 8 | 5 | 5 | 10 | 3 | 2 |
| 14 | 7 | 6 | 13 | 11 | 2 |
| 9 | 9 | 18 | 5 | 2 | 7 |
| 1 | 16 | 4 | 14 | 6 | 9 |
| 10 | 3 | 2 | 5 | 15 | 12 |

## Page 40

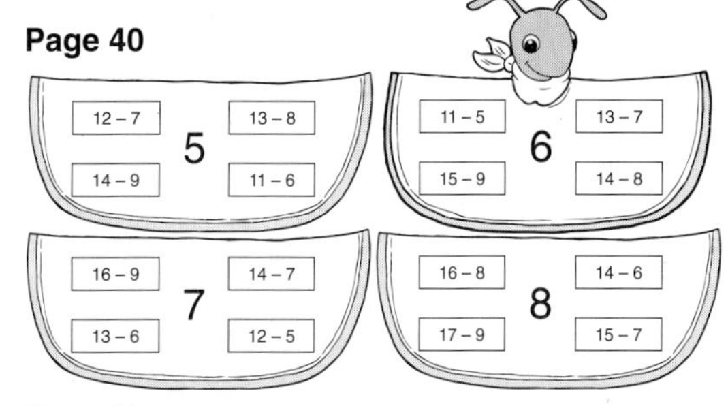


## Page 41

## Page 42

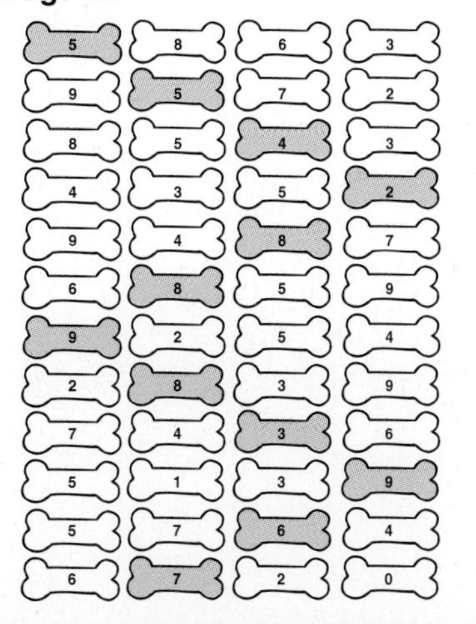

**Page 43**

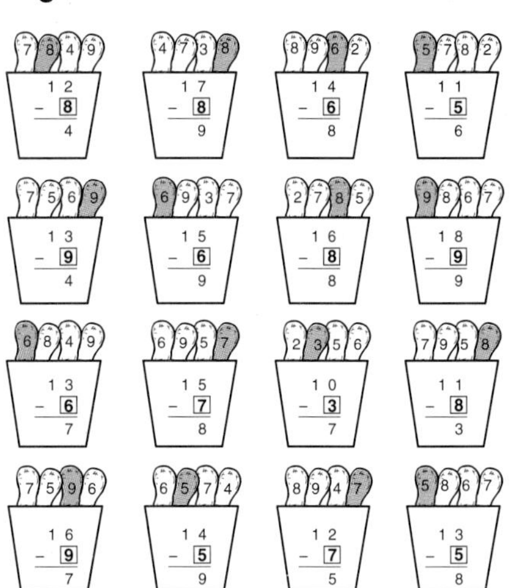


**Page 49**
Order of answers may vary.

**Page 44**

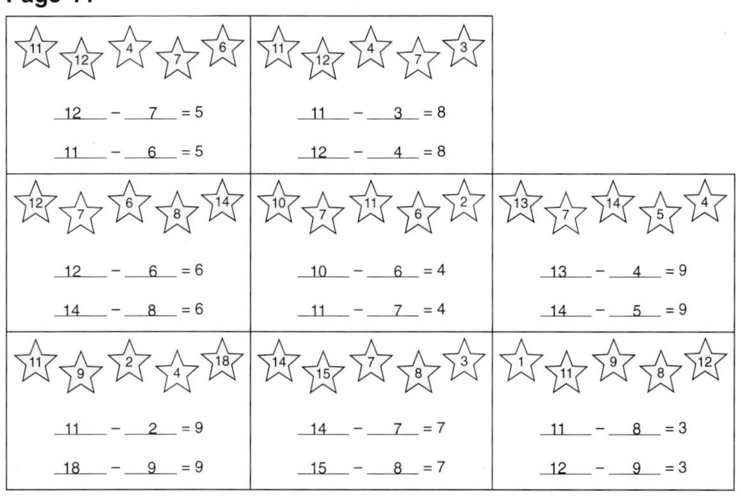


**Page 59**

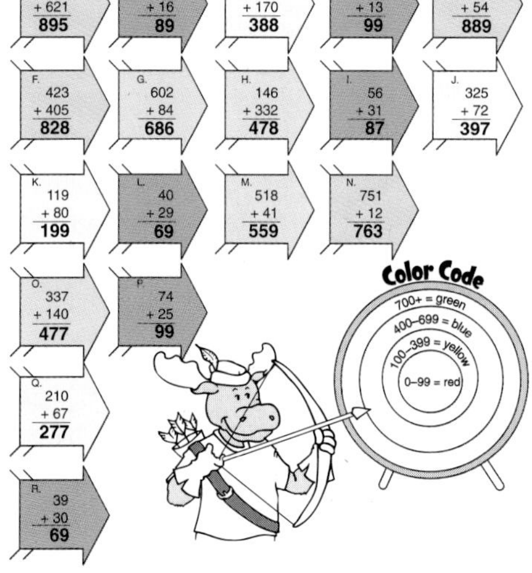


**Page 48**
Order of answers may vary.

A. 6 + 7 = 13
7 + 6 = 13
13 – 6 = 7
13 – 7 = 6

B. 8 + 9 = 17
9 + 8 = 17
17 – 8 = 9
17 – 9 = 8

C. 5 + 4 = 9
4 + 5 = 9
9 – 5 = 4
9 – 4 = 5

D. 3 + 9 = 12
9 + 3 = 12
12 – 3 = 9
12 – 9 = 3

E. 8 + 6 = 14
6 + 8 = 14
14 – 8 = 6
14 – 6 = 8

**Page 56**
a. 17
b. 69
c. 79
d. 67
e. 74
f. 65
g. 93
h. 78
i. 76
j. 68
k. 99
l. 96
m. 85
n. 44
o. 53

**Page 57**
A. 142
B. 50
C. 150
D. 90
E. 75
F. 132
G. 111
H. 113
I. 131
J. 124
K. 64
L. 102
M. 95
N. 191
O. 83
P. 86
Q. 63
R. 130

**Page 58**
Sundae 1
Across
A. 789
D. 844
E. 985

Down
A. 789
B. 848
C. 945

Sundae 2
Across
A. 554
D. 989
E. 647

Down
A. 596
B. 584
C. 497

Sundae 3
Across
A. 689
D. 877
E. 796

Down
A. 687
B. 879
C. 976

**Page 60**
O. 90
R. 344
U. 532
D. 1,125
H. 548
Y. 945
H. 335
V. 91
I. 917
E. 811
A. 141
S. 136
L. 1,037
T. 840
T. 72

<u>H A V E</u>  <u>Y O U</u>  <u>H E A R D</u>
<u>T H E</u>  <u>L A T E S T</u>  <u>D I R T</u>?

## Page 64

| | |
|---|---|
| A. 12 | K. 40 |
| B. 46 | L. 50 |
| C. 42 | M. 53 |
| D. 33 | N. 21 |
| E. 23 | O. 24 |
| F. 71 | P. 52 |
| G. 6 | Q. 30 |
| H. 61 | R. 7 |
| I. 34 | S. 51 |
| J. 23 | T. 28 |

## Page 65

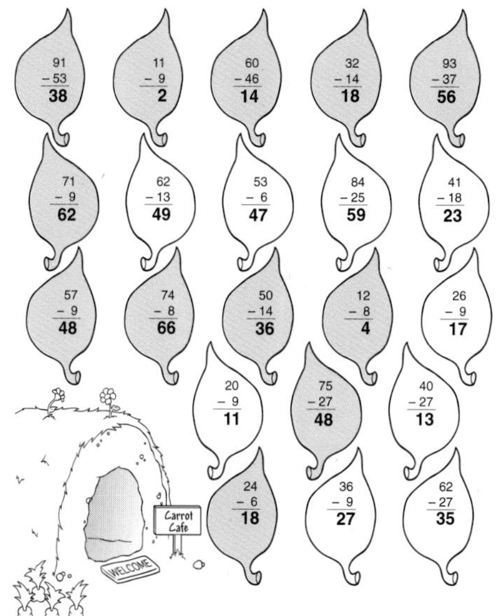


## Page 66

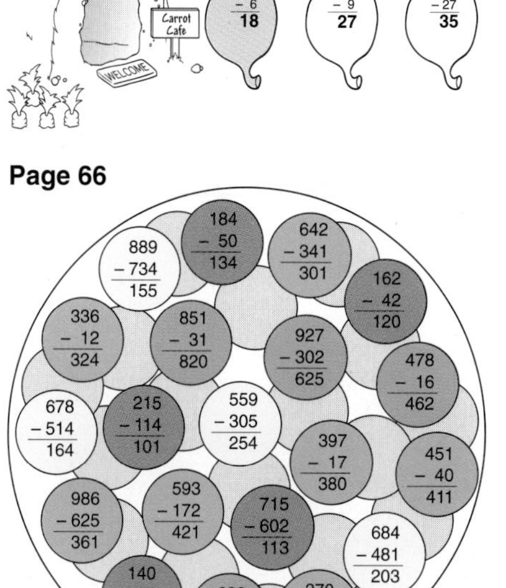


## Page 67

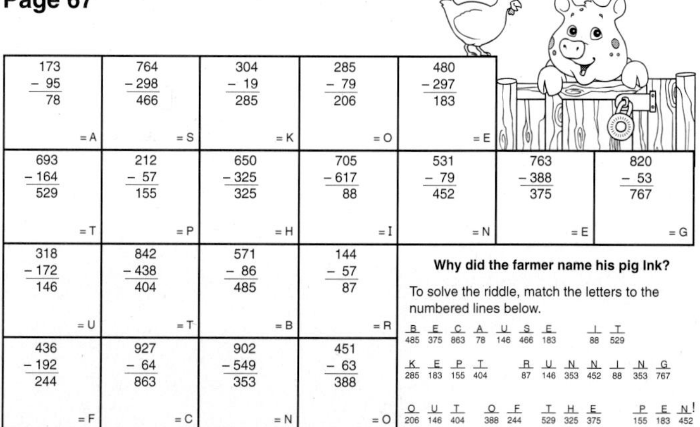

| | | | | | | |
|---|---|---|---|---|---|---|
| 173 − 95 = 78 = A | 764 − 298 = 466 = S | 304 − 19 = 285 = K | 285 − 79 = 206 = O | 480 − 297 = 183 = E | | |
| 693 − 164 = 529 = T | 212 − 57 = 155 = P | 650 − 325 = 325 = H | 705 − 617 = 88 = I | 531 − 79 = 452 = N | 763 − 388 = 375 = E | 820 − 53 = 767 = G |
| 318 − 172 = 146 = U | 842 − 438 = 404 = T | 571 − 86 = 485 = B | 144 − 57 = 87 = R | | | |
| 436 − 192 = 244 = F | 927 − 64 = 863 = C | 902 − 549 = 353 = N | 451 − 63 = 388 = O | | | |

**Why did the farmer name his pig Ink?**

To solve the riddle, match the letters to the numbered lines below.

B E C A U S E (485 375 863 78 146 466 183) I T (88 529)
K E P T (285 183 155 404) R U N N I N G (87 146 353 452 88 353 767)
O U T (206 146 404) O F (388 244) T H E (529 325 375) P E N! (155 183 452)

## Page 68

Across
A. 5,843
E. 4,312
F. 2,130
G. 1,242

Down
A. 5,421
B. 8,312
C. 4,134
D. 3,202

Across
A. 7,362
E. 4,616
F. 3,264
G. 5,251

Down
A. 7,435
B. 3,622
C. 6,165
D. 2,641

## Page 71

A. 4 + 4 = 8, 2 x 4 = 8
B. 2 + 2 + 2 = 6, 2 x 3 = 6
C. 5 + 5 = 10, 5 x 2 = 10
D. 3 + 3 + 3 = 9, 3 x 3 = 9
E. 6 + 6 + 6 = 18, 6 x 3 = 18
F. 2 + 2 + 2 + 2 = 8, 2 x 4 = 8
G. 4 + 4 + 4 + 4 = 16, 4 x 4 = 16
H. 4 + 4 + 4 + 4 + 4 = 20, 4 x 5 = 20

## Page 72

A. 2 x 3 = 6
B. 4 x 2 = 8
C. 3 x 4 = 12
D. 2 x 6 = 12
E. 5 x 2 = 10
F. 4 x 4 = 16
G. 6 x 3 = 18
H. 3 x 5 = 15

## Page 76

Code: △ = 0 □ = 1 ○ = 2

a. 3 x 2 = 6
b. 1 x 7 = 7
c. 1 x 6 = 6
d. 0 x 2 = 0
e. 1 x 0 = 0
f. 5 x 1 = 5
g. 4 x 0 = 0
h. 2 x 9 = 18
i. 4 x 2 = 8
j. 0 x 8 = 0
k. 2 x 1 = 2
l. 2 x 7 = 14
m. 5 x 0 = 0
n. 1 x 9 = 9
o. 0 x 7 = 0
p. 1 x 2 = 2
q. 2 x 6 = 12
r. 3 x 0 = 0
s. 1 x 8 = 8

## Page 77

## Page 78

| | |
|---|---|
| A. 2 x 3 = 6 | H. 1 x 3 = 3 |
| B. 5 x 4 = 20 | I. 4 x 4 = 16 |
| C. 7 x 3 = 21 | J. 9 x 3 = 27 |
| D. 3 x 1 = 3 | K. 5 x 1 = 5 |
| E. 9 x 4 = 36 | L. 2 x 4 = 8 |
| F. 4 x 2 = 8 | M. 6 x 2 = 12 |
| G. 6 x 4 = 24 | |

## Page 79

## Page 80

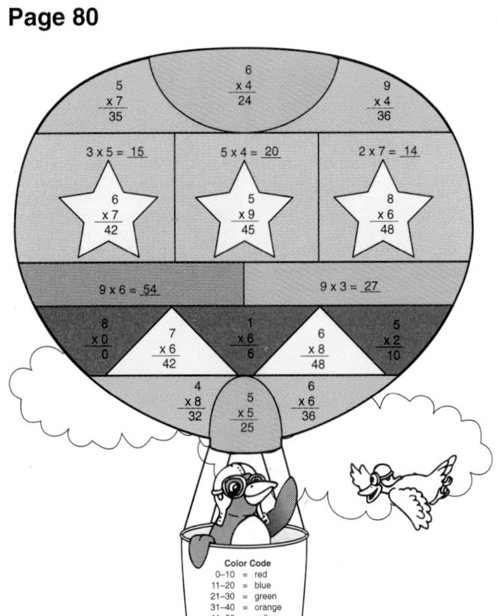


## Page 81

a. 6 x 3 = 18
b. 7 x 0 = 0
c. 4 x 5 = 20
d. 6 x 8 = 48
e. 9 x 2 = 18
f. 7 x 6 = 42
g. 6 x 1 = 6
h. 0 x 9 = 0
i. 5 x 3 = 15
j. 5 x 5 = 25
k. 6 x 4 = 24
l. 5 x 6 = 30
m. 2 x 8 = 16
n. 1 x 2 = 2
o. 8 x 5 = 40
p. 0 x 1 = 0

## Page 82

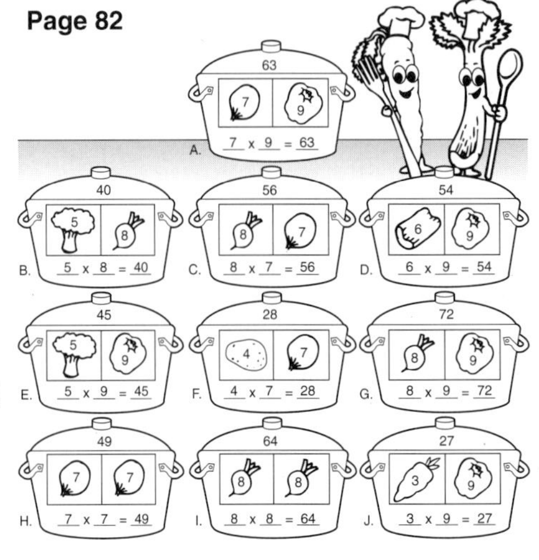


## Page 83

a. 9 x 8 = 72
b. 8 x 8 = 64
c. 1 x 8 = 8
d. 7 x 5 = 35
e. 9 x 9 = 81
f. 9 x 4 = 36
g. 6 x 7 = 42
h. 8 x 2 = 16
i. 9 x 6 = 54
j. 8 x 5 = 40
k. 2 x 9 = 18
l. 3 x 7 = 21
m. 6 x 8 = 48
n. 7 x 7 = 49
o. 1 x 7 = 7
p. 9 x 5 = 45
q. 8 x 8 = 64
r. 7 x 4 = 28
s. 9 x 3 = 27
t. 2 x 7 = 14
u. 8 x 3 = 24
v. 1 x 9 = 9
w. 8 x 7 = 56
x. 8 x 4 = 32

## Page 86

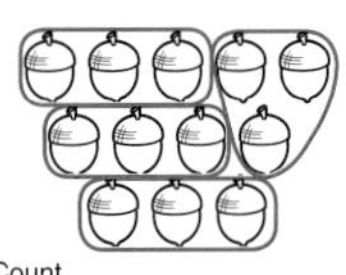

Count.
Circle groups of 3.
How many groups? 4
12 ÷ 3 = 4 groups

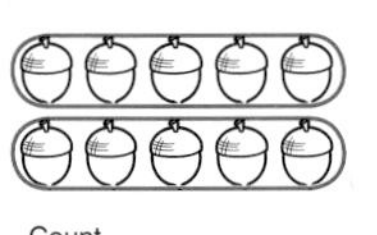

Count.
Circle groups of 5.
How many groups? 2
10 ÷ 5 = 2 groups

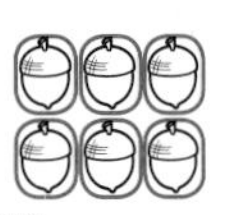

Count.
Circle groups of 1.
How many groups? 6
6 ÷ 1 = 6 groups

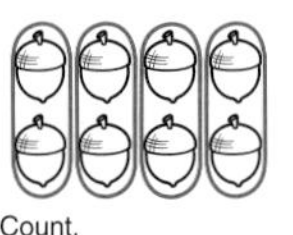

Count.
Circle groups of 2.
How many groups? 4
8 ÷ 2 = 4 groups

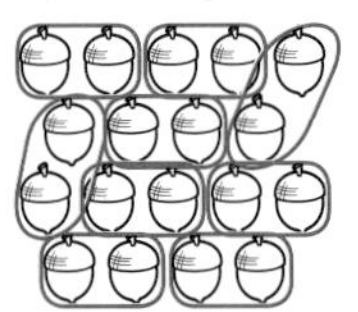

Count.
Circle groups of 2.
How many groups? 9
18 ÷ 2 = 9 groups

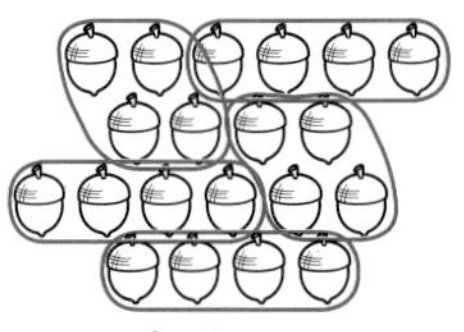

Count.
Circle groups of 4.
How many groups? 5
20 ÷ 4 = 5 groups

## Page 87

A. 20, 5, 4
20 ÷ 5 = 4

B. 6, 2, 3
6 ÷ 2 = 3

C. 15, 3, 5
15 ÷ 3 = 5

D. 27, 3, 9
27 ÷ 3 = 9

E. 24, 4, 6
24 ÷ 4 = 6

F. 10, 5, 2
10 ÷ 5 = 2

## Page 91

Letters A, C, D, E, F, I, J, K, L, N, T, V and W should be colored.

## Page 92

a. 1
b. 0
c. 1
d. 0
e. 2
f. 0
g. 1
h. 2
i. 0
j. 2
k. 1
l. 0
m. 2
n. 0
o. 2
p. 2
q. 1
r. 2
s. 0
t. 2
u. 1
v. 1
w. 1
x. 2
y. 0
z. 0

## Page 93

| Race | Total Number of Wheels | Problem | Number of Racers |
|---|---|---|---|
| A. unicycles | 4 | 4 ÷ 1 = | 4 |
| B. tricycles | 18 | 18 ÷ 3 = | 6 |
| C. wagons | 16 | 16 ÷ 4 = | 4 |
| D. scooters | 18 | 18 ÷ 2 = | 9 |
| E. wagons | 32 | 32 ÷ 4 = | 8 |
| F. tricycles | 15 | 15 ÷ 3 = | 5 |
| G. unicycles | 7 | 7 ÷ 1 = | 7 |
| H. scooters | 12 | 12 ÷ 2 = | 6 |
| I. tricycles | 27 | 27 ÷ 3 = | 9 |
| J. wagons | 20 | 20 ÷ 4 = | 5 |
| K. tricycles | 12 | 12 ÷ 3 = | 4 |
| L. wagons | 28 | 28 ÷ 4 = | 7 |
| M. tricycles | 21 | 21 ÷ 3 = | 7 |

## Page 94

## Page 95

A. 15 — 4 (3) 6 — 15 ÷ 3 = 5
B. 16 — 3 7 (4) — 16 ÷ 4 = 4
C. 18 — 4 5 (6) — 18 ÷ 6 = 3
D. 48 — (6) 7 9 — 48 ÷ 6 = 8
E. 10 — 3 4 (5) — 10 ÷ 5 = 2
F. 42 — 4 5 (6) — 42 ÷ 6 = 7
G. 21 — (3) 4 5 — 21 ÷ 3 = 7
H. 20 — 3 (4) 6 — 20 ÷ 4 = 5
I. 6 — 4 5 (6) — 6 ÷ 6 = 1
J. 32 — (4) 5 6 — 32 ÷ 4 = 8
K. 28 — 3 (4) 5 — 28 ÷ 4 = 7
L. 6 — (1) 7 8 — 6 ÷ 1 = 6
M. 25 — 3 4 (5) — 25 ÷ 5 = 5
N. 45 — 4 (5) 6 — 45 ÷ 5 = 9
O. 14 — (2) 3 4 — 14 ÷ 2 = 7

## Page 96

A. 2
B. 7
C. 4
D. 2
E. 9
F. 6
G. 4
H. 6
I. 9
J. 1
K. 4
L. 3
M. 8
N. 3
O. 1
P. 6
Q. 7
R. 5
S. 5
T. 9
U. 5
V. 7
W. 2
X. 3

## Page 97

A. 40
B. 2
C. 28
D. 64
E. 1
F. 36
G. 21
H. 0
I. 24
J. 49
K. 6
L. 45
M. 14
N. 8
O. 32
P. 7
Q. 3
R. 81
S. 4
T. 35
U. 27
V. 9
W. 16
X. 5

## Page 98

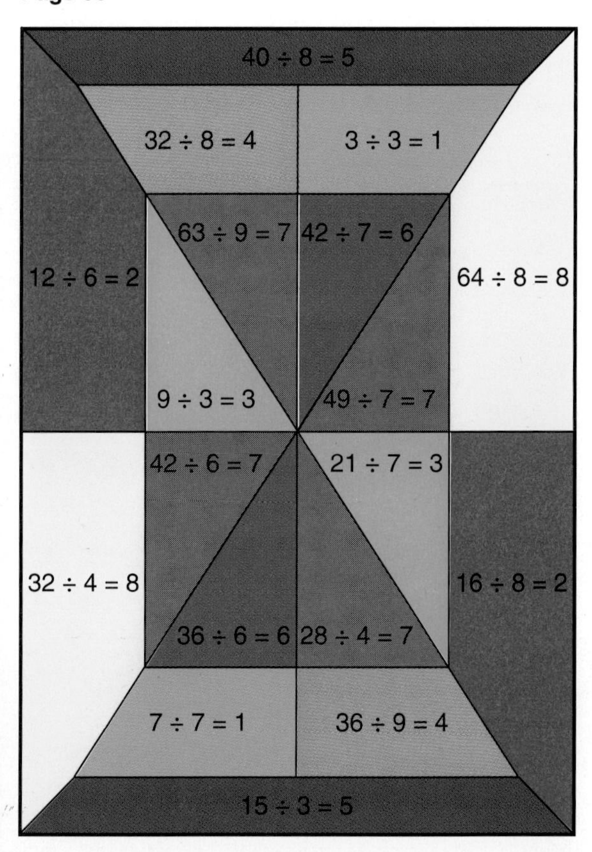


## Page 101

Order of answers may vary.

| A. | B. | C. | D. |
|---|---|---|---|
| 6 x 7 = 42 | 7 x 8 = 56 | 6 x 8 = 48 | 8 x 9 = 72 |
| 7 x 6 = 42 | 8 x 7 = 56 | 8 x 6 = 48 | 9 x 8 = 72 |
| 42 ÷ 6 = 7 | 56 ÷ 7 = 8 | 48 ÷ 6 = 8 | 72 ÷ 8 = 9 |
| 42 ÷ 7 = 6 | 56 ÷ 8 = 7 | 48 ÷ 8 = 6 | 72 ÷ 9 = 8 |

| E. | F. |
|---|---|
| 6 x 9 = 54 | 7 x 9 = 63 |
| 9 x 6 = 54 | 9 x 7 = 63 |
| 54 ÷ 6 = 9 | 63 ÷ 7 = 9 |
| 54 ÷ 9 = 6 | 63 ÷ 9 = 7 |

## Page 102

Order of answers may vary.

| A. | B. | C. | D. |
|---|---|---|---|
| 2 x 4 = 8 | 1 x 3 = 3 | 3 x 4 = 12 | 4 x 5 = 20 |
| 4 x 2 = 8 | 3 x 1 = 3 | 4 x 3 = 12 | 5 x 4 = 20 |
| 8 ÷ 2 = 4 | 3 ÷ 1 = 3 | 12 ÷ 3 = 4 | 20 ÷ 4 = 5 |
| 8 ÷ 4 = 2 | 3 ÷ 3 = 1 | 12 ÷ 4 = 3 | 20 ÷ 5 = 4 |

## Page 106

A. 3
B. 4
C. 0
D. 4
E. 1
F. 0
G. 2
H. 9
I. 1
J. 9

## Page 107

A. 744
B. 918
C. 854
D. 1,016
E. 1,295
F. 1,695
G. 2,176
H. 676
I. 1,311
J. 624
K. 528
L. 2,080
M. 3,090
N. 952
O. 831

They'd rather see him mow a lawn than MAKE A HOLE IN ONE!

## Page 108

A. 266
B. 1,413
C. 1,018
D. 310
E. 348
F. 1,096
G. 560
H. 1,364
I. 1,206
J. 405
K. 2,136
L. 416
M. 918
N. 340

## Page 111

1. –, 27 cars
2. +, 83 races
3. +, 118 miles
4. –, 22 cars
5. –, 9 cars
6. +, 39 gallons

## Page 112

1. –, 87 carrots
2. +, 262 pounds
3. –, 212 ears
4. +, 498 shoppers
5. +, 776 plants
6. –, 63 melons
7. –, 223 squashes
8. +, 275 onions

**Page 113**
1. 40 doughnuts
2. 27 eggs
3. 12 doughnuts
4. 54 doughnuts
5. 35 hours
6. 24 hours
7. 64 doughnuts
8. 45 doughnut holes

**Page 114**
1. 9
2. 8
3. 6
4. 6
5. 7
6. 7
7. 6
8. 6

**Page 119**

**Page 120**
1. $\frac{1}{4}$, $\frac{3}{4}$
2. $\frac{1}{2}$, $\frac{1}{2}$
3. $\frac{1}{3}$, $\frac{2}{3}$
4. $\frac{3}{8}$, $\frac{5}{8}$
5. $\frac{2}{3}$, $\frac{1}{3}$
6. $\frac{1}{8}$, $\frac{7}{8}$
7. $\frac{3}{4}$, $\frac{1}{4}$
8. $\frac{2}{2}$, $\frac{0}{2}$
9. $\frac{4}{8}$, $\frac{4}{8}$
10. $\frac{2}{4}$, $\frac{2}{4}$
11. $\frac{6}{8}$, $\frac{2}{8}$
12. $\frac{0}{3}$, $\frac{3}{3}$

**Page 121**

Tank 1
$\frac{2}{3}$
$\frac{1}{3}$
Tank 2
$\frac{5}{8}$
$\frac{3}{8}$
Tank 3
$\frac{3}{4}$
$\frac{1}{4}$
Tank 4
$\frac{1}{2}$
$\frac{1}{2}$
Tank 5
$\frac{2}{6}$
$\frac{4}{6}$
Tank 6
$\frac{2}{5}$
$\frac{3}{5}$

**Page 122**

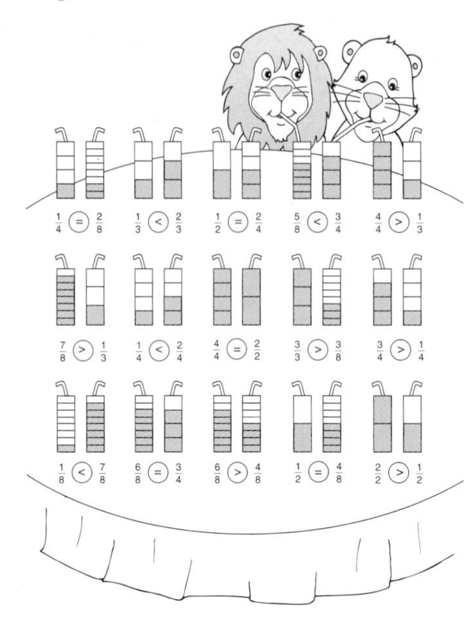


**Page 123**

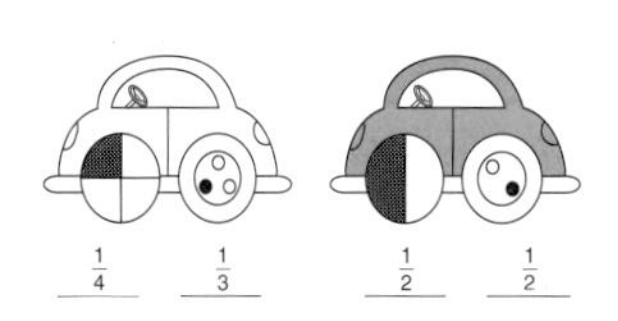


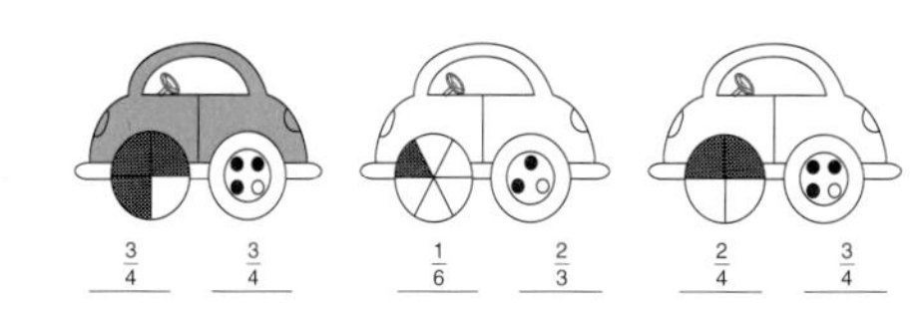


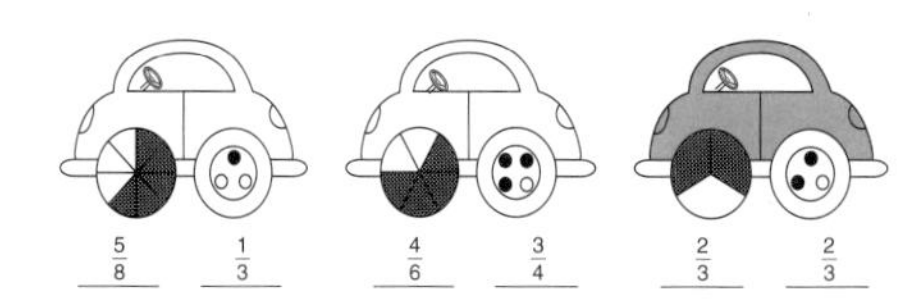


**Page 126**

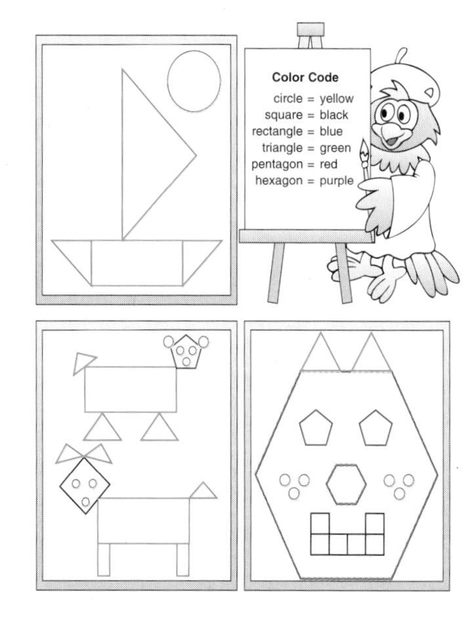


**Page 129**

**Page 130**
1. 6
   12
   8
2. 6
   12
   8
3. 5
   8
   5
4. 0
   0
   0
5. 1
   0
   0
6. 2
   0
   0

1. cube
2. rectangular prism
3. square pyramid
4. sphere
5. cone
6. cylinder

**Page 131**

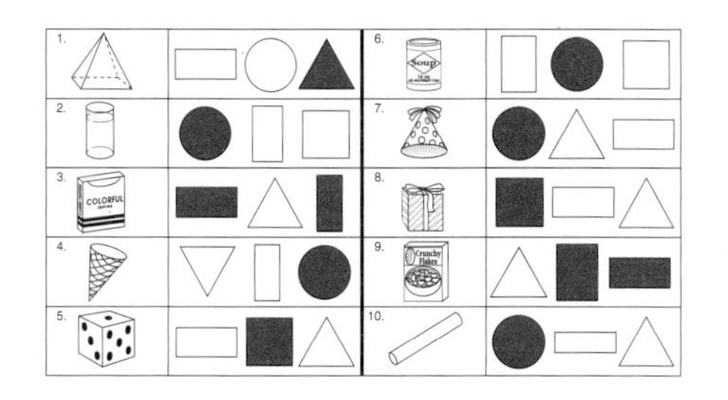

## Page 134

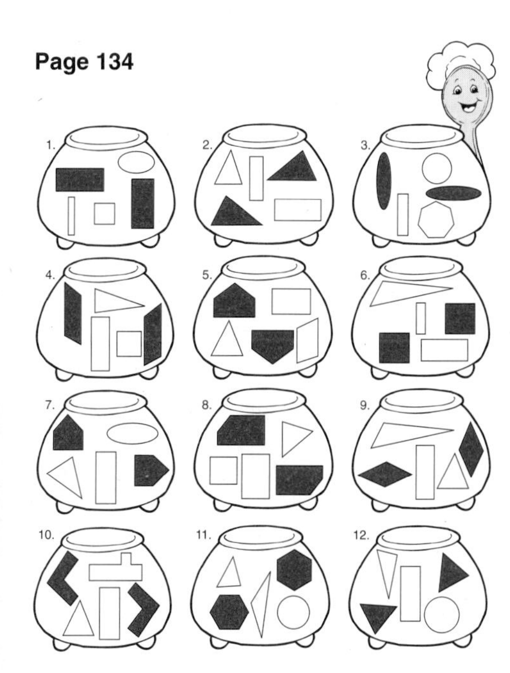

## Page 137

## Page 140

A. flip
B. slide
C. flip
D. turn
E. slide
F. slide
G. flip
H. turn
I. flip
J. turn

## Page 144

A. 2
B. 4
C. 5
D. 3½
E. 2½
F. 5½
G. 3
H. 4½
I. 1
J. 1½

## Page 145

| | |
|---|---|
| G | A |
| S | P |
| Y | M |
| A | O |
| N | L |
| D | Q |
| P | A |
| C | T |
| O | R |
| E | P |
| F | A |
| Z | E |

AT A YARD SALE

## Page 149

Tower 1: 4 + 4 + 20 + 20 = 48 cm
Flag: 5 + 5 + 2 + 2 = 14 cm
Tower 2: 3 + 3 + 18 + 18 = 42 cm
Window 1: 2 + 2 + 7 + 7 = 18 cm
Window 2: 2 + 2 + 5 + 5 = 14 cm
Door: 7 + 7 + 8 + 8 = 30 cm

## Page 150

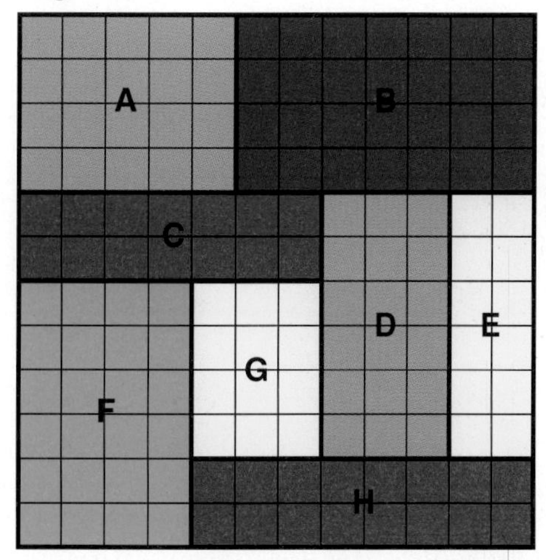


| Quilt Patch | Length | Width | Area |
|---|---|---|---|
| A | 4 cm | 5 cm | 20 sq. cm |
| B | 4 cm | 7 cm | 28 sq. cm |
| C | 2 cm | 7 cm | 14 sq. cm |
| D | 6 cm | 3 cm | 18 sq. cm |
| E | 6 cm | 2 cm | 12 sq. cm |
| F | 6 cm | 4 cm | 24 sq. cm |
| G | 4 cm | 3 cm | 12 sq. cm |
| H | 2 cm | 8 cm | 16 sq. cm |

## Page 153

A. red
B. blue
C. blue
D. red
E. blue
F. red
G. red
H. blue
I. blue
J. red
K. blue

## Page 154

A. ounces
B. pounds
C. ounces
D. ounces
E. ounces
F. ounces
G. ounces
H. pounds
I. ounces
J. ounces
K. pounds
L. pounds

## Page 155

## Page 159

1. ml
2. ml
3. L
4. ml
5. L
6. L
7. L
8. ml
9. L
10. L
11. ml
12. ml

## Page 160

| Item | Unit | |
|---|---|---|
| 1. sink | pint or **gallon** | |
| 2. pool | **gallon** or quart | |
| 3. dog bowl | **pint** or gallon | |
| 4. mug | quart or **cup** | |
| 5. baby bottle | **pint** or gallon | |
| 6. teakettle | **cup** or quart | |
| 7. small bowl | **pint** or gallon | |
| 8. barrel | quart or **gallon** | |
| 9. drinking glass | quart or **cup** | |
| 10. large fish tank | pint or **gallon** | |
| 11. bathtub | **gallon** or quart | |
| 12. pitcher | pint or **quart** | |

## Page 163

A. 25° F, building snowmen
B. 75° F, bike riding
C. 60° F, gardening
D. 95° F, surfing
E. 30° F, shoveling snow
F. 85° F, sailing
G. 65° F, hiking
H. 20° F, ice-skating

## Page 166

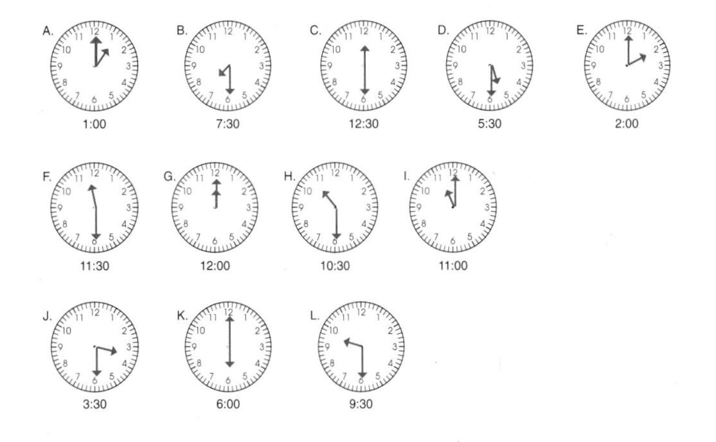


## Page 167

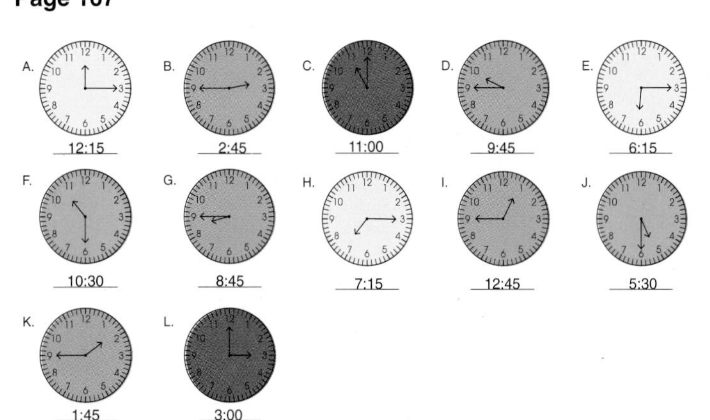


## Page 168

A. 6:05 A.M. We arrived at the dock.
B. 6:40 A.M. We set sail.
C. 8:10 A.M. We caught a fish.
D. 11:55 A.M. We ate lunch.
E. 12:35 P.M. We passed Feather Island.
F. 3:25 P.M. We played cards.
G. 5:50 P.M. We ate fish and chips for dinner.
H. 7:20 P.M. We arrived back at the dock.

## Page 169

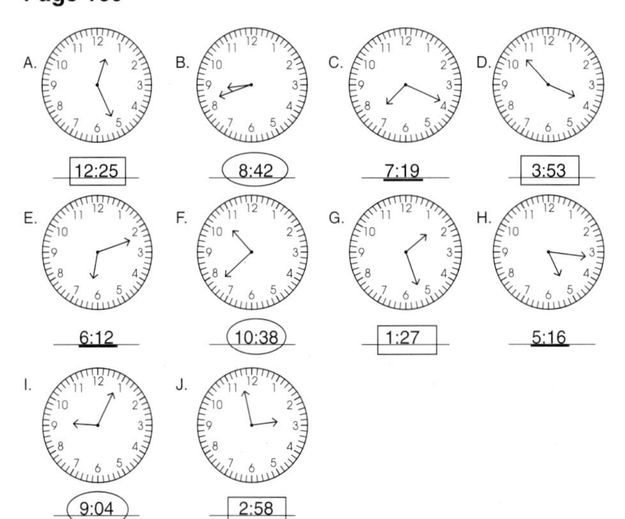


## Page 172

1. 1, 7, 8, 14, 15, 21, 22, 28, 29
2. 2, 9, 16, 23, 30
3. 11
4. Friday
5. 18
6. 3, 5, 10, 12, 17, 19, 24, 26, 31
7. Wednesday

The trip will be on Wednesday, May 4.

## Page 174

1. 30
2. May
3. 4
4. February
5. November 1
6. May 7
7. June 24
8. 7
9. May
10. Tuesday

## Page 173

| January | February | March | April |
|---|---|---|---|
| May | June | July | August |
| September | October | November | December |

## Page 177

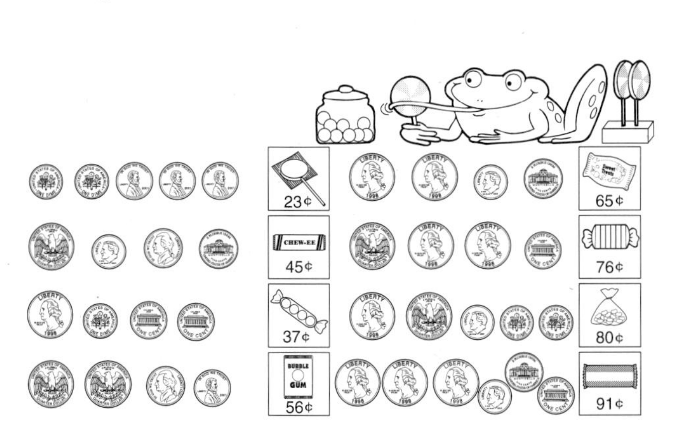

## Page 178

Order of coins in each row may vary.

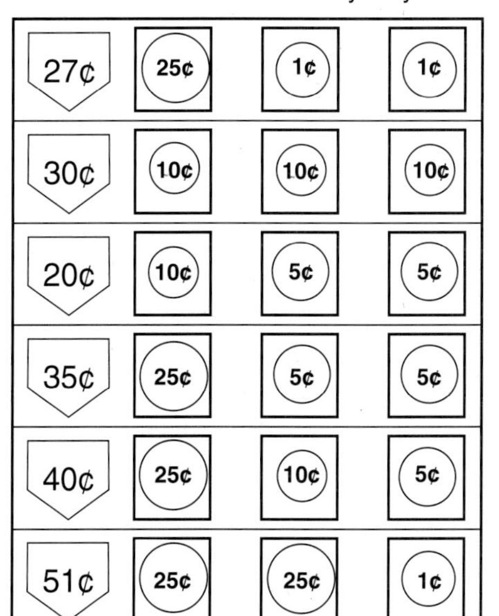


## Page 179

Coins colored may vary but should equal the amounts indicated on the bags.

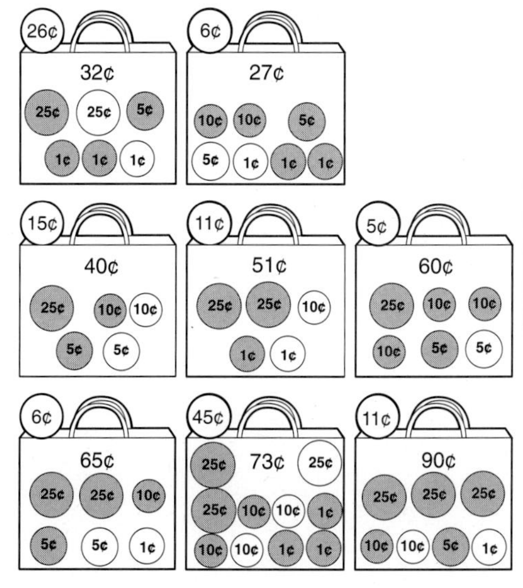


## Page 180

| Peggy | Percy | Porgy |
|---|---|---|
| \$ 1 . 2 5 | \$ 2 . 3 0 | \$ 1 . 4 5 |
| **Pauline** | **Plucky** | **Pam** |
| \$ 2 . 5 0 | \$ 3 . 2 2 | \$ 2 . 7 5 |
| **Penny** | **Perry** | **Pablo** |
| \$ 3 . 6 4 | \$ 4 . 8 5 | \$ 4 . 5 3 |

| Bank Slip | |
|---|---|
| (P) e r c y | \$2.30 |
| P a u l (i) n e | \$2.50 |
| P e (g) g y | \$1.25 |
| P o r (g) y | \$1.45 |
| P e r r (y) | \$4.85 |
| P a (b) l o | \$4.53 |
| P (a) m | \$2.75 |
| P e (n) n y | \$3.64 |
| P l u c (k) y | \$3.22 |

Because he worked in a piggy bank!

## Page 181

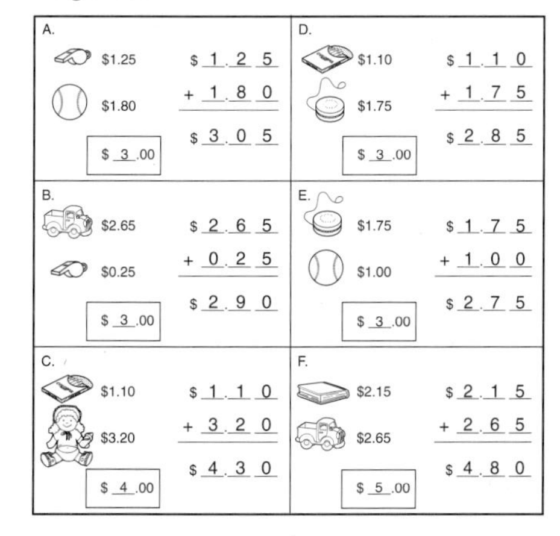


## Page 187

1. 7
2. cheddar, American
3. 5

| Cheese Sold | |
|---|---|
| **Type of Cheese** | **Number of Slices** |
| Swiss | △△△◺ |
| Cheddar | △△△△△△ |
| American | △△△△△△△ |

## Page 188

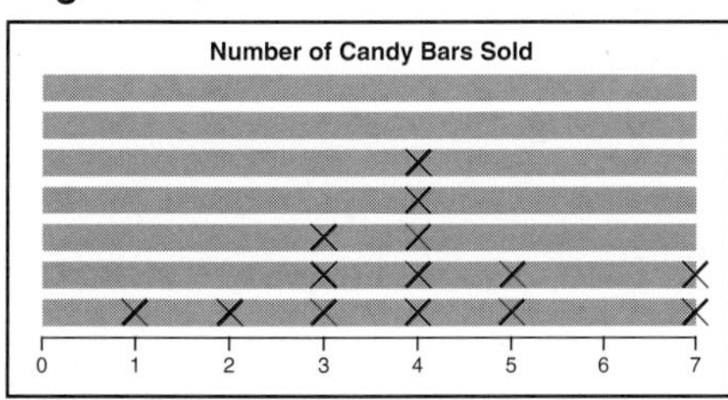


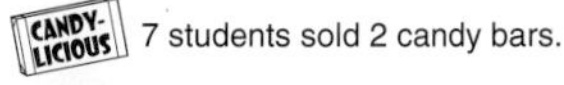

5 students sold 4 candy bars.

7 students sold 2 candy bars.

14 students sold candy bars.

Some students sold 6 candy bars.

The number of candy bars sold the most was 4.

## Page 189

| How Many Crackers Were Eaten? | | |
|---|---|---|
| **Type of Cracker** | **Tally Marks** | **Total** |
| (zebra) | 𝍸 II | 7 |
| (tiger) | 𝍸 IIII | 9 |
| (seal) | III | 3 |
| (giraffe) | 𝍸 | 5 |

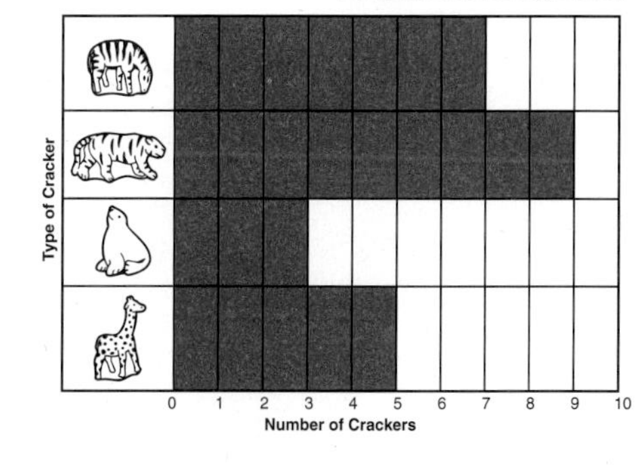


## Page 190

| Answer | Tally Marks | Total |
|---|---|---|
| apple | IIII | 4 |
| peach | III | 3 |
| cherry | III | 3 |

| Answer | Tally Marks | Total |
|---|---|---|
| coffee | IIII | 4 |
| tea | II | 2 |
| milk | IIII | 4 |

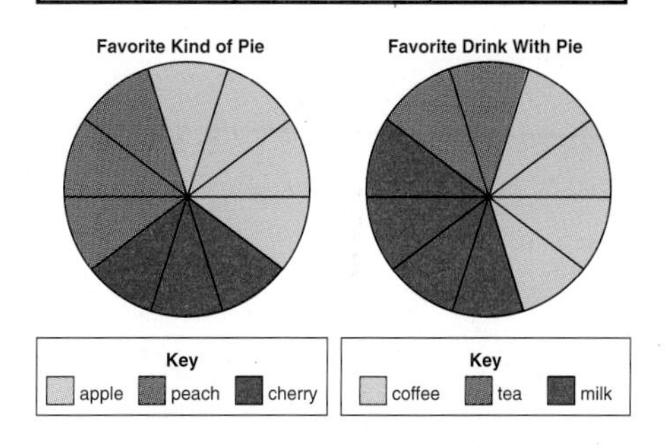


## Page 191

1. Wednesday
2. 60
3. Tuesday, Wednesday
4. Answers may vary. Possible responses include more people are off from work during the day to visit the store, and the store might be open longer hours that day.

**Page 195**

1. blue, red

2-4. answers will vary

**Page 197**

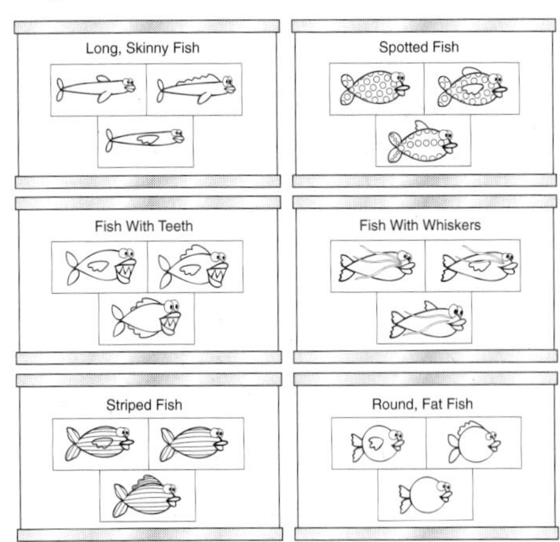


**Page 198**

Order of answers on each suitcase may vary.

Things to wear
shoes
socks
pants
shirts

Things to take to the pool
beach towel
raft
suntan lotion
pool toy

Things to keep in the bathroom
comb
toothbrush
soap
toothpaste

**Page 200**

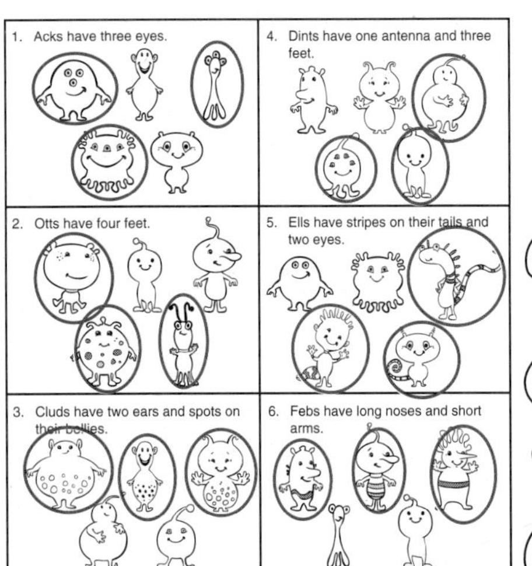


**Page 201**

| | True | False |
|---|---|---|
| 1. Sue plays a string instrument. | (C) | R |
| 2. Stan plays a percussion instrument. | T | (M) |
| 3. Simon plays a brass instrument. | (S) | W |
| 4. Tommy plays a percussion instrument. | (T) | P |
| 5. Sally plays a brass instrument. | J | (D) |
| 6. Sara plays a percussion instrument. | (U) | B |

He had the DRUMSTICKS!

**Page 204**

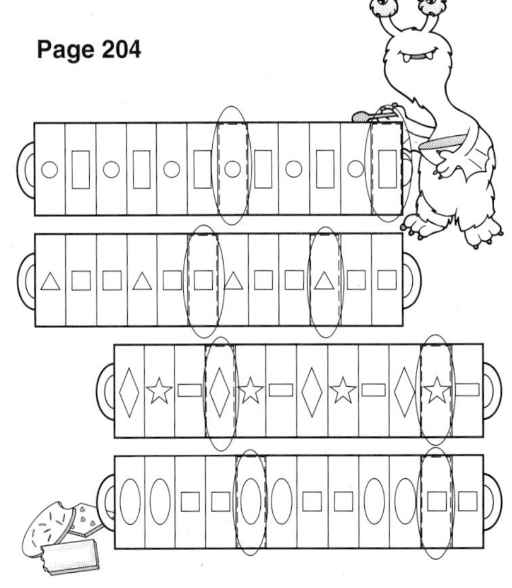

**Page 205**

**Page 208**

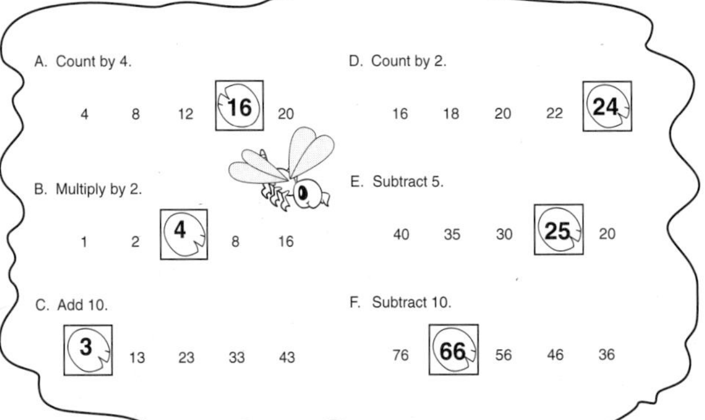


**Page 209**

A. Add 11.
11, 22, 33, 44, 55, 66, 77

B. Add 1 more than the number before.
1, 2, 4, 7, 11, 16, 22

C. Count by 5.
5, 10, 15, 20, 25, 30, 35

D. Subtract 2.
18, 16, 14, 12, 10, 8, 6

E. Multiply by 10.
1; 10; 100; 1,000; 10,000; 100,000; 1,000,000

F. Add 3.
4, 7, 10, 13, 16, 19, 22

**Page 212**

| | 25 | 32 | 33 | 50 |
|---|---|---|---|---|
| Al | ✔ | ✗ | ✗ | ✗ |
| Jim | ✗ | ✗ | ✔ | ✗ |
| Kate | ✗ | ✔ | ✗ | ✗ |
| Rita | ✗ | ✗ | ✗ | ✔ |

| | 27 | 53 | 74 | 95 |
|---|---|---|---|---|
| Al | ✔ | ✗ | ✗ | ✗ |
| Jim | ✗ | ✗ | ✔ | ✗ |
| Kate | ✗ | ✗ | ✗ | ✔ |
| Rita | ✗ | ✔ | ✗ | ✗ |

**Page 213**

| | baseball | drawing | in-line skating | reading | baking |
|---|---|---|---|---|---|
| Harry | ✔ | ✗ | ✗ | ✗ | ✗ |
| Fran | ✗ | ✗ | ✔ | ✗ | ✗ |
| Jerry | ✗ | ✔ | ✗ | ✗ | ✗ |
| Polly | ✗ | ✗ | ✗ | ✔ | ✗ |
| Frank | ✗ | ✗ | ✗ | ✗ | ✔ |

| | running | basketball | tennis | swimming | horseback riding |
|---|---|---|---|---|---|
| Chuck | ✗ | ✗ | ✔ | ✗ | ✗ |
| Patty | ✗ | ✗ | ✗ | ✔ | ✗ |
| John | ✗ | ✔ | ✗ | ✗ | ✗ |
| Tina | ✗ | ✗ | ✗ | ✗ | ✔ |
| Debbie | ✔ | ✗ | ✗ | ✗ | ✗ |